The Climate of China and Global Climate

Proceedings of
the Beijing International Symposium on Climate
Oct. 30 — Nov. 3, 1984 Beijing, China

Edited by
**Ye Duzheng, Fu Congbin,
Chao Jiping and M. Yoshino**

.With 357 Figures

China Ocean Press Beijing
Springer-Verlag Berlin Heidelberg New York
London Paris Tokyo

Edited by
Ye Duzheng, Fu Congbin,
Chao Jiping and M. Yoshino

Published by
China Ocean Press
Distributed by
Domestic — **China Ocean Press**
Foreign — **Springer-Verlag Berlin Heidelberg New York London**
Paris Tokyo

Springer-Verlag Berlin Heidelberg New York London Paris Tokyo is the exclusive distributor for all countries outside the People's Republic of China

ISBN 3–540–16718–8 Springer-Verlag Berlin Heidelberg New York
ISBN 0–387–16718–8 Springer-Verlag New York Berlin Heidelberg

2132/3140–54310

PREFACE

This volume contains most of the papers presented at the Beijing International Symposium on Climate (October 30 — November 3, 1984) sponsored by the Chinese Academy of Sciences and cosponsored by ICCL/IAMAP (International Committee of Climate/International Association of Meteorology and Atmospheric Physics) and INQUA Cl3 (International Quaternary Association). With the relationship between climate in China and global climate as the major subject of the Symposium, about 60 eminent international scientists presented their new contributions in this field and discussed common interesting scientific problems.

The climatic change is not only a challenging scientific problem to which climatologists in many countries are dedicating themselves, but also an economically important problem with which people all over the world are much concerned. Extensive scientific between scientists in different countries would be beneficial for promoting the further development of climate research.

The volume is edited in the following five parts:

(a) Climate history in the past 2000 years.

The history of climatic change in China and its comparison with that of other parts of the world are reconstructed by analysing large amounts of geological, geophysical, paleogeographical, and paleobiological evidence and historical literature. It is the first time that some of the unique data sets from China are published in such a volume as this.

(b) Air-sea interaction in the short-term climate variation.

The emphasis has been placed on the air-sea interaction in the tropical Pacific and the relevant monsoon variability.

(c) Land surface processes related to climate variation.

The discussion focuses on the role of the heat and water budgets of the earth surface and their physical properties in climate formation and climate variation.

(d) Impact of human activity and some other natural factors on climate.

Solar activities, volcanic activity, polar ice cover, CO_2 and desertification are included.

(e) Prediction methods of monthly and seasonal climate variation.

Several statistical, thermodynamic, dynamic and stochastic models and their experimental results are presented.

The present collection of papers is a valuable reference book for scientists working in the field of climate research.

Opening Speech by Prof. Tu-Cheng Yeh
Chairman of the Symposium

Ladies, Gentlemen, and Comrades

First of all please allow me on behalf of the Chinese Academy of Sciences to express our warm welcome to all the foreign friends and scientists coming from various parts of the world.

Today I am very happy to inaugurate formally the important conference on "Relationships between the climate of China and global climate: past, present and future" sponsored by the Chinese Academy of Sciences and cosponsored by ICCL/ IAMP & INQUA C13.

It is well known that the problem of world climate, especially the world climate change, its physical causes and predictions, has become a more and more important issue, not only in the atmospheric sciences, but also in the geophysical sciences in general and the biological sciences. It has aroused more and more concern, not only among the atmospheric scientists, but also among the geophysical scientists in general, and biological scientists. The world climate has indeed a very profound impact on many spheres of human activities. Needless to say, it is an economically important problem. It is also indeed scientifically challenging. In recent years numerous researches on climatic change, its physical causes, the possibility of its prediction and so on have been undertaken in various countries. Chinese scientists, as well as those in other countries, have paid and are still paying great attention to the problem of world climate, and have done research work on this problem. But almost all the research papers are written in Chinese. Due to the language barrier, few of these researches have been known to the foreign world. Although for some Chinese scientists foreign literature is not unfamiliar, not all Chinese scientists are good enough in foreign languages. Therefore, foreign literature cannot be said to be very well known in China either. This meeting will give a unique opportunity for the Chinese and foreign scientists to communicate directly with each other the interesting achievements in the field of climatic change and variability, and their causes.

China is a large country with its territory covering roughly 9,600 thousand km^2. If the climatic change of this large part of the world is not known to the outside world and not connected to the climatic change of other parts of the world, the knowledge of world climate change cannot be said to be complete. Since the atmosphere, also the climate, has no boundary, world climate should be studied as a whole. Climatic change and variability in one region should be connected with those of the other parts of the world. It is difficult, even impossible, to study the climate, climatic change and especially its prediction in an isolated region. In this sense the Chinese scientists, studying climatic change and variability in China, need the knowledge of the outside world. Similarly, the outside scientists, studying the climate of the other parts of the world, also need to know what is going on in China. Through mutual discussions this conference will undoubtedly provide a good chance for the scientists to link the climatic change and variability outside to those within China. That is the reason why we suggest that the general goal of this symposium is to make a comparison of climatic change between China and other regions in the world in the past, present and future, and to further explore the physical causes leading to these differences and similarities.

Some 70 eminent scientists have gathered together here to present their latest research words and to discuss further problems of common interest. I believe that through the lectures and helpful discussions and extensive exchange of ideas, mutual understanding and friendship will be strengthened and this will definitely promote further development of research in this field. I earnestly hope this meeting can further encourage the exchange of scientific research results in this field between the Chinses and scientists from the outside world. Now, may I wish this conference all success and a happy stay in China to all the foreign scientists and friends.

Thank you very much.

Summary Speaking at the Symposium

M. Yoshino

I. Foreword

The "Beijing International Symposium on Climate" with the title of "Relationships between the Climate of China and Global Climate: Past, Present and Future" was very successful. A large number of participants came and discussed the theme at the every session of the Symposium. The general goal of the symposium was to make a diagnostic comparison of climatic change between China and other regions in the world in the past, present and future. The physical causes for the differences, similarities and parallelism were also main points in the discussion.

The presentation was grouped into the following five sessions:

1. Climatic fluctuations over the past 2000 years or more.
2. Air-sea interaction with particular reference to the West Pacific.
3. Land surface-climate interaction.
4. Prediction methods for monthly and seasonal climate variations.
5. Impacts of human and natural activity on climate.

I, as one of the co-chairmen of the symposium, would like to summarize the results of Symposium..

II. Discussions

On the last day morning we had a time for discussion grouping the participants into three groups: Modelling Group, SST Group and Past Climate Group, with special reference to the climates of China. The following is the summary of discussion:

(a) Modeling Group (Chairperson: D. L. Hartmann)

There can be no doubt that the circulation driven by topography and latent heating in and around China plays a central role in the general circulation of the Earth's atmosphere and influences climate and weather in many distant parts of the world. The proper simulation of these circulations involve several very difficult modeling problems. For example, the proper treatment of mountains and accurate simulation of clouds and precipitation are essential. Discussion centered around the following main topics, and was stimulated by the presentations given at this meeting.

(i) There is need for better physical understanding of the physical processes linking and surface conditions such as soil moisture, snowcover and vegetation with the climate of the air. It was suggested that simplified models, designed to investigate selected physical mechanisms, can provide some of the needed physical insight.

(ii) Soil moisture is an important quantity for which we should be developing prediction models. Before an accurate soil moisture simulation can be attained, however, we must be able to simulate precipitation much more accurately and we must have a database of soil moisture observations to be used for analysis and model verification. An adequate database does not at present appear to exist.

(iii) The coupling of atmosphere and ocean plays an important role in climate variability. The development of coupled models presents an important scientific challenge which should be rigorously pursued. Here again, models of varying degrees of complexity are needed. It may be possible to illustrate much of the fundamental physics with simplified models. General circulation models will be needed to integrate all of the important processes and feedbacks. Observational studies suggest that modeling of the effect of extratropical SST anomalies on climate needs to be studied further.

(b) SST Group (Chairperson: M. Yoshino)

Many facts have been made clear between SST over the North Pacific and China's climate and the related circulation patterns: for example, tropical easterly jet, 150–200 hPa level and 500 hPa topography patterns, Tibetan high, North Pacific subtropical high, summer and winter monsoons, cyclogenesis over the East China Sea, frequencies of typhoon formation and occurence, precipitation, and air temperature both in winter and summer.

We have concentrated the discussion to the early summer or summer precipitation over the Changjiang River region, Mai-Yu, and summarized the following relationships. There is a positive correlation to the SST of the Eastern Equatorial Pacific (EEP) in the preceding 6–8 months. Namely, the warmer SST over EEP causes floods in the region. But when the subtropical high develops too strongly, the region suffers from drought, because the North Pacific high develops strongly and extends to East China through strong Hadley Circulation.

The subtropical high over the North Pacific, especially in the western part, has a positive correlation to the SST over EEP in the preceding 2–6 months (mostly 3–4 months). The warmer Tibetan Plateau (corresponding to the cold SST over the Indian Ocean) encourages the downward stream in the western part of the North Pacific high.

Emphasis was placed upon further studies on the physical processes between the effects of SST on the climate on the various scales, because the response will be different in accordance with their scales. Also, lag correlation between them and case studies for the individual year should be analyzed in detail. The Plateau's effects on climate over East Asia should be examined in association with SST change over EEP.

(c) Past Climates Group (Chairperson: F. A. Street-Perrott)

The following items were concluded:

1) Importance of paleoclimatic data: Instrumental records too short to provide adequate samples of late Holocene climatic variability. Application to practical problems: Distribution of climatic hazards, magnitude/frequency data for engineers, water resources, effects of man on climate, and improved statistical forecasting. Research applications: Data on climatic forcing (sunspot, dust . . .), and sensitivity experiments.

2) Interdisciplinary and international cooperation: Data collection, statistical analysis, and numerical modeling.

3) Research Priorities: Improve space/time data coverage, identify sensitive

areas (arid zones, high mountains etc.), apply new techniques (geomorphology, paleolimnology and zones, etc.), obtain long time series, quantify and standardize data, set up systematic archives/data bases, and compare results of different techniques.

4) International Multidisciplinary Symposium on Past Climates of Chian. This was strongly supported by the participants at the discussion. It was considered that many scientific organizations will support this Symposium.

III. Closing Remarks

The papers presented at the Symposium are classified as (a) regions concerned, (b) periods (timescales) studied, and (c) methods or techniques of studies and data analyzed.

(i) The regions studied are grouped into six and the numbers of papers are shown as follows:

Regions studied	Number of papers concerned
Global or hemispheric	15
East Asia	10
China	16
Other parts of East Asia	4
Other continents or oceanic regions	17
Comparison between China and other regions	12

This figure shows that the numbers of the treated regions are almost the same between global or hemispheric, China and other continents or oceanic regions. However, the comparative studies on the phenomena in China and other regions show a somewhat lesser figure.

(ii) The studied periods time-scale treated, are thus summarized as follows:

Periods studied	Number of papers concerned
Geological period	8
Historical period since 2,000 yr B. P.	9
Instrumental observation period since 19 century	48

Based on this table, it is our opinion that we should encourage more studies or the geological and historical periods. Also, it is worthy of note that there were very few papers which treated the future climate, even though this was a part of the subtitle of the Symposium.

(iii) The methods or techniques of studies are grouped roughly into three:

Methods o r techniques of studies	Number of papers concerned
Analysis of the instrumentally observed data	30
Simulation model	19
Analysis paleoclimatic evidence and past climate records	18
Chronicles or diaries	8
Tree ring	3
Pollen analysis	1
Prehistoric evidence	1
Glacier or glaciation	3
Sand and soil sediments	1
Deep sea or lake bottom core	1

This table makes clear that the meteorologists and climatologists should analyze the past climates more intensively, using paleoclimatic evidence and past climate records. On the other hand, because the models may indicate tendencies of the future climate, this should be encouraged also.

It was clear to the participants that the topics on SST and ENSO have interested many researchers. Actually, the number of papers concerning to airsea interaction, SST or ENSO was 23, which is about one third of the total papers presented. The climatological studies on these topics will follow at coming meetings and symposiums.

(A detailed report of this Symposium will appear in the Bulletin of the American Meteorological Society, September issue, 1985)

CONTENTS

Section I: Climate history in past 2000 years and beyond

Section II: Ocean-atmospheric interaction in short-period climatic variation

Section I

Climate History in Past 2000 Years and Beyond

Long-Term Climatic Change in the Loess-Plateau of China

An Zhisheng[1], Liu Tungsheng[2]

Abstract — Studies on magneticestratigraphy, climatistratigraphy and biostratigraphy indicate that the Lochuan loess-paleosol sequence in Shaanxi province has recorded entirely the histroy of long-term climatic change in the past 2.4 m. y. on the loess plateau. The loess plateau has experienced 11 periods of dry-cold climate and 11 periods of warm-humid climate that constitute 10 complete cycles of climatic fluctuation. The estimated difference of annual mean temperature between the dry-cold and warm-humid periods is less then $13°C$ and the maximum difference of annual precipitation is about 550 mm.

The climatic record in Lochuan loess-palaeosal sequence is comparable to that from the deep sea sediments.

Introduction

Loess is widely distributed at $30°-40°$ N in northern China. In the loess-paleosol series saved were a number of paleoclimatic records which may by and large reflect the paleoclimatic changes in northern China. The loess-paleosol series of Lochuan in Shaanxi Province, 130 meters thick and located in the midreach of the Yellow River has preserved the information about the major geological and climatic events happened since 2.4 m.y. B. P. It is recognized as one of the most complete pieces of geological-climatic records in the Northern Hemisphere (Fig. 1).

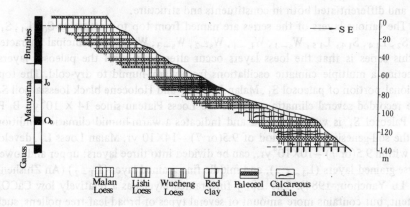

Fig. 1. Lochuan loess-paleosol stratigraphical series

The grain size distribution, $CaCO_3$ content, heavy mineral assemblage, micromorphological characteristics, susceptibility and other geochemical indices

1. Xi'an Laboratory of Loess and Quaternary Geology, Academia Sinica, Xi'an, China
2. Institute of Geology, Academia Sinica, Beijing, China

(FeO/Fe$_2$O$_3$ and SiO$_2$/Al$_2$O$_3$ ratios, and amino-acid content and composition) of loess-paleosols are the geological evidences of paleoclimate. The fossil assemblages of land snail, pollen and mammalian fauna buried in the loess-paleosols are the biological evidences of paleoclimate. For example, the Metodontia assemblage of fossil snail in paleosols reflects a warm-humid environment (Chen Deniu et al., 1982), a species of which is still alive and spreads southward down to the Changjiang River Reaches; the Cathaica assemblage in loess reflects a dry-cold condition and stretches in the south not beyond the Yellow River Reaches, but extends northwestward up to Inner Mongolia, Xinjiang and Tibet. Therefore, it is possible to turn the geological and biological evidences recorded by the loess paleosol series into indirect indices reflecting the paleoclimatic conditions.

Magnetostratigraphical investigations have provided a fundamental scale for setting up the time sequence of loess-paleosol series (F. Heller and Liu Tungsheng, 1982; Liu Tungsheng and An zhisheng, 1984). By calculating the depositional rate of loess with reference to the known magnetic boundary ages, a time scale for Lochuan loess-paleosol series was established (Table 1).

The time sequence and individual indirect climatic indices of Lochuan loess-paleosol sereis exhibit a variational nature, thus making it possible to reconstruct the sequence of paleoclimatic changes with different time scales over the last 2.4 m.y. in Lochuan, or even northern China.

Paleoclimatic Changes Since 14 × 10^4 yr. B. P.

The loess contains buried paleosols of different developing extent, which constructed a loess-paleosol series more or less consecutive both in time and deposition and differentiated both in constituents and structure.

The various layers of the series are named from top to bottom as S$_0$, L$_1$, S$_1$, L$_2$, S$_2$, L$_{14}$, S$_{14}$, L$_{15}$, W$_{s-1}$, W$_{L-1}$, W$_{s-2}$, W$_{s-3}$, W$_{L-3}$. A principal character of this series is that the loess layers occur alternately with the paleosol layers, reflecting a multiple climatic oscillations from warm-humid to dry-cold. The top-sectional portion of paleosol S$_1$, Malan Loess S$_2$ and Holocene black loessial soil S$_0$ have recorded several climatic stages of the Loess Plateau since 14 × 10^{4}yr. B. P.

Paleosol S$_1$ is well-developed and indicates a warm-humid climatic condition for the soil-genesis over a period of 9.5(or 7) −14×10 yr. Malan Loess L$_1$, developed within 9.5(or 7) −10×10^{4}yr., can be divided into three layers: upper and lower coarse-grained layers (L$_{1-1.3}$), and middle fine-grained layer (L$_{1-2}$) (An Zhisheng and Lu Yanchou, 1984). The L$_{1-2}$ finegrained layer has a relatively low CaCO$_3$ content, but contains more amount of several types of broad-leaf-tree pollens, such as *Alnus Betula Quercus* and *Ulmaceae*, indicating a temperately cool and humid environment as contrasted to the dry-cold climate recorded by the L$_{1-1.3}$ coarse-graind layers (Fig. 2).

The climatic events that have happened since 14 × 10^4 yr. B. P. were also recorded in itermountainous basins, cave and plain deposits and land/sea alternate sediments (Fig. 2). During this period there were 5 climatic events in northern China, which are measured with the time scale to be more than 1 × 10^4 yr., i.e. a

Table 1.   The time sequence of Lochuan loess-paleosol series

Sequence	Age ($\times 10^4$ yr.)
S_0	1
L_1	
S_1	14
L_2	
S_2	25
L_3	
S_3	33
L_4	
S_4	41
L_5	
S_5	56
L_6	
S_6	66
L_7	
L_7	72
L_8	
S_8	77
L_9	
S_9	90
L_{10}	
S_{10}	
L_{11}	
S_{11}	97
L_{12}	
S_{12}	
L_{13}	
S_{13}	
L_{14}	
S_{14}	109
L_{15}	
W_{s-1}	148
W_{1-1}	167
W_{s-2}	187
W_{1-2}	
W_{s-3}	222
W_{1-3}	240

dry-cold one of 9.5 (or 7) -5×10^4 yr. B. P., a temperately cold one of $5-2.5 \times 10^4$ yr. B. P., a dry-cold one of $2.5-1 \times 10^4$ yr. B. P. and a warm-humid one of after 1×10^4 yr. B. P. It should be noted that there was a severely dry-cold episode about 1.5×10^4 yr. B. P. during the last glacial period of China, when the dust accumulation was greatly enhanced and the sea level in the east dropped down to -155 meters.

6

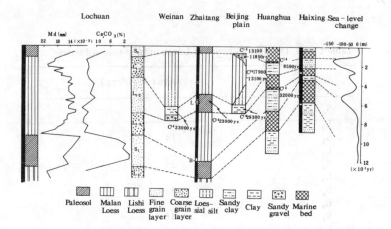

Fig. 2. Comparison of paleoclimatic records of of Northern China since 12×10^4 yr. B.P. (after An Zhisheng and Lu Yangchou, 1984)

Variational Sequence of Climatic Elements (Annual Mean Temperature and Annual Mean Precipitation) of Lochuan over the Period Since 1.1 m.y. B. P.

From the loess-paleosol series ($S_{14}-S_0$) developed since 1.1×10^4 yr. B. P., we identified three types of loesses (weakly, moderately and strongly weathered), and five different paleosol categories, namely, black loessial soil, calcareous cinnamon soil, cinnamon soil, leached cinnamon soil and brown cinnamon soil. This series has recorded an environmental alteration sequence from desert steppe to forested steppe. The climatic elements (annual mean temperature and annual mean precipitation) represented by the five paleosol categories can be inferred from those pertinent to their recent geographical distribution zone. The annual mean precipitation during the deposition of weakly weathered loess may be represented by that (~200 mm) of modern transitional zone between black loessial soil and desert soil. From the consideration that the temperature during the climate-deterioration period in northern China which occured about $1.5-1.8 \times 10^4$ yr. B. P., decreased at least by 8–10°C, it is estimated that the annual mean temperature during the deposition of weakly weathered loess was 0°C, or slightly below 0°C. According to the temperature and precipitation range from weakly weathered loess to black loessial soil, the climatic elements recorded by the moderately and strongly weathered loesses were evaluated through interpolation. Thus, a variational sequence of the climatic elements of Lochuan during the last 1.1 m.y. was obtained in accordance with the time sequence of individual loess and paleosol layers as well as their climatic element values (Fig. 3).

The maximum amplitude of the annual precipitations during this period is about 600 mm and that of the annual mean temperatures is about 14°C. Around 0.5 m.y. B. P., when the paleosol S_5 was formed, the annual mean precipitation was

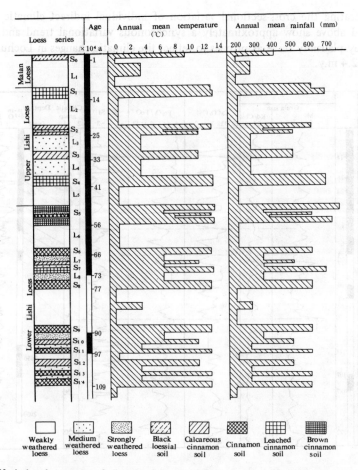

Fig. 3. Variational sequence of climatic elements of Lochuan over the last 1.1 m.y

about 800 mm and the annual mean temperature was 12–14°C (An Zhisheng and Wei Langying 1980).

Paleoclimate During the Past 2.4 m.y.

Curves were plotted for particle size distribution (represented by medium particle diamenter M and coarse silt content to clay content ratio Kd/CI), percentage content of $CaCO_3$, Fe_2O_3/FeO ratio and susceptibility K of the Lochuan loess-paleosol series (Fig. 4). The big particle size, high $CaCO_3$ content, small Fe_2O_3/FeO ratio and low susceptibility, as shown on the left side of the horizontal axis in Fig. 4, all indicate that the loess and paleosol experienced rather weak biochemical weathering, but suffered relatively strong wind-blow and a dry-cold climate, while the opposite conditions of these elements reflect a warm-humid climate. The curves

8

demonstrating the cross-sectional variation of individual indirect climatic indices indicated above show approximately a synchronous variational trend and, therefore, may plausibly serve as proxies to reflect the climatic changes in Lochuan over the last 2.4 m.y.

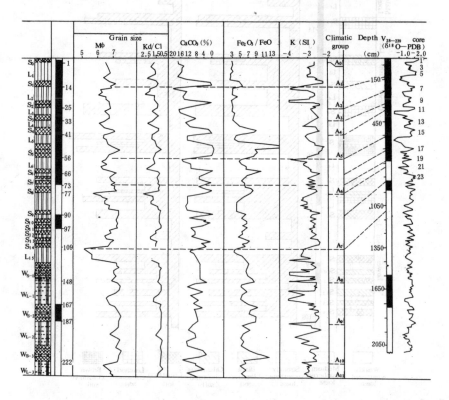

Fig. 4. Curves showing variations of indirect climatic indices of Lochuan loess-paleosol series over the last 2.4 m.y.

An analysis of the curves has yielded the following understandings: The quasi-Matuyana Chron (0.73–2.4 m.y.) is characterized by a relairvely frequent fluctuation in climate with an average warm — humid climatic condition while the Brunhes Chron (since 0.73 m.y. B. P.). features a dry-cold average condition and clear periodic climatic oscillations with some larger amplitudes.

Judging from the alternate occurences of thick-layered (> 2 meters in thickness), weakly or moderately weathered loesses and paleosols or loess-paleosol assemblages in Lochuan, there were at least 11 prolonged dry-cold episodes related to wind-borne dust accumulations, which built up 10 climatic change cycles in a large time scale with each oscillating from warm-humid to dry-cold (A_1, A_2 ...... A_{10}) and two semi-cycles of incomplete climatic changes (warm-humid A_0 and dry-cold A_{11}).

Since 2.4 m.y. B.P., the Loess Plateau, as the representative of northern China, has experienced multi-cycled climatic changes characterized by clear oscillations from warm-humid to dry-cold with a general tendency of becoming dry. The W_{s-2} paleosol assemblage developed about 1.8 m.y. B. P. and the S_5 paleosol developed about 0.5 m.y. B.P. have recorded two climatic optimum periods. The Malan Loess of about 9.5 (or 7) -1×10^4 yr. B. P. and the upper and lower sandy loesses (L_1 and L_{15}) developed about 0.8 and 1.15 m.y. B.P., respectively, have retained a record of a severely dry-cold climate.

Comparison Between Pleistocene Land/Sea Paleoclimatic Records

Detailed studies have already been made on the records of deep sea sediments accumulated since the late Jaramillo Subchron (about 0.9 m.y. B. P.). Here, we intend to make a mere comparison of the Lochuan climatic change sequence in this period with the climatic records of V_{28-238} (Shackleton and Opdyke, 1973, 1976). From the portion of S_9-S_0 paleosols 11 climatic change cycles have been recorded, which oscillated from warm-humid to dry-cold and are denoted by C_I, C_{II} ...C_{XI}, respectively (Fig. 5). The S_0 paleosol-genesis period (1×10^4 yr. B. P.) is compared with the oxygen isotope stage 1, while S_1 and L_1 (C_{II}) ($14-1 \times 10^4$ yr. B. P.) with the stages 2–5 (S_1 vs. stage 5 and $L_{1-3}, L_{1-2}, L_{1-1}$ vs. stages 4, 3, 2, respectively). Normally, each paleosol and loess layers between S_1 and S_9 are compared with the odd and even-numbered stages, respectively, except for paleosol S_5. S_5, consisting of 3 layers of paleosols and 2 thin layers of loesses, spans an interval of about 10×10^4 yr., and reflects a climatic optimum period. So it is obviously comparable to the oxygen isotope stages 13, 14 and 15, which feature an average warm-humid climatic condition.

Comparisons between the climatic records of Lochuan loess-paleosol series and core V_{28-238} have justified that the paleoclimatic changes in Lochuan were also of global significance.

Autospectral Analysis of Loess-Paleosol Series

Variations of several Lochuan loess-paleosol indices with time, such as Kd/Cl (coarse silt content/clay content), $CaCO_3$ content, Fe_2O_3/FeO ratio and susceptibility, reflect at different angles the characteristics of paleoclimatic events. For a better understanding of the time sequence of the geological-climatic events, we conducted a multi-dimensional spectral analysis for the time sequence of the indices mentioned above. First, two time models were designed (Table 2). The age of model 1 was estimated based on the depositional amount and the magnetic boundary ages. For model 2, individual ages were calculated by determining the ratio of the time lengths spanned by individual loess and paleosol layers of each assemblage from the fractions taken by the cold and warm periods of each cycle recorded in cores V_{28-238} and V_{28-239} with reference to the initial ages previously calculated for individual loess-paleosol assemblages. Spectral analysis of the time sequence of various indices within the interval of the past 1 and 0.77 m.y. were performed through the use of the two time models. The results reveal that, for the

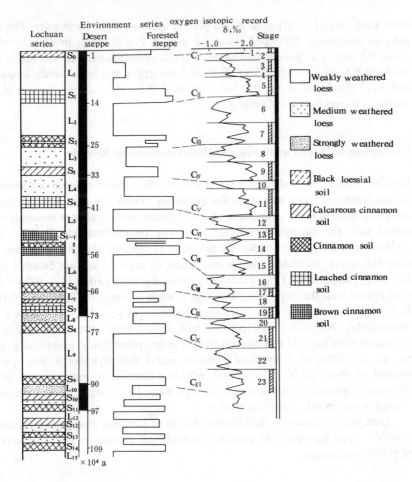

Fig. 5. Comparison between Lochuan loess-paleosol series and V_{28-238} oxygen isotope records

time sequence of the loess-paleosol series, there exist several dominant oscillation cycles, such as those of quasi-8 $\times$ 10^4 yr. and quasi -2 $\times$ 10^4 yr. (Fig. 6,7). The dominant quasi-8 $\times$ 10^4 yr. cycle indicates an approximate average time length of the geological-climatic cycles represented by the loess paleosol assemblages. Assuming the elemental change cycle is 2 $\times$ 10^4 yr. long and counting from 1.5 $\times$ 10^4 yr. B. P. (when the most severely dry-cold event took place in northern China), we anticipate that there will possibly reappear a severely dry-cold glacial climate 5,000 yr. later in northern China.

In short, since 2.4 m.y. B.P. there have been multiple climatic oscillations from warm-humid to dry-cold with the climate becoming dry and dry in the Loess Plateau, or even northern China. The climatic changes are governed primarily by the mechanism of the global climatic variations, and are also closely related to the uplift of the Tibetan Plateau since late Cenozoic. With the uplift of the Tibetan

Table 2. The time sequence of Lochuan loess and paleosols since 1 m.y. B. P.

Strata	Model I ($\times 10^4$ yr.)	Model II ($\times 10^4$ yr.)
S_0	1.0	1.0
L_1	9.5	7.0
S_1	14.0	14.0
L_2	20.0	20.0
S_2	25.0	25.0
L_3	30.0	29.0
S_3	33.0	33.0
L_4	37.5	35.0
S_4	41.0	41.0
L_5	46.0	44.6
S_5	55.5	55.5
L_6	4.2	3
S_6	66.0	66.0
L_7	70.0	70.5
S_7	72.0	72.0
L_8	74.2	74.0
S_8	77.0	77.0
L_9	87.0	82.2
S_9	90.0	90.0
L_{10}	92.0	92.0
S_{10}	94.0	94.0
L_{11}	96.0	95.7
S_{11}	97.0	97.0
L_{12}		

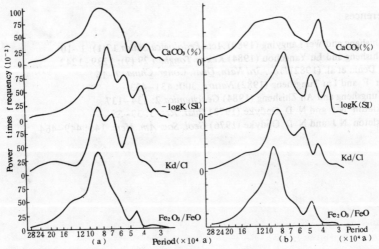

Fig. 6. Spectral characteristics of loess-paleosol time sequence within an interval of 1 m.y

12

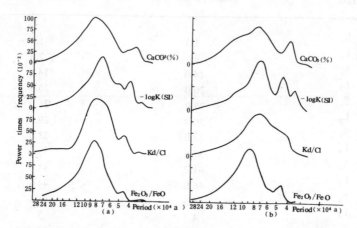

Fig. 7. Spectral characteristics of loess-paleosoi time sequence within an interval of 0.77 m.y

Plateau, the development and changes of desert and loess accumulations in China have been affecting the radiation equilibrium at the mid-latitude zone through albedo feed-back and the dust produced, which was of prime importance to the paleoclimatic evolution in northern China, and might in some extent affect the global climate.

Acknowledgement. The authors wish to thank Prof. Wen Qizhong and Mr. Lu Longhua for their help in this study.

References

An Zhisheng and Wei Langying (1980) *Acta Pedologica Sinica* 17(1): 1−10
An Zhisheng and Lu Yangchou (1984) *Kexue Tongbao* 29 (9): 1239−1242
Chen Deniu et al. (1982) *Proc. 3rd Natn. Quat. Congr. China,* 7−15
Heller F and Liu Tungsheng (1982) *Nature* 300: 431−433
Liu Tungsheng and An Zhisheng (1984) *Geochemica* 2: 134−137
Shackleton N J and N D Opdyke (1973) *Quat. Res.* 3: 39−55
Shackleton N J and N D Opdyke (1976) *Geol. Soc. Am. Mem.* 145: 449−464

An Analysis of Long-Term Variation of Storm Damage in Japan

Koichiro Takahashi[1]

Abstract — The storm damage variation in Japan over the long period of 1650–1983 has been analyzed by examining historical and other relevant data. The 70-year cycle observed in storm damage variation is probably the result of the 70-year cycle in volcanic activity.

Introduction

Every year Japan suffers considerable damage from typhoons and localized downpours during the Baiu season, the rainy season in June and July. The actual extent or amount of damage varies from year to year due to a number of elements such as climatic, social, and technological change. In this paper, I have tried to analyze storm damage variation over long period of time, 1650–1983, examining historical records and other relevant data.

Estimation of Time Series of Storm Damage in Japan

The degree of damage from storms cannot possibly be expressed by only one parameter, since there are many kinds of damage — loss of life, destruction of buildings, landslides, loss of crops, to cite a few major examples — and since the kinds of damage do not necessarily reflect the type of intensity of a storm. Nevertheless, here I have defined the degree of storm damage by the geometric mean of yearly loss of life and residential buildings due to storms, because the estimation of these figures is possible as far back as 1650, using existing historical data.

Understandably, the records of the twentieth century are accurate and reliable, but estimated values before 1880 need some adjustment, for Japan underwent drastic social changes and modernization around that time and many of the records before then were lost or not up to the modern standards. Thus, I multiplied the values obtained from historical records by a factor of 3, because in estimating the number of severe earthquakes in ancient Japan, I discovered that the figures in historical record correspond well with those of the present age when tripled.

As stated before, old records of storm damage are imperfect and sometimes only one of the two kinds of damage is listed. When this occurs, the other is estimated by the average ratio between the loss of life and loss of building.

The amount of storm damage, however, depends not only on the intensity of storms but also on the size of population and the number of buildings. For the sake of easy comparison, therefore, the degree of storm damage is calculated for the

---

1. Japan Meteorological Agency

standard population of 100 million. The results, however, still vary considerably year by year and occasionally extraordinarily large values might appear. This is inconvenient for statistical analysis, so I took the common logarithm of the amount of damage plus 1. Table 1 shows the degree of storm damage calculated in this manner for the period 1650–1983. The major sources of data are as follows:

1. Okajima, K.: Nihon Saiishi (The Chronology of Japanese Disasters), 1894
2. Marine Meteorological Observatory at Kobe: Nihon Kisho Shiryo (Historical Records of Japanese Meteorological Phenomena), 1939
3. Arakawa, H. et al.: Takashio Shiryo (Historical Records of Storm Surges), 1961
4. Japan Meteorological Agency: Kisho Yoran (Geophysical Review), Monthly journal.

Analysis of Storm Damage Variation

Figure 1 shows the degree of annual storm damage in this century, the yearly minimum atmospheric pressure on the mainland of Japan and anomalies in the northern hemisphere. We can see that the storm damage fluctuates considerably year by year. The main cause for this fluctuation is the difference in storm intensity. The minimum atmospheric pressure is an index of storm intensity, especially the intensity of typhoons. On the whole, storm damage is severe when the minimum atmospheric pressure is low. In some years however, it is severe when the minimum pressure is not low. This is because the damage is caused by localized heavy rain during the Baiu season.

Another noticeable fact is that the damage is generally heavier between 1930 and 1960. The increased damage during this period may be attributed to two causes: a climatic change in storm intensities and the deterioration of counter-measures against storms because of World War II. It is not easy to separate the two causes, but I have made the following attempt to analyze them. The equation below indicates an empirical relation between the annual amount of storm damage I and the annual minimum atmospheric pressure on Japan's mainland P, based on the observed values during 1930–1960.

$$I = 0.0144 \, (1010 - P_m)^{3.5}$$

Here I is the geometric mean of lost lives and buildings on the mainland.

This empirical formula based on minimum pressure is valid when anti-storm measures are not advanced. Therefore, the ratio between the actual amount of storm damage and the estimation derived from the formula is an index of the effectiveness of the measures taken against storms.

Figure 2 shows the 5-year mean values of this ratio. Clearly, the ratio is very high around World War II (WW II), dropping considerably in recent years. In fact, the recent figures are approximately 1/7 of the WW II figures. This proves that better anti-storm measures reduce damage. Figure 2 also shows the 5-year mean values of annual storm damage calculated by the empirical formula, which represent the amount of damage when countermesures remain the same. The curve, therefore, shows the influence of storm intensity change due to climatic change.

Table 1. Annual degree of storm damage (in logarithm)

Decade	0	1	2	3	4	5	6	7	8	9
1650	5.1	2.9	4.0	5.2	3.9	4.3	3.4	2.7	3.1	2.7
1660	2.9	1.3	3.5	2.7	2.5	2.7	3.8	2.7	3.3	4.5
1670	4.2	3.3	2.5	4.3	3.7	3.4	3.4	4.0	3.7	3.1
1680	4.6	4.0	2.5	1.6	2.5	2.5	2.5	3.0	1.6	3.3
1690	2.5	2.5	2.5	2.8	2.5	2.0	2.4	2.4	2.4	3.3
1700	2.2	2.4	4.2	3.4	2.8	2.9	2.9	3.6	2.4	1.9
1710	1.2	1.2	3.2	4.0	2.0	3.5	3.4	1.8	1.8	1.8
1720	3.3	3.6	4.6	3.0	4.1	1.2	3.2	1.6	5.0	3.0
1730	1.6	1.8	1.8	1.8	1.8	1.6	3.2	1.8	3.3	2.7
1740	4.1	4.6	4.2	3.4	3.9	3.9	2.9	1.8	4.0	4.8
1750	2.9	3.5	1.8	2.9	3.2	3.1	2.5	4.3	3.5	2.5
1760	0.0	2.5	3.3	4.2	1.9	2.5	2.1	2.0	1.9	3.1
1770	1.9	2.5	3.7	2.5	3.1	1.3	2.0	3.6	4.4	2.5
1780	2.0	3.0	3.8	34.3	0.0	2.5	2.2	1.3	3.0	2.4
1790	1.9	3.9	3.7	2.5	2.5	3.7	3.9	3.9	1.9	1.9
1800	2.0	2.5	3.7	2.5	3.8	1.4	1.3	2.1	2.4	1.9
1810	1.6	1.6	3.5	1.9	2.6	4.2	2.9	2.6	2.0	2.5
1820	2.6	2.8	3.1	3.6	3.3	3.2	2.0	2.9	5.5	2.5
1830	2.2	3.2	1.6	2.2	2.9	4.2	3.8	2.5	2.4	2.5
1840	3.9	2.5	0.0	3.1	3.2	3.3	3.2	2.5	2.8	3.9
1850	3.5	2.8	3.3	1.2	1.2	2.8	2.5	2.7	2.5	3.1
1860	3.1	1.9	1.6	1.8	1.2	2.9	3.4	1.2	4.4	1.2
1870	4.0	4.3	0.0	2.5	3.6	2.7	2.1	2.5	1.8	2.1
1880	3.1	2.4	2.8	3.4	4.4	3.1	3.2	1.4	2.2	3.7
1890	2.5	2.3	3.6	3.7	1.4	2.7	3.6	2.6	3.5	3.2
1900	2.2	2.0	3.0	2.4	2.5	2.9	3.0	2.8	2.3	2.4
1910	3.4	2.6	3.0	2.0	2.8	2.4	3.0	4.1	2.8	2.9
1920	3.0	3.0	2.6	2.2	2.8	3.0	3.6	3.2	2.8	3.2
1930	3.4	3.2	3.6	3.0	4.3	3.3	3.2	3.4	3.8	2.9
1940	3.1	3.4	4.0	3.8	3.1	4.3	3.2	3.8	3.5	3.5
1950	3.7	3.9	2.8	4.2	3.8	3.3	3.2	3.4	3.4	4.3
1960	2.6	3.6	2.8	3.2	3.2	3.2	3.2	3.1	2.9	2.6
1970	2.8	2.8	3.1	2.2	2.6	2.5	2.8	2.5	2.4	2.4
1980	2.2	2.0	2.7							

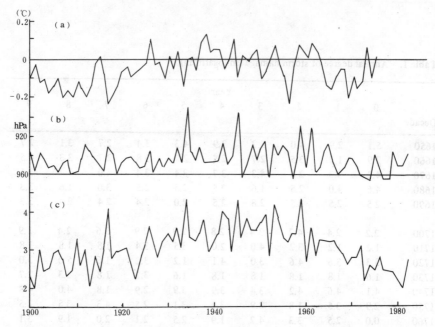

Fig. 1. Secular change of annual anomalies of northern hemispheric temperatures (a), minimum pressure (b), and degree of storm damage in Japan (c)

We see that these values are also high around WW II and recent values are approximately 1/3 of those of that period. Thus, we can conclude that the severe storm damage around WW II was the result of two factors: the changed measures against storms and changed storm intensities, the former having were influence than the latter.

70-Year Cycle in Storm Damage, Climate, and Volcanic Activity

Let us now look at storm damage variations over a long period of time. Figure 3 shows the 10—year mean values of storm damage degrees, middle latitude temperatures, temperatures in central England, and number of large volcanic erruptions worldwide. Here, the middle latitude temperatures are estimated based on observed temperatures in Japan, England, and the eastern part of the U.S.A., on tree rings in California and Germany, on historical records of famines and the condition of ice on lakes and along the seacoast. The main sources for these data are the writings of Lamb, Humphrey, and R. Yamamoto.

Looking at the graph in Fig. 3, we notice that the middle latitude temperature, temperature over England, and storm damage vary in parallel on the whole and that the volcanic activity also varies similarly, except that it is in opposite phase occurring 10—20 years in advance. Judging from the graph, we may say that the storm damage in Japan increases when the climate is warm. This tendency is also

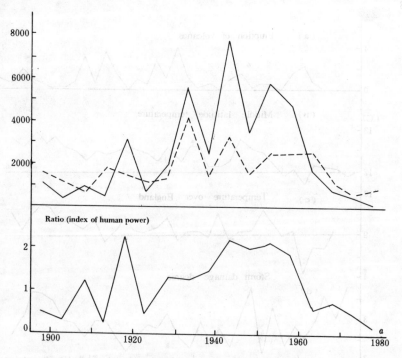

Fig. 2. Secular change in storm damage. ——— Observed storm damage; ––– Expected storm damage if no change occurs in anti-storm measures (upper figure)

observable in Fig. 1, in which changes of storm damage and temperature anomalies in the northern hemisphere are shown. Such a relationship may be explained in the following manner.

Major storm damage in Japan is caused mainly by attacks of strong typhoons and local torrential downpours during the Baiu season. The energy for both is supplied by the Pacific Ocean in the tropical zone. Hence, storm intensities will increase in a warm period, resulting in increased storm damage. It is also a generally accepted fact that a world-wide tempearature drop occurs after a great volcanic erruption, because of the umbrella effect of volcanic dust. Thus, increased volcanic activity will bring about a drop in temperature, resulting in decreased storm damage.

Examining the graph further, we also notice a 70-year cycle in the changes. In order to verify this observation, I made a 70-year periodgram analysis of volcanic activity, represented by the number of large-scale volcanic erruptions, middle latitude temperatures, temperatures of England, and degree of storm damage. The results are shown in Table 2.

Here, the corresponding year of phase is the first year of each decade, a is the standard deviation in the analysis and the analyzed period is between 1650–1980. The fluctuation amplitude of the 70-year periodgram values is fairly large compared with the standard deviation indicating that 70-year cycle has statistical significance. Also, the middle latitude temperature, temperature over

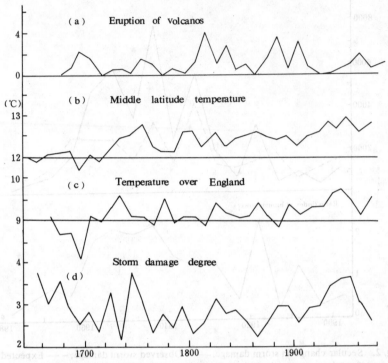

Fig. 3. Secular change of the number of large volcanic eruption over the world in a decade (a), the 10-year mean value of middle latitude temperature (b), temperature over England (c), and storm damage degree in Japan (d)

Table 2. The 70-year periodgram analysis of volcanic activity, middle latitude temperature, temperature of England, and storm damage degree

Phase, corresponding year	1900	1910	1920	1930	1940	1950	1960	ε	mean
Volcanic activity	1.8	−0.8	−1.3	−1.6	−0.4	1.6	0.4	0.6	1.6
Middle latitude temperature	−0.14	−0.12	0.06	0.10	0.08	0.10	−0.10	0.10	12.39
Temperature over England	−0.18	0.09	−0.01	0.19	0.16	−0.18	0.02	0.15	9.16
Storm damage degree	−0.32	−0.14	−0.30	0.24	0.03	0.39	0.11	0.11	2.89

England, and storm damage are in the same phase, while the volcanic activity is in the opposite phase, 10–20 years in advance of the other factors.

Conclusions

The result of my study led me to believing that there are two causes for the long-term variation in storm damage: the change in anti-storm measures and the changes of storm intensities due to climatic change; and that a 70-year cycle exists in the variations of storm damage, which tends to increase, on the whole, when the climate is warm. The 70-year cycle observable in the storm damage variation is probably the result of the 70-year cycle in volcanic activity, as indicated in my periodgram analysis, although the exact mechanism is not yet clear. Of course, there are other elements of climatic change, such as solar activity and ocean temperature, and no doubt they too influence storm damage variation. Still, the present analysis indicates that the influence of volcanic activity on climatic change, hence on storm damage in Japan, is considerable over a time scale of several decades.

Drought/Flood Variations for the Last 2000 Years in China and Comparison with Global Climatic Change

Wang Shaowu [1], Zhao Zongci [1], Chen Zhenhua [1] and Tang Zhongxin [2]

Abstract — The climatic change is reviewed for the last 2000-year period. The historical documentations are analyzed, extending the drought/flood type series back to 951 B.C. The serious droughts and floods in Changjiang River and Huanghe River areas are examined. The climatic trends in the Little Optimum and in Little Ice Age are compared.

Introduction

Mary historical documentations in China describe the disasters and damages caused by droughts and floods. Some of them were used in reconstructing the rainfall regime and studying the climatic change for the period from 1470 to 1970 (Wang Shaowu and Zhao Zongci 1981; Wang Shaowu et al. 1981). The publication of drought/flood maps for the last 500 years is a good example of how one can understand the climate of several hundred year age in detail (Meteorological Institute 1981). Further study of the historical data indicated that it is possible to extend the scope of research to about 2000 years ago.

Table 1. Examplex of historical documentations of meteorological disasters in ancient times

---

180 B.C.

  The Changjiang River and Han River overflowed, more than 10,000 homes were devastated.

29 B.C.

  Heavy rain persisted more than ten days, the dyke of the Huanghe River was breached About 97,000 people were moved to the hills by 500 boats and ships.

309 A.D.

  There was serious drought in June, the Changjiang River and Huanghe River were drained. People can wade across the river.

---

1. Peking University, Beijing, China
2. Meteorological Bureau of Hebei Province, China

Table 1 gives three examples, which are quoted from the "Tian zai ren huo biao" (Table of natural and man-made calamities), which contains the description of all kinds of calamities registered in the period starting from 246 BC, that is the first years of the Qin Dynasty, the first dynasty that unified nearly the whole of China. Of course, these data are not the earliest record of the climatic anomalies in the ancient times. In the "Gu jin tu shu ji cheng" (The anthology of ancient and modern books), a special volume is dedicated to astronomical and meteorological phenomena, with an account of the catastrophic droughts for the 6-year period from 1765 B.C. to 1760 B.C. in the Shang dynasty, the dynasty before Qin. But the aim of the present paper is to compile a successive series of droughts and floods, so 200 B.C. is adopted as the beginning of the research period, because since then more or less successive data are available. In this research more than ten data sets of droughts and floods for local provinces are extracted, which were compiled by the local meteorological bureaus based on local chronicles.

The present paper is a preliminary report of the research program aimed at reconstructing the climate in historical times and revealing long-term climatic variations. This kind of work, we believe, will also benefit the understanding of the interactions between climate and boundary surface such as snowcover, sea-surface temperature, and the transmissibility of the atmosphere. It will facilitate the design of climatic models and improve the theory of climate.

Drought and Flood Types up to 950 A.D.

The chronology of drought and flood types for the last 500 years was constructed in 1979. It was presented in the Climate and History Conference held in East Anglia University in July 1979 (Wang Shaowwu and Zhao Zongci 1984). The characters of drought/flood types are outlined briefly in Table 2.

Table 2. Characters of droughts and floods by types

1a	Floods over all China, but mainly in the Changjiang River region
1b	Floods in the Changjiang River, droughts to the north and the south of it
2	Floods in the south and droughts in the north
3	Droughts in the Changjiang River region, floods to the north and the south of it
4	Floods in the north and droughts in the south
5	Droughts over almost all China

Table 3. Choronology of drought/flood types for 950–1980 A.D.

	0	1	2	3	4	5	6	7	8	9
950	1_b	3	4	4	3	4	2	2	4	1
960	2	1_a	1_b	1_b	2	1_b	1_a	4	4	4
970	1_b	1_a	4	4	1_b	2	2	1_b	1_b	
980	1_b	2	1_a	1_a	2	3	2	1_a	2	
990	2	2	5	4	1_a	3	2	3	3	2
1000	3	3	4	2	5	3	2	1_a	2	3
1010	5	4	1_a	1_b	1_b	5	2	4		
1020	4	3	1_b	1_a	1_b	2	1_a	1_a	1_b	4
1030	2	4	5	5	1_a	1_b	1_b	2	2	3
1040	2	5	2	1_b	2	1_b	2	1_a	2	
1050	1_a	2	4	2	4	3	1_a	4	1_a	1_b
1060	1_b	1_a	3	4	1_b	3	2	2	1_b	3
1070	5	2	1_b	5	5	5	2	4	3	2
1080	2	4	4	1_b	3	1_b	4	2	5	2
1090	1_b	1_a	2	4	1_a	3	3	2	3	1_a
1100	4	5	5	2	2	2	1_b	4	3	5
1110	3	3	2	2	3	4	1_b	1_a	1_b	
1120	4	4	1_b	5	4	4	4	4	2	1_b
1130	3	5	2	2	3	3	4	3		
1140	5	5	5	4	1_b	4	3	5		
1150	4	1_a	1_b	1_b	3	1_b	3	1_a	4	1_a
1160	3	2	2	5	1_b	4	4	1_a	4	3
1170	1_b	3	2	3	2	3	2	1_a	3	3
1180	3	3	3	5	4	2	4	4	1_a	1_a
1190	3	2	1_a	4	3	3	2	5	2	2
1200	3	5	2	5	5	3	5	5	3	
1210	5	2	2	2	2	5	1_b	2	5	1_b
1220	4	1_b	2	1_a	1_a	2	2	2	4	4
1230	1_b	1_b	1_a	4	1_b	2	1_a	3	2	5
1240	5	3	1_b	2	5	3	3	3	5	2
1250	1_a	1_b	3	5	1_b	1_a	3	3	3	1_b
1260	3	1_b	2	2	3	4	1_b	1_a	1_a	3
1270	2	1_b	1_a	4	4	2	5	3	2	2
1280	2	2	2	4	4	1_a	4	4	1_a	1_a

Table 3 (*contd*)

1290	1_a	3	1_b	2	1_a	1_b	1_b	2	4	3
1300	1_b	1_a	3	2	3	2	5	3	4	1_a
1310	1_a	1_a	1_b	4	4	1_a	1_b	2	4	1_a
1320	3	2	3	4	4	1_a	1_b	1_b	2	5
1330	2	2	2	3	3	4	5	1_a	1_a	1_a
1340	1b	1_b	2	4	4	2	2	2	1_a	1_a
1350	2	4	5	4	4	2	1_a	3	2	5
1360	3	2	3	1_b	4	1_a	3	2	1_a	2
1370	2	2	2	2	5	5	1_a	1_a	4	4
1380	1_a	4	4	4	1_a	1_b	1_a	4	2	1_b
1390	4	3	4	2	1_a	1_a	4	4	2	4
1400	4	3	3	1_b	1_b	1_a	2	4	1_b	1_a
1410	1_a	1_a	1_b	1_b	4	3	1_a	2	1_a	4
1420	3	2	1_a	1_a	1_a	1_a	3	2	1_a	2
1430	4	4	4	5	5	5	1_a	2	3	4
1440	3	5	2	2	4	1_a	3	5	3	1_a
1450	2	5	4	1_b	2	3	3	4	3	3
1460	1_b	4	1_b	2	2	1_a	3	2	5	1_a
1470	2	1_a	2	2	1_a	1_a	4	3	3	2
1480	2	2	3	2	2	2	1_b	5	3	3
1490	2	1_b	2	2	1_a	2	1_b	2	1_b	1_b
1500	1_b	1_a	1_a	3	2	2	1_b	5	3	5
1510	1_a	1_a	5	5	4	1_b	2	1_a	1_a	1_a
1520	3	2	2	3	2	5	4	2	5	5
1530	3	1_b	3	2	3	3	4	1_a	4	2
1540	2	5	3	4	4	5	3	3	1_a	3
1550	1_b	3	1_a	3	3	1_b	3	1_a	1_b	3
1560	1_b	1_b	1_a	1_a	1_a	2	1_b	1_a	2	1_a
1570	1_a	1_a	2	2	1_a	1_a	3	1_b	1_b	1_b
1580	1_a	2	2	3	2	5	2	2	5	3
1590	5	1_b	4	4	4	4	2	1_a	4	2
1600	2	1_b	1_a	4	4	1_b	3	1_a	1_b	2
1610	1_b	2	1_b	1_a	1_a	2	2	2	5	2
1620	2	3	3	4	4	5	4	2	5	5
1630	3	4	3	2	2	2	5	2	2	2
1640	2	5	5	5	3	4	4	3	4	1_b
1650	2	1_a	3	4	4	1_b	1_a	4	1_a	4

Table 3 (*contd*)

1660	4	2	4	1_a	4	5	5	3	1a	1b
1670	1_b	3	3	1_b	2	1_a	1_a	3	3	3
1680	1_b	4	1_b	1_b	4	4	3	1_b	3	2
1690	2	5	3	3	3	1_a	1_b	4	1_b	3
1700	4	3	4	4	2	1_a	2	3	1_b	1_a
1710	1_b	5	2	2	2	1_a	1_b	3	1_a	2
1720	2	5	2	2	4	1_a	1_a	1_a	4	4
1730	1_a	1_a	1_a	3	1_a	3	1_a	3	5	4
1740	3	1_b	1_b	1_b	3	4	2	4	5	1_a
1750	3	4	3	4	1_a	1_a	1_a	4	1_b	2
1760	1_b	1_a	1_b	1_b	2	2	1_a	1_a	3	1_b
1770	3	3	3	1_a	2	3	1_a	1_b	5	4
1780	3	4	3	2	2	5	5	1_b	1_b	4
1790	4	1_a	2	2	2	1_a	3	3	4	1_a
1800	1_a	1_a	5	1_b	2	1_b	3	5	4	3
1810	4	5	2	2	5	4	4	2	3	3
1820	4	3	4	1_a	3	2	1_b	1_b	4	3
1830	4	1_a	1_b	1_b	2	5	5	2	1_b	2
1840	1_a	1_a	2	4	1_a	3	2	2	1_a	1_a
1850	1_a	4	4	1_a	3	3	3	5	2	2
1860	1_a	2	2	3	3	1_b	1_b	5	4	1_b
1870	1_b	3	3	3	3	1_b	2	2	2	3
1880	2	2	1_b	4	4	1_a	1_a	4	3	1_a
1890	3	3	3	4	4	4	4	4	3	2
1900	5	1_a	5	1_a	3	1_b	1_a	5	2	1_b
1910	4	1_a	1_b	3	3	1_a	1_b	4	3	2
1920	2	4	3	4	2	5	1_b	5	5	5
1930	5	1_b	3	4	5	2	2	3	1_b	2
1940	4	5	2	2	3	5	2	3	2	3
1950	3	5	2	3	1_a	2	4	1_b	4	3
1960	5	5	1_a	4	4	5	5	4	2	1_b
1970	2	4	5	3	5	2	3	3	3	1_b
1980	1_b	3	1_b	1_b	1_b					

To extend the type series from 1470 backwards it is necessary, first of all, to decide the drought/flood degree by stations; but the scarcity of early data forced us to use the regional degree instead of the stational one. Therefore, 10 regions are incentified and illustrated in Fig. 1, covering nearly the same area as the 25 station covered.

Fig. 1. Regionalizaiton of drought/flood analysis

According to the regional degree of drought/flood, the chronology of the types is extended from 1469 A.D. back to 950 A.D. At the same time, the types after 1470 A.D. are reviewed with the additional data and some revision is made for a few years. The chronology for the whole 1000 years is given in Table 3. The frequency of types is shown in Table 4, separately for 950–1469 A.D. and 1470–1984 A.D., the lower referring to the first and second period. Before 950 A.D. the data availability is too poor to give a definite indication of the drought/flood type, so the type series extended only to 950 A.D. Table 4 gives flood types, so the type series extended only to 950 A.D. Table 4 gives the frequency of types separately for the two periods. It shows that the frequency for the first period is generally similar to that for the second period, though some discrepancy can be found; for example, the frequency of type 2 and 4 in the first period is higher than in the second, but lower for types 3 and 5.

Table 4. Frequency of drought/flood type

	1_a	1_b	2	3	4	5
950–1469 A.D.	88	73	121	93	94	51
	(16.9)	(14.0)	(23.3)	(17.9)	(18.1)	(9.8)
1470–1984 A.D.	83	77	112	104	81	58
	(16.1)	(15.0)	(21.7)	(20.2)	(15.7)	(11.3)
950–1984 A.D.	171	150	233	197	175	109
	(16.5)	(14.5)	(22.5)	(19.1)	(16.9)	(10.5)

26

It has been found (Wang Shaowu et al. 1981) that the areas with positive anomaly of precipitation usually migrate from north China to the south in a cycle related to the solar activity. Examination of the 1000-year series of types shows that this kind of variation of drought and flood holds true for both the first period and the second. In Fig. 2 the 50-year frequency of type is given. The upper curve is the frequency combined type 3 and 4, both imply a flood in the north China; the middle and lower curves are the frequency of type 1 and type 2, which represent the flood along the changjiang River and the flood in the south China respectively. The broken lines join the maxima of curves, indicating that the maxima are more or less linked to each other. The tilt of broken lines suggested that the flood usually appeared early in the north (as seen from curve), then along the Changjiang River and in the south, so the migration of the positive anomalies of precipitation from north to south seems to have occurred in the first period.

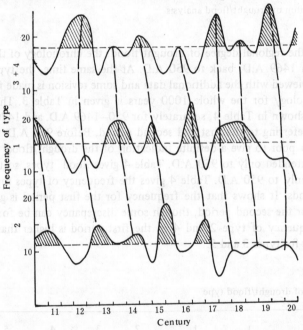

Fig. 2. Half-century frequency of types 3 and 4 (upper), type 1 (middle) and type 2 (lower)

Serious Droughts and Floods from 200 B.C. to 1984 A.D.

Before 950 A.D. the drought/flood degree series can be followed only for the 4th and 7th regions (Fig. 1). These two series are used as indicators of drought/flood along the Huanghe and Changjiang Rivers. Even so, it is difficult to obtain a successive series. Therefore, only the serious drought/floods are investigated, nearly equal to the 5th degree (strong drought) and lst degree (severe flood), but with greater area coverage. In the five-degree system as indicated in the previous paper

(Wang Shaowu and Zhao Zongci 1981), for any single station both the frquency of 1st and 5th degree is about 12.5%, so their sum is about 25%. However, the frequency of serious drought and flood defined here is much less, because only the drought/flood occupying broad areas rather than in a single station is considered as serious. As a result, 196 drought and 147 flood years are identified from 200 B.C. to 1984 A.D. in the Huanghe River, and 112 droughts and 178 floods in the Changjiang River for a shorter period, 200 A.D. to 1984 A.D. In Table 5 the numbers of serious droughts and floods are given in the row with letter D and F. It shows that the drought is more frequently observed in the Huanghe than in the Changjiang River, and on the contrary, the frequency of flood is greater in the Changjiang than in the Huanghe River. It corresponds well to the climatological characters for these regions.

Table 5. Frequency of serious droughts (D) and flood (F)

	D	F	D+F	D/(D+F)	(D+F)/Σ	Σ
Huanghe River	196	147	343	0.57	15.7%	2184
Changjiang River	112	178	290	0.39	15.4%	1884

The frequency of serious drought and flood as a whole is 0.157 and 0.154 for the Huanghe and Changjiang Rivers respectively. This means that a serious drought or flood, can generally be found once in 6 years. Of course, the droughts and floods mentioned above contain some most serious droughts or floods, that may occur once in a century or in 1000 years. But in the present stage of research they are not distinguished from the "general" serious droughts or floods.

Figure 3 gives the relative frequency of serious droughts, denoted by D/(D+F). It is used instead of D itself, because we are not sure that the influence of war and society was fully eliminated in a regional analysis. This kind of influence can sometimes significantly reduce the number of record, so some serious drought or flood can be missed. Therefore, the retiao D/(D+F) is used. Figure 3 shows the long-term variation of drought/flood for about 2000 years. Both curves obviously show the dry and wet period. Generally, there are four dry periods 4—6th, 11—13th, 15th, and 19—20th centuries in the Huanghe River, and four dry periods, 4th, 9th, 11—13th and 15—17th centuries in the Changjiang River.

Comparison of Climatic Change in China with Global Climatic Trend

Climatic change in historical times is a subject to which a great number of famous climatologists have contributed. However, the discussion has long been focused on temperature; only a few research works, such as the excellent monograph "Climate, present, past and future" (Lamb 1977), are concerned with precipitation. In this book Lamb indicated that the Little Optimum in the Middle Ages was a dry period over England, but a moist period in the Mediterranean. These climatic character-

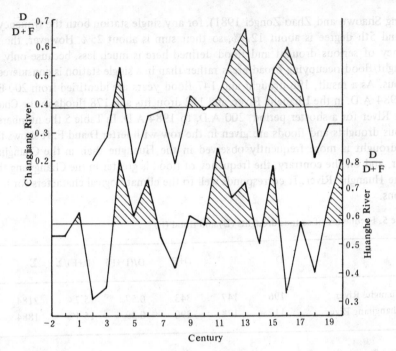

Fig. 3. Relative frequency of drought/flood D/(D+F) for Changjiang River (upper) and Huanghe River (lower)

istics were explained by the northward movement of depression tracks in north Europe and eastward progress of the European trough with a lengthening of the wave. On the contrary, in the period of the Little Ice Ages, the climate in England was cold and wet, which was, as suggested by Lamb, formed under the prevailing circulation with opposite characteristics.

So, it is of interest to examine the climatic characters in China for these two time intervals. Therefore, two epochs are compared, one covering the 11th and 12th centuries, that corresponds to the Little Optimum; the other, which lasted from 1550 A.D. to 1750 A.D. roughly represents the Little Ice Ages. Table 6 gives the frequency of drought/flood types in these two epochs. It shows that in epoch 2 type 1_a (flood over whole country) occurred with a frequency of 12 years more than in epoch 1, but the frequency of type 5 (drought over whole country) was higher in epoch 1. This difference is obviously outlined in Table 7, in which the frequency of serious drought and flood for these two epochs is compared.

It is clear that epoch 2 was characterized with the wet summer in China, and epoch 1 was a dry period. It is well konwn that most of the summers with serious floods along the Changjiang River are cold. 1755 A.D. was a typical example (type 1_a in Table 3). It was learned from local chronicles that in this summer the people wore cotton padded jackets. It is interesting to note that 1755 A.D. was one of the four damage years in Japan. As Yoshino (1978) indicated, in the Little Ice Ages Japan suffered from cold-wet summers. He suggested that the Okhotsk high may

Table 6. Frequency of drought/flood types in the Little Optimum (epoch 1) and in the Little Ice Ages (epoch 2)

	1_a	1_b	2	3	4	5
Epoch 1(1000–1199 A.D.)	25	30	44	42	36	23
Epoch 2(1550–1750 A.D.)	37	32	42	36	34	18
Normal	33	29	45	38	34	21

Table 7. Frequency of serious drought and flood in the Little Optimum (epoch 1) and in the Little Ice Ages (epoch 2)

	Huanghe River		Changjiang River	
	D	F	D	F
Epoch 1(1000–1199 A.D.)	26	8	13	13
Epoch 2(1550–1750 A.D.)	17	20	10	18
Normal	18	13	12	18

be well developed in this epoch. This is in full agreement with our result. Okhotsk high is the main feature of the May-u situation, the situation with flood along Changjiang River. The evidence both in Japan and in China proves that within the Little Ice Ages the trough in East Asia may be placed near Lake Baikal and the blocking high occupied the east coast of Asia.

References

Lamb HH (1977) Climate: present, past and future. vol. 2. Methuen London, 835

Meteorological Institute (ed) (1981) Drought/flood maps for the last 500 years. State Meteorological Administration China. Cartography Ditu Chubanshe, p 332

Wang Shaowu, Zhao Zongci (1981) In Wigley et al. (eds) Climate and History. Cambridge University Press, 271–288

Wang Shaowu, Zhao Zongci, Chen Zhenhua (1981) Geojournal 5: 117–122

Yoshino MM (1978) Climatic change and food production Univ Tokyo Press, pp 331–342

A Preliminary Study on Long-Term Variation of Unusual Climate Phenomena During the Past 1000 Years in Korea

Kim Gum Suk[1] and Choi Ik Sen[1]

Abstract — In this paper, the variations in climate conditions for spring drought and warm or cold weather during the past 1000 years in Korea are discussed, using the unusual climate data included in historical records.

Introduction

In Korea, there are many descriptions of unusual climate phenomena in various historical documents (Social Science Academy 1978; Tamura 1960). Among these, records after 1000 A.D. are particularly abundant and in regular sequence. The records on drought and unusual temperature phenomena from many such records on unusual climate phenomena were used in this paper.

The records which involve descriptions of drought, frozen sea during spring and winter season, snowfall and frost occurring during the summer season and the effects of unusual climate on plant growth are valuable, and can be compared with recent climate data.

Drought in Spring During the Past 1000 Years

Records on drought can almost be used for identifying the period of nonprecipitation. In the present paper, intensity of drought in spring is characterized by the length of the nonprecipitation period from April to June. An index on drought during some decades is made by decade summation of the months (in 0.5 month unit) for a continuous drought year.

Both rainfall data and drought records from historical documents are available for the period 1771 A.D.–1910 A.D. and thus the inter-relation between these was investigated. Figure 1 shows a relation between the square root of decade mean rainfall in Seoul from April to June and the summation of drought months during the same period. As shown in Figure 1, the inter-relation is in good linear relation Therefore we could transform rainfall data during 1911–1980 into data of summation of drought months by using the above relation.

The decade summation of months of spring drought during 1011–1980 is shown in Fig. 2.

As shown in Fig. 2, the 15th and 16th centuries are the periods during which droughts occurred most frequently in spring over the last 1000 years. In parti-

---

1. Hydro-Meteorological Research Institute of Hydro-Meteorological Service, Pyongyan, D.P.R. Korea

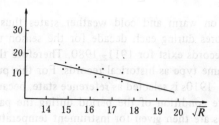

Fig. 1. Relation between decade mean rainfall and decade summation of months of spring drought

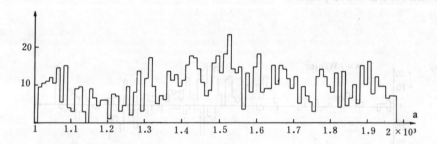

Fig. 2. Decade summation of spring drought months.

cular, the 16th century shows spring droughts over 30 years continuing for 3 months and in the 1520, 9 years (except 1524) are years of drought persisting for 2–3 months. Drought persisting for 3 months in the spring occupy 16% of the 1000 years, while this means one occurrence per 6 years.

The period 1881–1910 is a severe spring drought period during the last 100 years and the period 1901–1910 especially is the most severe spring drought decade during the last 400 years. However, the 20th century has not been a severe drought period over the past 1000 years, but a period of rather frequent spring droughts.

Variation of Unusual Warm and Cold Climate During the Past 1000 Years

It is characteristic of historical records that descriptions of unusual warm weather phenomena are more from the cold season, while those of unusual cold weather phenomena are from the warm season. Therefore the estimation of average temperature conditions from such records is not easy. However, we could make of an index on warm and cold weather states based on historical records. At first, each month is classified into 4 classes according to grades of warm (+) and cold (−)

32

states. Index values on warm and cold weather states thus are determined by summation of the scores during each decade for the season under consideration.

Only instrument records exist for 1911–1980. Therefore these records must be transformed to the same type as historical records. For this purpose, the temperature condition in the 1910s is selected as reference state, because it corresponds to the mean temperature condition of the period during the past 1000 years (Chu Kochen 1973). Scores are then given for instrument temperature data. The scores of each month are thus determined by the same evaluation method consistently over the past 1000 years.

Figure 3 (a, b) show the decade summation of scores only for winter seasons and spring-summer-autumn seasons whose months have a score of more than 2 points (or less than 2 points).

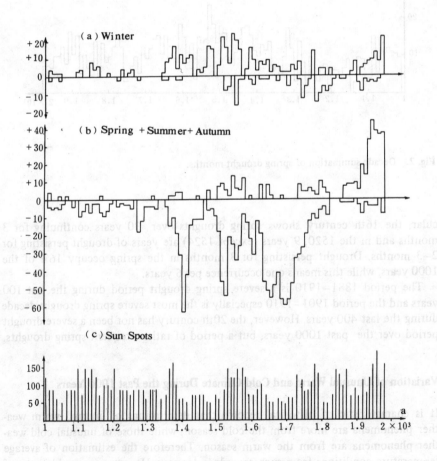

Fig. 3. (a) and (b) are the decade totals of scores. (c) is maximum values of relative sun spot number for 11-year periods after Shove (Lamb 1972)

Historical records on unusual warm and cold weather from the end of the 14th century to the 18th century are abundant, since the period was an age when astronomical and meteorological works of the Li dynasty were being carried out actively and systematically. However, records in the 19th century are rather poor for various reasons. Such differences depending on ages is, however, comparatively small in severe unusual weather phenomena having scores of ±2. We therefore think that the severe weather conditions of all ages can be compared to each other.

We also referred to complementary data of solar variability, annual tree rings, and historical climate records of neighboring countries.

The relation between solar variability and unusual temperature condition

The histogram of Figure 3(c) shows maximum values of relative sun spot number for 11- year periods given by Shove (Lamb 1972). If Fig. 3(c) is compared with Fig. 3(a,b), it is obvious that the climate in Korea was also cold in the Mounder Minimum (1645–1715) and the Supeler Minimum (about 1500). Then the cold period before and after 1900 with smaller sun spots can be identified by mordern instrumental records.

Winter seasons in the early 19th century corresponding to periods of weak sun activity were cold, whereas it is not obvious whether summer seasons too were cool or not.

Cool weather phenomenon in summer seasons in the Korean East Sea shore region

This phenomenon occurs owing to intensifving of the Okhotsk high toward south or south-west directions in spring-summer seasons and is a common weather phenomenon seen at Hokkaido and Tohoku in Japan as well as the northern region of Korean East Sea shore.

Wonsan is situated in the shore region of the Korean East Sea and the observation there has been carried out since 1905. We could select 17 years in which in summer seasons the mean temperatures in Wonsan and Nemuro in Hokkaido, Japan were lower than the normal temperature by the amount of the standard deviation or a value larger than 1.Sixteen years except one year (0.0) were years in which both stations showed negative deviation in summer seasons. This indicates that the cool weather phenomenon mentioned above possesses simultaneously over wide region.

If we select a year in which severe damages were caused to the harvest of agricultural products by cool summer weather from the historical records (Takahasi 1977) on cold weather damages at Tohoku and Hokkaido in Japan for various periods, the results are as in Table 1. Table 1 also shows the numbers of years when Suwa lake situated north-west of Tokyo did not freeze fully during the winter season (Arakawa 1975).

Correlation of season mean temperature between Pyongyang and Tokyo is considerably high and particularly so in winter seasons ($r_{win} = 0.781, r_{sp} = 0.616, r_{sum} = 0.711, r_{at} = 0.744, n = 5$). Therefore it is considered that a significant relationship exists between temperature in winter in Korea and the freezing of

Table 1. Cool weather damage in summer seasons at Tohoku and Hokkaido of Japan and number of years when Suwa lake did not freeze throughout the (year/10 year mean)

	1631–1740	1741–1840	1841–1880	1881–1930	1931–1970
Duration (years)	110	100	40	50	40
Number of years of cool weather damage	13/1.2	20/2.0	2/0.5	6/1.2	5/1.2
Number of years Suwa lake not frozen	5/0.5	9/0.9	8/2.0	3/0.6	3/1.3[a]

a Statistics untiil 1954

Suwa lake. As shown in Table 1, cool weather phenomena in the summer of the period 1841–1880 were not frequent when compared with the cases before and after the period and in this period, warm winters were more frequent.

Little Ice Age in Korea

Many climate characteristics of the last Little Ice Age have been described for various places of the world. The growth of glaciers in Little Ice Age also suggests cool summers as well as severe cold winters.

No record of orographic glaciers exists in historical records of Korea. Nevertheless unusual cold weather phenomena such as freezing of the Korean East Sea and frost and snow during warm seasons had occurred frequently in the Little Ice Age, while these phenomena have not happened once during modern instrumental registration.

Table 2 indicates the number of years in which frost and snow phenomena occurred during each period.

Features of warm and cold states will be considered according to each period as follows:

Before 1380 A. D.

According to Lamb (1972), the period 1000–1200 was warm over all the Northern Hemisphere and records on icing of Iceland show that winter seasons especially in this period were warm. It is likely that the winter seasons were relatively warm during 1000–1200, although records with respect to this period in historical documents of Korea are scarce .

Table 2. Years in which frost and snow phenomena occurred in warm seasons in middle and south Korea (frost/snow)

Period		1381– 1420	1421– 1520	1521– 1630	1631– 1740	1741– 1840	1841– 1880
Spring	Number of years	16/4	8/10	18/19	24/4	6/6	0
	Decade mean	4.0/1.0	0.8/1.0	1.6/1.7	2.1/3.7	0.6/0.6	0.0
Summer	Number of year	8/2	5/4	13/3	34/23	4/3	0
	Decade mean	2.0/0.5	0.5/0.4	1.2/0.3	4.3/2.1	0.4/0.3	0.0
Autumn	Number of years	2/4	8/12	5/11	11/15	0/1	0
	Decade mean	0.5/1.0	0.8/1.2	0.4/1.0	1.0/1.4	0.0/0.1	0.0

1381 –1420

In this 40-year period, warm seasons were cold and snow and frost frequent, whereas winter seasons were relatively warm and severe cold winter did not occur. However, the warm half year was very cool, and this indicates features of the Little Ice Age.

1421–1520

As shown in Fig. 3(a,b), this period was a relatively warm one in winter seasons as well as in summer seasons.

1521–1630

Cold winters and warm winters occurred repeatedly, severe cold winters also happened often and Korean East Sea froze even 5 times (winter, spring of 1555, spring of 1565, winter, spring of 1599). In late spring and even summer, frost and snow phenomena were frequent and features of the Little Ice Age are noticeable.

1631–1740

The warm seasons of this period were the coolest of the past 1000 years. Occurrence of 10 years mean frosts and snows are 4.3 years and 2.1 years respectively in summer seasons and the Korean East Sea froze 4 times (spring of 1654, 1659, spring and winter of 1707). This period seems to be the peak of the Little Ice Age in Korea.

1741–1840

Freezing of sea occurred once and severe cold winters were few, although unusual low temperature phenomena were frequent according to records of unusual warm and cold weather phenomena. The summer cimate condition also recovered considerably and the temperatures rose. However, frost phenomena in summer seasons did not vanish fully and cold weather damage phenomena were also still frequent. Therefore this period is one in which some features of the Little Ice Age still remained partially.

1841–1880

This period is identified as a warm one. In this period, years when the Suwa lake did not freeze occurred often, and records of severe cold winters do not exist. Cold weather damage in summer in the Korean East Sea region took place rarely and rainfall in Seoul was relatively heavy. Records of frost and snow in summer seasons could also not be found. This period can be identified as a warm and wet age also from data of annual tree rings. Winters in this period were not so warm as after the 1950s of the 20th century, but this period can be regarded as a warm age.

1881–1930

It is obvious from instrumental data of Korea and its neighboring countries that both summer and winter during this period were cold. However, this period is identified as a cold age because features of the Little Ice Age like snow and frost in summer are not found.

1931–

Winter was very warm and annual mean temperature was also high for 30 years from the 1950s, entering a global cold period to 1970s. Temperature of summer and autumn was temporarily low during the period from the latter half of the 1960s to the first half of 1970s and then recovered. On the whole, this period is regarded as the warmest one during the past 1000 years.

Defining warm and cold features of winter and summer seasons according to various periods, the result is as in Table 3.

Table 3. Cold periods and warm periods in Korea

Periods	Classification	Features of winter/summer
1381–1420	Cold period	w/c
1421–1520	Warm period	w/w
1521–1630	First period of Little Ice Age	c(w)/c
1631–1740	Second period of Little Ice Age	c/cc
1741–1840	Third period of Little Ice Age	c/c
1841–1880	Warm period	w/w
1881–1930	Cold period	c/c
1931–	Warm period	w/w

w, warm; c, cold or cool; cc: very cold or very cool

As mentioned above, we defined as the latest Little Ice Age in Korea the period of 320 years from 1521 to 1840, based on extremely cold and warm weather records from historical documents. The Little Ice Age was then divided into three periods, depending on cold and warm features of winter and summer seasons.

We may define interglacial periods, but in future this must be proved by more abundant data.

This paper is a preliminary study on the Little Ice Age and spring drought in Korea, and will be complemented by more elaborate future study.

References

Arakawa H (1975) Climate change. Hydrometeorology Press. Leningrad (in Russian)

Chu Kochen (1973) A preliminary study on the climatic fluctuation during the last 5000 years in China. *Sci Sin* 16(2)

Lamb H (1972) Climate present, past and future. *vol 1*. Fundamenentals and climate now, Methuen & Co Ltd, London.

Social Science Academy (1978) Unusual climate data book during the past 2000 years in Korea. *vol 1, 2* (in Korean)

Takahasi K (1977) Climate change and food. Daimeido Press Tokyo. (in Japanese)

Tamura S (1960) Stuay on history of meteorology in Korea. Science Academy Press. (in Korean, translated by Li Yong Tae)

Three Cold Episodes in the Climatic History of China

Zhang Peiyuan[1] and Gong Gaofa[1]

Abstract — An attempt is made to identify some of the main characteristics of the previous three cold episodes in China in different time scales and in comparison with other parts over the world to show some specific features of climatic changes in the monsoon area.

Late Pleistocene

One way of reconstructing former climates and therefore of recognizing climatic change is the reconstruction of former biota. This can be done by analyzing those parts of these biota that fossilized in suitable environments.

Woolly rhinoceros (*Coelodonta antiquitatis* Blumenbach) and woolly mammoth (*Mammuthus primigenius* Blumenbach) were believed to be typical animals of the ice age. They lived in 12 – 14 × 1000 years BP over the tundra or plateau. Their activities were restricted according to the extension or retreat of the ice sheet or glacier.

During the Late Pleistocene in Europe, the woolly rhinoceros distributed as far south as 36°N (for example: near the Black Sea), but in the eastern part of China, the farthest southern extension is 33°N (Aba, 33°N, 102°E and Su Shien, 33°N, 117°E).

The southern limit of woolly mammoth in Europe is about 40°N, and in China the fossil sites can be expected to occur as far as 38° – 39°N (Chow 1978).

The presence of these two specific animals of glacial climate in more southern parts of China than in the western part of Eurasia probably indicates the more severe condition during the ice age.

Sub-Boreal–Sub-atlantic

The time dividing the Sub-boreal and Sub-Atlantic period is believed to be 3000–2000 B.P. (Lamb 1981, p 135) wrote that by about 500–200 B.C. the long-term average temperature in Europe was about 1°C lower, and in the south-western part of North America about 0.5°C lower, than it had been in the warmest post glacial period.

Based on the archeological excavations, there were large areas near Huanghe (Yellow River) heavily vegetated by bamboo about 5000 years ago. Now, only the Changjiang valley is vegetated by bamboo. The difference in annual mean temperature between these two areas is about 2°C, which shows that the range in

---

1. Institute of Geography, Academia Sinica, Beijing, China

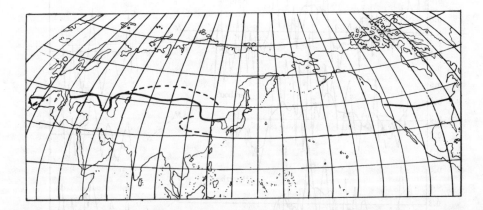

Fig. 1. South limit of *Mammuth primigenius* (bold line) and *Coelodonta antiquitatis* [broken line; modified after Chow (1978)]

temperature is larger in China than in North America and Europe (Zhang 1982).

A Warm Epoch in the Little Ice Age

The middle and later seventeenth century provides the earliest instrument observation records. These, like the evidence of glaciers in many parts of the world and of Arctic sea ice, indicate a colder climate than that of the twentieth century.

In China, many ancient weather records were incorporated in personal or official diaries and so have been preserved.

The data were daily records of notes of clear sky and hours of rainfall or snowfall, which were taken in three cities in the lower Changjiang (Yangtze River) valley namely Nanjing (1722–1785), Suzhou (1725–1782), Hangzhou (1723–1769), and Beijing (1724–1903).

Bearing in mind that the number of snow-days depends on both atmospheric moisture and temperature, in humid regions, because of the abundance of moisture, the snow-days may be determined to a large degree by the temperature.

The ratio of snow-days to the sum of snow-days and rain-days is correlated to the temperature in winter (December-February). The ratio can eliminate the effects of variation in moisture conditions. Modern instrument observations were used to evaluate the relationships between these two variables.

The same tests were made in Beijing and there are no significant relations, which can be interpreted by the following facts. The days with precipitation in winter (Dec. – Feb.) over the lower Changjiang valley are 25 (in Suzhou), 26 (Nanjing) and 38 (Hangzhou) respectively, almost 1 precipitation-day in every 3 days. The ratio of snow to precipitation days provides information about the temperature. But in Beijing, because of the semi-arid climate, there are about 9.4

40

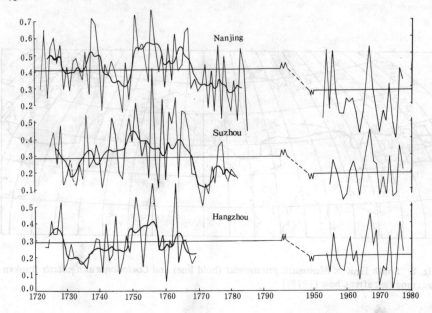

Fig. 2. Changes in ratio of snow and precipitation in Hangzhou, Suzhou and Nanjing.

Table 1. Correlative coefficient amongst ratio of snow-days and precipitation-days and average temperature in three cities.

Stations	Duration	Years	Coefficient
Hangzhou	1954–1978	24	−0.6675
Suzhou	1955–1978	23	−0.7510
Nanjing	1953–1978	25	−0.7201

snow-days in 3 months, that is only 1 precipitation day in every 10 days, and no evident relationship could be found between the ratio and temperature.

Regression equations between this ratio and the temperature of Hangzhou, Suzhou and Nanjing have been developed on the data of the last 30 years and the evidence given in Table 1. The temperature in winter (December-February). The

Changes in ratio of snow and precipitation are shown in Fig. 1. The general trend for these three places appears that the ratio in 20's–70's in the 18th century was 10–15% higher than that of today. This means there were more snow-days during that time than at present. Based on regression equations, ancient winter temperatures have been derived from the ratio of snow and precipitation (see Table 2).

Table 2 suggests that the winter temperature of the 18th century is more than 1°C higher than those now considered normal over the lower Changjiang valley.

Table 2. Ratio of snow and precipitation, derived from winter temperature in Hangzhou (1723–1769), Suzhou (1725–1782) and Nanjing (1722–1785).

Station	18th Century				Present			
	Duration	Yrs	Ratio	Temp.°C	Duration	Yrs.	Ratio	Temp.°C
Hangzhou	1723–1769	46	0.289	3.7	1954–1978	24	0.207	4.7
Suzhou	1725–1782	57	0.289	3.6	1955–1978	23	0.190	4.6
Nanjing	1722–1785	63	0.417	1.6	1953–1978	25	0.280	3.1

Analyzing the freezing years of the lakes and rivers located at the lower Changjiang valley, which are due to the severe cold waves coming from Siberia in winter, Zhu Kezhen (1973) discovered that the 18th century was the warmest period during the last 500 years. This study also shows that even in this warm period the winter temperature was lower than that the present by 1.0 −1.5°C over the lower Changjiang valley.

The severity of winter in the 18th century in Nanjing are shown in Table 3, based on the data mentioned above. It appears that the warm and cold winters often occur in groups of consecutive years. By the ratio of snow to precipitation and derived temperature, the severity of winter are divided into six groups (Table 3), showing that even in the warm episodes the winter temperature was lower than today by 0.5 −1.6°C; and in the cold episodes the differences will range from 2.1 to 2.4°C.

Table 3. Severity of winter in the 18th century in Nanjing

Duralion	−1728	1729–35	1736–41	1742–49	1750–67	1768–
Yrs duration	7	7	6	8	18	>18
Regime	cold	warm	cold	warm	cold	warm
Ratio	0.4827	0.4092	0.4845	0.3567	0.5100	0.3030
Temp.winter°C	0.9	1.5	1.0	2.1	0.7	2.6
Departure from today,°C	−2.2	−1.6	−2.1	−1.0	−2.4	−0.5

The dates of bloom of some kinds of flowers were often recorded by numerous ancient writers, especially in their diaries. The blooming dates of *Malus micromalus* Makins and *Prunus persica* have been collected over the lower Changjiang valley, and are summarized in Table 4.

Table 4 shows that the phenodate was later in the 20′s–70′s of the 18th century by 6 days than that of today. From an empirical formula of the distribution of *Malus micromalus* Makins in China, it appears that the difference of 4 days in phenophase corresponds to 1° shift in latitude, so that the variation of 6 days

42

Table 4. Days-lag from present blooming dates of *Malus micromalus* Makins and *Prunus persica*

Year	1778	1780	1787	1790	1791	1792	1796	1800	1808
Days-lag	15	20	4	10	9	6	5	−1	−6

in the 18th century implies a shift of 1.5° in climate belt.

Wind observations were available in the mid-18th century, recorded in eight directions once a day in Hangzhou (1723–1768). The summary is shown in Fig. 3 (upper panel), and suggests that the prevailing wind during that time is NW for November-January, NE for February to June and September, and SE for July and August. Comparing with that of the instrumental records of 1953–1979 (lower panel), there are evident differences between these two periods, the frequency of NE was much more prevailant during the mid-18th century. It is known that in the lower Changjiang valley the northeast wind is often associated with wet weather, and the 18th century is known as a wet period during the last 500 years (Zhang 1981).

It is now possible to argue that the climatic history of the 18th century from different sources exhibits a good deal of agreement with the reality of climatic pattern.

Fig. 3. Distribution of wind direction frequency during 1723–1769 and 1953 – 1979 in Hangzhou

All temperature and proxy temperature records in England, central and northern Europe, north Greenland, California, and New Zealand indicate that a sharp change to much warmer conditions in the early and middle 18th century lasted to the end of this century. (Lamb 1981, p 232). These trends are the same as in China.

The seven coldest and seven mildest winters in central England were chosen between 1659 and 1979 and shown in Table 5.

Table 5. Average temperature over Dec. Jan. and Feb. in the seven coldest and seven mildest winters in central England between 1659 – 1979. (After Lamb, 1981)

Winter	1683–4	1739–40	1962–3	1813–4	1794–5	1694–5	1878–9
°C	−1.2	−0.4	−0.3	+0.4	+0.5	+0.7	+0.7

Winter	1868–9	1833–4	1974–5	1685–6	1795–6	1733–4	1934–5
°C	6.8	6.5	6.3	6.3	6.2	6.1	6.1

The long-term average for winters 1850–1950 is 4.0°C, so the temperature of the warmest winter of the 18th century is 2.1°C higher than that of present, but in China, the warmest period in the 18th century is 2.2°C lower than it is at present. This evidence implies the severity of cold waves during the Little Ice Age in China.

This phenomenon may be attributed to the existence of the Qinghai-Xizang Plateau, which restricts part of the westerlies in a more southern latitude and delivers enough space in the eastern plain for the invasion of cold air from the north in winter.

The relationships between the Siberian High and the cold winters in Shanghai (representing Hangzhou and Nanjing) have been summarized by Yen (1981) in Table 6.

Table 6 suggests that when the Siberian High is located to the east (about 103°−108°E), the lower Changjiang valley experiences cold winters.

Table 6. Relationships between the Siberian High and cold winters in Shanghai (After Yen)

	Location of S. H. long. E,	lat. N	No. of cold winter	No. of warm winter
1871–80	100.5	49.5	4	
1881–90	108.0	51.0	3	
1891–00	107.5	52.0	3	
1901–10	101.5	49.5	3	
1911–20	108.0	53.0	3	
1921–30	103.0	52.0	1	
1931–40	100.5	53.0		4
1941–50	102.0	52.0		5
1951–60	100.0	48.5		3
1961–70	97.0	46.5		3

When this evidence is added to the evidence presented before, the cases of cold climate is considerably strengthened.

References

Chow Benshun (1978) *Vertebrata PalAsiatica* 16(1): 47–58

Lamb, H H (1981) Climatic history and the modern world. Methuen, London

Yen Jiyun (1981) *Sym Climat Change Stud* Science Press Beijing, pp 71–77

Zhang Peiyuan (1982) Editorial Committee of Physical Geography(ed), Historical physical geography. Academia Sinica. Science Press, Beijing, pp 6–18

Zhang Siengong (1981) Symp Climat Change Stud. Science Press, Beijing, pp 46–51

Zhu Kezhen (Chu Ko-chen) (1973) *Sci Sin* 14: 226

A Historical Survey of Arid Areas of China

Sheng Chenyu[1]

Abstract — The paper consists of six parts. In part 1 it is shown that the arid and semi-arid areas cover the provinces and autonomous regions of Xinjiang, Inner Mongolia, Gansu, Ningxia, and most parts of Shaanxi, Shanxi, and Qinghai. Its extent ranges approximately from 35°. North in Shanxi to 50°. North in Inner Mongolia in latitude. The comparison between this distribution and the global geographical model of arid and semi-arid areas proposed by the renowned climatologist W. Koeppen is illustrated and their coincidence is evidenced. Some opinions that the climate of Northwest China in early days was much more humid than now, are quoted in part 2. In part 3, the paper gives historic evidence documented from various historical sources. These materials may serve as proof to refute those inferences that the Northwest of China was formerly more humid. The genetic causes of the *Silk Road* in the Northwest are discussed in part 4. The characteristics of climatic fluctuation in arid areas are also mentioned in part 5. Finally, in part 6, some instructive ideas are expressed as the conclusion of the whole work.

Geographical Model of Arid Areas and Contemporary Distribution of Arid Regions in China

A wide zone of arid climate extending from the western side of the landmass in subtropical latitudes, across the interior of these latitudes, and diverting to the hinterland of temperate latitudes is proposed in any kind of climatic classification. Figure 1 illustrates the geographical distribution of the climatic model proposed by the German climatologist W. Koeppen (Koeppen and Geiger 1936). It is clearly seen, that both hemispheres have similar arid zones, but the northern one, especially the Eurasia landmass, extends nearly to the eastern side of the continent. The cause of the arid zone is closely correlated with the subsiding aircurrent in the subtropic latitudes and paucity of moisture accompanied with high mountains encircling within temperate latitudes. The arid zone of China results mainly from the latter condition.

The climatic atlas published recently by the State Bureau of Meteorology divides the wet/dry condition into four grades, namely humid, sub-humid, sub-arid and arid (State Meteorological Administration 1966). Supposing we combine the first two grades and designate then as arid region, then besides the immense cold-arid area of the Tibetan Plateau and scattered localities along the Yuanjiang River valley in the Yunnan province, the area covers Xinjiang, Inner Mongolia, Gansu, Ningxia, and most parts of Shaanxi, Shanxi, and the Chaidamu Basin in Qinghai, that is, the conterminous arid areas are located in the Northwest of China, whose southern limit reaches 35°N in Shaanxi and the northernmost extends to 50°N in eastern Inner Mongolia. All of the desert and steppe in China are situated in this

---

1. Department of Meteorology, Nanjing University, Nanjing, China

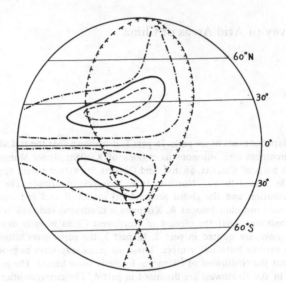

Fig. 1. Ideal distribution of arid and semi-arid climates on the Earth. *Barbed line,* extent of landmass; *dotted line,* area existing dry season in the year; *solid line,* semi-arid climate, and *dashed line,* arid climate.

area, which occupies one third of the total area of China; therefore, their rational utilization is of great importance for state economy (Fig. 2).

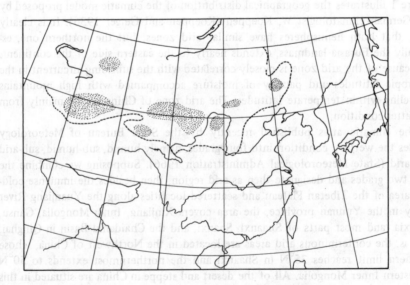

Fig. 2. Arid and semi-arid climates in China (stippled areas—sandy desert)

Discussion

Historic evidence can cite many geo-scientists and historians to recall that the climate of the Northwest have been desiccated in the past thousand years (Stein 1912, Hedin Sven 1933; Huntington 1907), although the immense Northwest area is climatically arid and semi-arid and is occupied by continuous desert and gobi-desert at present. This historic evidence may be listed as follows: (1) the cradle of ancient Chinese civilization originated in the Huanghe River Basin; (2) the renowned *Silk Road* known abroad, which connected with Arabia, India, and ancient Rome, ran across the Northwest from the capital of the Empire of China during early centuries. Many of the cities, documented in history are ruined or completely destroyed; (3) the Hetao (i.e., the great bend of the Huanghe River in Inner Mongolia) area was once farmland, but is now a sandy region with little agriculture. So we can read in the records. But we can also find historical data to prove that the climate of the Northwest has been dry for as long as recorded history. In other words, the climate of Northwest China in historic time was never moister in condition than now.

Historical Evidence in the Arid Area

Recorded history, the earlier and reliable records should be considered from the Zhou Dynasty, that was from the eleventh to third century B.C., and lasted approximately 800 years. In that time, the agricultural society of the Zhou Dynasty and its kingdoms of dukes occupied the mid-and lower reaches of the Huanghe and Changjiang Rivers. Minorities of pastural races settled in the remaining part of the territory, for example the Hun, Qiang, Di and others. The territory of Zhou was mainly limited to the Central Shaanxi Plain at that time, when folklore poems proclaimed, "they are prolific in the grains of millet and wheat" (Shi Nianhai 1963). This indicated that the Central Shaanxi Plain had less rain and was suitable for drought-resistant crops. Another example was the excavation of an irrigation channel. The first emperor of the Qin Dynasty adopted the proposal of a water worker called Zheng Guo, who submitted a program to dig a channel more than 150 km in length, connecting the two tributaries of the Weihe River in 246 B.C., so that the agricultural condition of the Central Shaanxi Plain, where the land was alkaline and liable to drought, could be much improved. These two examples dearly expressed that, although the surrounding region of Northwest China the climate was dry, yet the hinterland of Xinjiang, Gansu, and Inner Mongolia must have been without doubt much drier.

It is known that the Northwest of China was an arena for tribal rivalry in history. During the period of less rain, the pastoral tribes invaded the arable land of Hans. They fought each other in the area of the Northwest. For this reason, we traditionally called the battle field "sandy field" and those killed in the war "the corpse packed with horse skin". The origin of these terms stemmed from the fact that the ancient war occurred in the desert area and that the cavalry was the main force in fighting.

In the late period of warring states (ca. 475–221 B.C.), the kingdom of Qin had built a section of the Great Wall from Min county in Gansu to Zhunger county in Inner Mongolia to repel the invasion of Huns. It is worthy of note that this section of the Great Wall made a detour to the Northern Shaanxi Highland because of the severely salinized soil to the north of the Liupan Mts, as has been commented by Prof. Shi Nianhai of Shaanxi Normal University by his own field survey in recent years. Similarly, the southward route of Huns and Turks took the longer way to the Central Shaanxi Plain instead of a shorter way from the time of Qin and Han Dynasties down to Sui and Tang Dynasties (ca. 200 B.C. to 900 A.D.). The cause of this measure was that spotted sandy desert and barren land existed with little grazing grass, the nickname "the sea of drought" being age-old. (Shi Nianhai 1981).

In the Han Dynasty, the sixth emperor (ca. 140–87 B.C.) decided to extend his sovereignty to the western part of Gansu and set up four prefectures, as then he could establish the passway to the western Territory (now called Xinjiang and part of Central Asia) (Xiang Da 1957). In 138 B.C., the emperor ordered a expedition of more than 100 persons under the command of Zhang Qian to develop relations between the Han Dynasty and minorities of the West. Due to the hardship of the natural environment and to political causes, they failed to complete the task and came back with only two survivors after 13 years abroad. In 119 B.C., a diplomatic mission of 300 persons was re-organized and sent to Wusen (SE of Lake Balkhash) and Sogdiana (SE of Aral Sea), and the communication between the East and the West eventually inaugurated (Chang Renxia 1981). The Emperor Liu Che (i.e., the sixth emperor of the Han Dynasty) also nominated Lee Guangli " General of Sutrishna" with the mission to bring back the famous bread of horse from Sutrishna (SE of Tashkent in Central Asia). According to the historic document, the military corps amounting to several thousands of soldiers suffered from hunger and were tired when they arrived Yucheng (east of Tashkent) (Anonymous[1]). They had spent 2 years on the journey and only one to two tenths of the corps was alive when they came back to Dunhuang. The details of hardship en route to the West Territory was recorded in *History of Zhou Annals of Qoco* (the kingdom of Qoco is now called Turfan): "Sandy desert abounds, it is hard to accurately count the mileage from Dunhuang to Qoco, other than measured by the skeletons of men and domestic animal and dung of horses. There are a lot of demons and monsters on the way."

Desert occupied a wide area in our country as early as in Han Dynasty, there were some counties named with the word of "sand" (pronounced "sha" in Chinese), for examples the Shaling county (now called Tuktu county in Inner Mongolia) and the Shanan county (now to the south of Baotou) (Shi Nianhai 1981). .The name of a place often implies the natural environment, hence, these two counties located nearby the Hetao area in Inner Mongolia obviously indicated that a broad area of drifting sand had been spread over there since the Han Dynasty. Other evidence was found in the *Historic Records: Native Products*, written by Si Maqian in the Western Han Dynasty, in which the dividing line between agriculture and animal husbandry ran from Jieshi to Longmen, that was from the northern coast of the Gulf of Bohai in Hebei province through Beijing, Taiyuan, thence along the southern part of the Luliang Mts., and to somewhere on the Huanghe River in the southwest of Shanxi province. The *Historic Records* plainly pointed out that north of the line abounded with "horse, cow, sheep and its products" in the period of warring states (ca. 475–221 B.C.) (Si Maqian).

In 78 A.D., a troop of 10,000 soldiers commanded by Ban Chao attacked Gumo (now called Akesu) and Shichen. Ban Chao had been stationed for many years in the West Territory and had submitted a letter to the Emperor, in which he said, "I fortunately protect the security of West Territory. I have no regret if I die on the borderland. But, I should wish to go back Yumenkuai when I am alive. I am afraid of the later generations complaining about the loss of this territory." (Anonymous[2]) Afterwards, the younger sister of Ban Chao, named Ban Zhao, also submitted a letter to Emperor, in which she asked the Emperor to promise to call her brother back to the homeland. The letter reads, ". . . my brother has been here 30 years. Although he is healthy at present . . . most of his subordinate have already died there. It is a great pity that my brother must die in the wild area as though he had kept the loyalty when he was young . . . ". (Anonymous[2]). It is clearly proved from this literature that Xinjiang and Central Asia were a broad area of dry land even in the Han Dynasty. The so-called "kingdom" of so and so actually was these oases scattered over the immense arid area. Therefore, Ban Yong, the son of Ban Chao frankly said, "The West Territory had been neglected by those former rulers, because it was not advantageous for us, and hard to maintain the supplies . . . " (Anonymous[2]).

It has been pointed out that the West Territory was the main channel of cultural exchange between the East and the West. As early as in the Han Dynasty, Buddhism was introduced to China through this channel from India. In the fourth century, several eminent Indian monks took this way to China, and the Chinese monks took the same way to India. All of them left us valuable travel writings, by which we can infer the natural environment at that time. Fa Xia, one of the eminent monks who went India in 399 A.D. by the land route and came back in 412 A.D. by sea, wrote a booken titled *A Trip to the Buddhism State.* It recorded as follows: "*There* are monsters and ghosts and hot wind in the sandy river (i.e., desert); any man will definitely die if encounters them. There is no bird in the sky and no animal on the ground. It is hard to find out where the way is, but the bleached bone is the only mark to indicate." " . . . Go over the Gobi and cross over the desert." *The Biograph of Fa Xia* said, "He spent more than one month across the desert from Yanqi to Khotan . . . There is no inhabitant on the way, walking is difficult in the sand, the arduous struggle confronting the course is beyond of imagination by man."

The Sui and Tang Dynasties were other peak times in the conquest of the West Territory, the border extending westward to the Aral Sea. Communication between the Tang Dynasty and the West reached a new level when the high tide of spreading Buddhism culture occurred. The cave temple culture prevailing in Buddhism had begun to be introduced to China before the Tang Dynasty. For example, most of the cave temples at Dunhuang were established in the Northern Wei Dynasty (386 –

534 A.D.). These stone inscriptions, wall paintings and Buddhist Sutra leave now become a historic treasure in the cultural arts. But it should be pointed out that the environmental condition at Dunhuang was very cruel. It had been called "Shazhou" (i.e., sandy state) and "Guazhou" (i.e., melon state) in history, because it was full of sand and merely suitable for melon-planting. *The Biography of Chang Shouguai* said, " . . . there is plenty of sand, it is not good for planting, with little rainfall, the thawed water from mountains irrigates the field." A poem written by another writer described: "the Yumenguan is so isolated, yellowish sand covers with withered grass. The neighbors are savage peoples. The commander meets with unexpected aggression." (Xiang Da 1957).

In 629 A.D., 230 years after the Buddhist master Fa Xia, the Central Plain of Shaanxi suffered from hunger, an eminent monk Xuan Zhuang left for India via the corridor of western Gansu and Xinjiang. His diary reported, " . . . from Liangzhou (now called Wuwei) westward by a little north to Ganzhou (now called Zhangye) runs 470 li (235 km) in distance, further 400 li (200 km) westward to Suzhou (now called Jiuquan), another nearly 7000 li (?) (3500 km) to Yumenguan . . . from Shazhou southwestward crossing the desert more than 700 li (ca. 350 km) in distance to Kroraina, or called Charklik . . ." *The Biograph of Buddhist Master of "Great Kindness Temple"* reported, " . . . travel at night and take rest in daylight, eventually arrived Guazhou . . . There are five beacon towers to the northwest of Yumenkuai, each tower at 100 li (50 km) distance from the other. There is no water and grass. The territory of Hami lies beyond the range of five beacon towers." The report, describing the return trip from India by Xuan Zhuang said, " . . . taking more than 400 li (nearly 200 km) in distance to Tukhara (i.e., the area west of Pamir), all its towns are desolate and uninhabitated. Thence eastward more than 600 li (300 km) to Calmadana (now called Charchan), the castle is reserved, but without any inhabitant." All these descriptions express that the southern part of Xinjiang was already a wide desert in the early stage of the Tang Dynasty.

To sum up these comments, cited from various literature, none of then could prove that the vast area of Xinjiang and western Gansu was much moister in historic time, even since the very beginning of the *Silk Road*. As to the desert of the Hetao area (i.e., the great bend of the Huanghe River in Inner Mongolia), several geographers claimed that, it was a "man-made desert" of the past 1000 years, and considered it as indicative of moister condition in early history (Hou Renzhi 1965). But the opposite idea is suggested by Zhao Yunfu, who claims that there was desert as early as in the Tang Dynasty (Zhao Yunfu 1981). It may be proved by the writing on Shen Yazhi (c. 781−823 A.D.), who said: " . . . the land set up as a prefecture called Xiazhou (now in the northern part of Shaanxi province), . . . there is a lot of drifting sand blocking the river, the climate is extreme hot or prolonged winter for the seasons, the people are rigid and the wind is severe." He further said: "The territory of Xiazhou extends several thousand li in width, full of drifting sand." All this historic evidence shows that the northernmost part of Shaanxi had always maintained its arid condition, desolate and with little agriculture. On the other hand, it has been declared that this area has become desiccated within the past 1000 years, because it had been as county, called Sheyan in the Han Dynasty, and the kingdom of Xia had been set up during 407−431 A.D.. The kingdom of Xia was governed by a descendant of the Huns (named Helianpopo). The city Tongwan, the capital of Xia, was situated somewhere between the Sheyan River (now called the Wuting River) and the He River (now called the Hailiutu River). The reason for the insistance that the climate was moister in earlier times is the beautiful landscape which had been praised by Helianpopo. Meanwhile, as Zhao Yunfu pointed out, even if small lakes and tributaries of clear water appeared in the low land of a desert area, it should not be reckoned as signifying wetter conditions, Building up the city of Tongwan was a military requirement rather than an economic need. The author agrees with the idea of Zhao. As the documents of the Ming Dynasty pointed out, the city of Yulin had been encircled by drifting sand, and the accumulated sand outside of the city already reached the top of the city wall in the late sixteenth century. It is impossible that the great change could have occurred during the millenium from Helianpopo to Ming Dynasty.

Social Factors and Natural Factors Forming the *Silk Road*

In the time before the sea-route was fully utilized, the main channel of exchange between the East and the West was dependent on the land route, which was maintained for more than 2000 years in history. As everybody knows, this is the was called the *Silk Road*. The very beginning of the *Silk Road* should be credited to Zhang Qian, who first found the way to link culture between China and the countries of central and western Asia. We should note the words " to chisel through the channel" are used to describe the process of building the route in ancient literature. It implied the deep meaning, which reflected the stubbornness and courage of these pioneers to overcome countless difficulties. Numerous verses or poems about seeing their relatives and friends go to the West Territory in ancient Chinese literature reveal the feelings of sadness and despair. How high a price the ancestors of the Chinese people paid to maintain the *Silk Road* unimpeded and unblocked!

It should be emphasized that to build the channel to the West was mainly due to the development of social economics. For example, the invention of silk production technology was at least early in the seventeenth century B.C. So that the ancient western peoples called China "Seres" (i.e., silk country). Later, the paper production was invented as well. These technologies were gradually transmitted to central Aisa, northern Africa, and Europe. After the mid seventh century, caravans frequently traveled across the great desert, shuttling between China and the countries in the West. The social factor was of great importance for the development of *Silk Road*.

Secondly, the origin of *Silk Road* and its choice of site, in some measure, implied the adaptation and struggle of men against the natural environment. For example, in the first section of the Road, the people chose the main course along the western Gansu corridor; in mid-section, there were two sub-routes diverging westward from Liulan: one was along the northern side of the Taklamaken Desert; another was along the southern side of it. It clearly showed that the early people were good at utilizing the natural environment. They selected the course along the inclined plain in front of a high mountain or along a wide valley where the

52

ground water level was not too deep (Zhao Xingchen 1983). Hence, it should be understood that the *Silk Road* was a result not of the moister climate at that time, but on the contrary, was forced to choose a relatively better course due to the prevatent conditions. These facts might be regarded as the natural factors of the *Silk Road*.

A few sub-routes were built in historic times. One route was by way of Xining, then either along the Tatung River across the Qilian Mts. to link with the Western Gansu Corridor, or crossing the Chaidamu Basin and the Altun Mts. to join the main track at Ruoqiang. Choosing the best course was essentially dependent upon actual circumstance when possible; however, it did not correspond with a variation of wetness, in which there was no significant change at all.

Climatic Variation of Arid Regions

This paper does not deny the fact that climatic fluctuations also exist in arid and semi-arid regions. However, it should be emphasized that the nature of drought or semi-drought in present desert and steppe zone did not change in the historic period, while climatic fluctuation merely resulted in less quantitative variation. In other words, this kind of fluctuation only affected the geographic displacement of the transitional zone, and did not change the qualitative characteristics of the entire arid area. A few years ago, the author discussed variation in arid areas in the past 10 to 20 years on the basis of criteria of Koeppen (Sheng Chenyu, in press). Figure 3 is the result of that work, and shows that the delimitation line between wet and arid vibrated rapidly in the North China Plain and Loess Plateau, it vibrated 4° in longitude in the eastern North China Plain and 8°–9° in latitude in the western Loess Plateau during the period of 1951–60. The west and north of the line, which connects Hailar, Linxi, Huateh, Dongsen, Yinchuan, and Chungwei, is the area of arid climate. The east and south of the line, which connects Nenjiang, Harbin, Zhanglin, Shenyang, Zhunghe, Xinyang, Nanyang, and the Qinling Mts., is the area of wet climate. Between these two lines is the transitional zone of arid/wet climates.

Figure 4 is another result worked out recently by the author (Sheng Chenyu et al., in press). It shows that the region encircled by the line from the border between China and the People's Republic of Mongolia (east of Hami) through Kuerlei, south of Kucha, Kashgar, north of Khotan, Ruoqiang, Dunhuang, Jiuquan, Laodongmiao to Dugulike is 100% of desert climate, another minor center of desert climate is located near the lower Manus River in northern Xinjiang. Furthermore, two frequency isopleths on Fig. 4 are noted, one is 0 % of desert climate, and runs from north of Xilinhot, through west of Beijing, Taiyuan, Yan'an, Lanzhou, south of Chaidamu Basin, and to the north slope of Mts. Kara Kunlun and Altun. Another is 0% of the steppe climate, which runs from the northern end of Mts. Da Hinggan through Nenjiang, Harbin, Changchun, Shenyang, Dalian, Qingdao, Xuzhou, Nanyang, Hanzhong to near Chengdu. The wide zone between these two isopleths is the intermediate climate of semi-arid and semi-wet. Certainly, it is hard to be sure whether a more violent variation has occurred during the last 2000 years

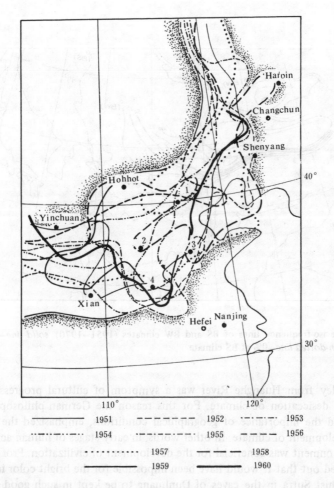

Fig. 3. Displacement of arid/wet delimiting lines in East China. 1 – Beijing; 2 – Taiyuan; 3 – Jinan; 4 – Zhengzhou

since the Han Dynasty. But the author believes that the Northwest of China was without doubt an arid climate, with at most a variation in the degree of dryness.

Regarding the questions mentioned in Section 2, for example, why did the upper and mid reaches of Huanghe River become the cradle of Chinese culture? According to the view points of archeologists, human society always originated from a drier climatic condition with possibility of irrigation. Under these circumstances, it was not only possible to solve the problem of food and drink, but also easter to over- come the lesser rivality of nature. All ancient cultural sources in the world—Egypt, Babylon, Greece, and India—were situated in arid or relatively dry climates. The Inca and Maya civilizations originated from the much drier highland in lower latitudes of the western hemisphere. Our ancestors—Peking man (*Sinanthropus pekinensis*), Hetao Man, Dingcun Man etc. —also lived in a similar environment (Shi Nianhai 1963). .Therefore, that Chinese culture gradually pushed to Changjiang

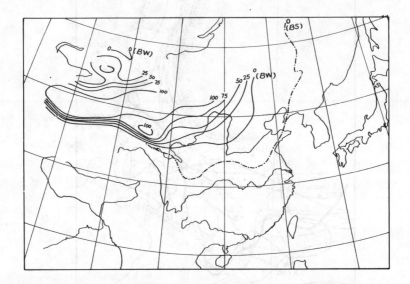

Fig. 4. The iso-frequency lines of BS and BW climates (1951–1970). *solid line*—% of BW climate; *dash-dotted line*—0% of BS climate

River Valley from Huanghe River was a symptom of cultural progress, but not caused by desiccation of climate. For this reason the German philosopher Hegel emphasized the importance of geographical conditions, emphasized the effect on social development of climate. In other words, in early stages of human activity, the drier environment was beneficial for the development of civilization. Prof. Lu Jiong had pointed out that it would have been impossible for the bright color in painting and Buddhist Sutra in the caves of Dunhuang to be kept in such good condition over so long a period, if the climate had been very moist (Lu Jioug 1942).

As to the rise and decline of *Silk Road* and the disappearence of cities on the way, its causes were complicated, the variation of natural condition including climate was of less importance, instead, its social causes, political, economic, and others, were more essential. For example, the alteration of the traffic route from land course into sea course would seriously have affected the decline of the route and its cities. Moreover, numerous incidents between rival nationalities would severely damage the construction of the cities.

The loss of Liulan should be ascribed to its own weakness. According to literature, it was "full of sand and saline, lack of arable land." Later (330 A.D.), the lower reach of the Tarim River was diverted by being silted up by drifting sand. Finally, the city became a ruin because of failure to irrigate (Zhao Xingchen 1983). In the arid area, owing to greater varibility of thaw water from the high mountains and possible silting up of the river basin, an alteration in the water system frequently occurred, hence the loss of a certain city cannot be regarded as evidence of desiccation.

Regarding the origin of deserts in the Ordus highland, Hetao area, and the upper valley of the Xiliaohe River, the author has spoken in Section 3, but will make some further comment in this section. Those areas actually belonged to the transitional zone of desert climate and steppe climate, they potentially had the possibility to be a desert, when the natural vegetation cover was once destroyed by human activity. Consequently, human activity must not be confused with climatic desiccation.

In short, in the past 1000 — 2000 years of history, there was no tremendous variation in climate in the Northwest, but pulsatory changes were not excluded. For example, the literature of the Zhou Dynasty recorded, in the ninth to eighth centuries B.C., "frequent drought, naked hill and dried-up river", "turmoil comes from Heaven, hunger occurs everywhere"; sometimes even the rivers in the Central Plain of Shaanxi ran dry. However, the degree of environmental variation in the Central Plain of Shaanxi was still insignificant (Shi Nianhai 1981).

Conclusions

Some enlightment can be gained from the characteristics of the Northwest arid region for the historic period. Firstly, the progress of desertification affected by natural factors, which includes the climatic factor, was relatively slow and prolonged. Climate is the most changeable among natural facters, but the extent of its variability is still limited by geographical environment. It is impossible that the qualitative variation of climate can have occurred within a short historic period. On the contrary, the destructive effects on environment of human activity are significant, which plays an important role in the progress of desertification. But the main influence of human activity on desertification will not take place unless the environment itself potentially possesses the possibility of desertification. So human activity is merely an act of acceleration or triggering-off in the progress of desertification.

It should be noted that neglecting the triggering action of human activity would have serious result in the neighboring regions of arid climate. Ecological equilibrium should be emphasized whenever the arid or semi-arid areas are to be exploited. According to the survey of the settlement in arid area, a family with five persons requires 16 000 kg. of firewood per year, that is equivalent to the vegetation covered on 1.34 ha. . According to the census taken in 1980, the population of six counties in northern Shaanxi province totaled 1 290 000. On an verage, there were 38 persons per sq. km. Hence, the limit of 20 persons in semi-arid region, which was proposed by the Congress of World Desertification in 1977, was exceeded. Moreover, simply raising the production indices for animal husbandry without considering the ecological equilibrium should be checked. In the early 1960's, we found deterioration of pasture land in the eastern part of Inner Mongolia. This was an example of neglecting the ecological equilibrium. Owing to surplus livestock, the pasture was usually grazed before the grass was mature. Year by year, the grazing fields had degenerated.

In conclusion, the author confirms that the Northwest of China did not have moister condition in its long history. But now and then, we should pay more attention to the effects of climatic variation due to human activity in the arid vicinities. Rationing of agricultural and grazing production, with the major prerequisite of maintaining ecological equilibrium, is of great importance. We should closely monitor the harmful influences arising from human activity, but also are against exaggerating its effect and randomly to extend the limit of desertification into wetter climate environment.

References

Anonymous [1]. The Biographies of Chang Kian and Lee Guangli. *Annals of Han Dynasty*

Anonymous [2]. The Biographies of Ban Chao and others. *Annals of Later Han Dynasty*

Chang Renxia (1981) *The Silk Road and Cultural Arts of West Territory* (in Chinese). The Arts Shanghai

Hedin Sven (1933) *Across the Gobi Desert*

Hou Renzhi (1965) *Geogr*

Huntington E, (1907) *The Pulse of Asia*

Koeppen W, Geiger R (Eds) (1936) *Handbuch der Klimatologie*, Bd 1, Teil C.

Lu Jiong (1942) *Acta Meteorol* 16: (3)

Sheng Chenyu (in press) *Variation of Wet/Dry Division Line in the Decade* 1951–1960

Sheng Chenyu et al. (in press) *The Secular Variation of Wet/Dry division Line in China*

Shi Nianhai (1963). *The Land of China* (in Chinese). Sanlian Book

Shi Nianhai (1981) *The Land of China* Vol. 2 Sanlian Book

Shi Nianhai (1981) In: philosophy and social science (Ed) *Jour. of Shaanxi Normal Univ*

Si Maqian *Historic Records—Native Products*

Stein A (1912) *Ruins of Desert Cathay*

State meteorological Administration (1966) *The Climatic Atlas of China*. Cartography press Beijing

Xiang Da (1957) *The Civilization of Changan and Xian in Tang Dynasty* (in Chinese), Salian Book

Zhao Yunfu (1981) *Histor Geogr*, 1st issue

Zhao Xingchen (1983) *The China Desert* 3 (2)

The Climatic Changes of Drought-Wet in Ancient Chang-an Region of China During the Last 1604 Years

Li Zhaoyuan[1] Quan Xiaowei[2]

ABSTRACT

According to the historic records of the ancient Chang-an (Xi-an today) region, a climatic series of drought-wet indices of this region for the last 1604 years was reconstructed. The method of data processing is the same as explained in Zhang De-er (1983) and the Ref.under Atlas of ...(1981).

By method of optimum classification (classifying by statistic variances), power spectrum analysis, and harmonic analysis, some characteristics of the drought-wet climate in Chang-an and its relationships with those in other places on the northern hemisphere were explored. The main findings are as follows:

The climatic change of drought-rainfall in Chang-an can be divided into six periods: I. ca. 380–680 A.D., dry; II. 690–990, wet; III. 1000–1220, dry; IV. 1230–1410, wet; V. 1420–1630, dry; VI. 1640–now, wet. On the average each period ranged about 267 years. The wet or dry periods were compared with the mean value of the series over all the 1604 years (see Table 1) which presents a dry feature of the climate in Chang-an comparing with the mean level of whole China.

On the secondary time scale of about a century, the drought-wet climate in Chang-an may be divided into 18 episodes, each of which ranged about 89 years on average. There were three rather dry episodes respectively in 460 – 500, 1140–1220, 1420–1490, and several wet episodes appeared in 690 – 750, 1230–1250, 1380–1410, 1640–1700, 1810–now.

Since 380 A.D. the duration of the dry periods has become shorter and the latest wet period of these two time scales are the longest of all. By calculating the ratio of dry or wet years during each period, we discovered that even in the wet periods in Chang-an, dry years still occurred frequently.

The result of harmonic analysis to the drought-rainfall series, each of which are obtained respectively from 11,31, 51years weighted running average to the original series, presents two main periods of about 71 years and 518–531 years. The power spectrum analysis shows that the main periods are of about 74–94 years (maximum lag M=533). This means the climate in Chang-an varied, on the background of changing with the period of about 520 years, with the notable periods of 71–75 years.

In order to understand the atmospheric circulation associated with the variation of wet or dry years in Chang-an, the mean (May to September) fields of 500 hPa geopotential height from years 1951 to 1980 were investigated. In the 30

1. Meteorological Institute of Shaanxi Province, Xi-an, China
2. Institute of Atmospheric Physics, Academia Sinica, Beijing, China

58

Table 1. Mean features (a), period (b) and episode (c) of the wet/dry climate in Chang-an

(a).

	Mean feature of the wet/dry climate in Chang-an				
Index	1	2	3	4	5
Number of yrs.	107	275	566	529	127
%	6.67	17.14	35.29	32.98	7.92
Total.	wet year 23.8% normal 35.3% dry year 40.9%				
Mean value of the indices series 3.2					

(b)

Period (A.D)	Range (Years)	Mean value of indices	Ratio of drought years (%)	Ratio of rainy years (%)
380 – 680	310	3.4	44.8	15.2
690 – 990	310	3.1	41.6	28.7
1000 – 1220	230	3.4	50.4	15.2
1230 – 1410	190	3.1	34.2	25.3
1420 – 1630	220	3.3	45.0	15.9
1640 – now	344	2.9	32.0	34.0

(c)

Episode (A.D)	Range (Years)	Mean value of indices	Ratio of drought years (%)	Ratio of rainy years (%)
380 – 450	80	3.3	36.3	16.3
460 – 500	50	3.7	62.0	10.0
510 – 680	180	3.3	43.9	16.1
690 – 750	70	2.9	32.9	35.7
760 – 800	50	3.3	52.0	28.0
810 – 870	70	3.1	41.4	32.9
880 – 990	12	3.2	42.5	22.5
1000 – 1080	90	3.4	60.0	16.7
1090 – 1130	50	3.0	30.0	24.0
1140 – 1220	90	3.5	52.2	8.9
1230 – 1250	30	2.8	10.0	30.0
1260 – 1370	120	3.3	44.7	22.5
1380 – 1410	40	2.9	22.5	30.0
1420 – 1490	80	3.5	55.0	15.0
1500 – 1630	140	3.3	29.3	16.4
1640 – 1700	70	2.8	25.7	38.6
1710 – 1800	100	3.1	35.0	25.0
1810 – now	174	2.9	32.8	37.4

Table 2. Main Periods of The Wet/Dry Climatic Change in Chang-an

Method	Running average	Main periods
Harmonic analysis	11-years weighted	71, 531, 27, 26, 797, 41
	31-years weighted	71, 524, 787, 50
	51-years weighted	70, 74, 518, 777, 141
Spectrum analysis	50-years (maximum lag M=533)	148, 129, 94, 74, 69, 32

years, serious dry years occurred four times (1953, 1957, 1958, 1964) and a very wet year occurred twice (1969, 1978). The composite mean fields of summer half year (May–September) in these years are given in Fig. 1.

From Fig. 1 we see that wet or dry years in Chang-an were obviously associated with the positions of both the trough over East Asia and the Indian low pressure. When the Indian low becomes lower and its center is located in a more south-west position, Chang-an will be at the west side of the trough, a dry year will very likely occur. Otherwise, when the Indian low is weak and in the north-east position, Chang-an will be under the trough or sometimes at the east side of it, a wet year would occur. The trough above West European also differs between the dry years and wet years in Chang-an. These features should be tested carefully over a longer period, such as the periods and episodes mentioned above.

The statistical correlations of the wet/dry climatic indices series between Chang-an and the other places show results similar to those obtained from the study on atmospheric circulation. Table 3 lists several correlation coefficients between the drought-rainfall indices of Chang-an and the indices of a place in northern hemisphere. Some other coefficients are omitted because they are too small to approach the significant level (only two coefficients with Russia and California are given as examples). In comparison, the correlations between Chang-an and Calcutta, Germany, and the Nile River region in north Africa are not too inaccurate. It is interesting that the correlation coefficient between Chang-an drought-rainfall indices and the dates of Indian monsoon onset have such a high value. This means that the earlier the Indian monsoon comes, the drier climate of Chang-an will be and vice versa. The early Indian monsoon implies that the south-west flow is active and strong, which would induce the earlier Mai-U in the mid-low reaches of the Yangtze River and then the early onset of the rainy season in north China. At the side of the prevailing monsoon belt, in this case, Chang-an would perhaps be affected by the monsoon for a shorter period. Also the westerly trough would be located at a more east position. Such a pattern would be favorable to the development of drought in the Chang-an region.

The above hypothesis will be examined further in the analysis of global atmospheric circulation associated with the rainfall distribution pattern in Asia or on an even larger scale.

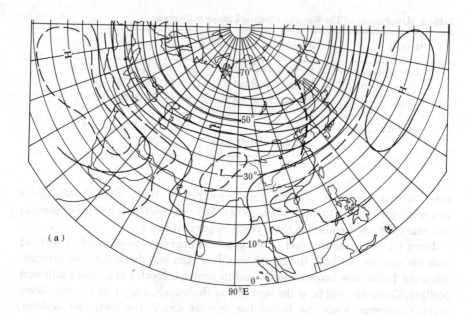

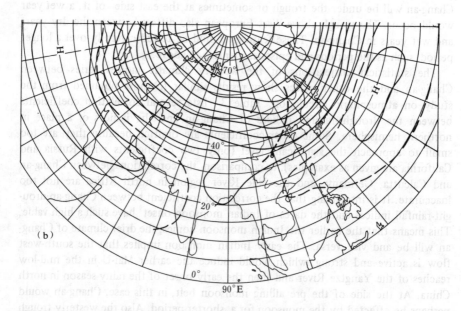

Fig. 1.(a) The composite mean field of summerhalf-year (500hPa H, May–Sept.) associated with a very dry year in Chang-an; (b) The composite mean field of summerhalf-year (500hPa H, May–Sept.) associated with a very wet year in Chang-an

Table 3. Correlation coefficients between the climatic indices of Chang-an and those of other places on the Northern Hemisphere

	Drought/rainfall indices of the mid-reach of the Yellow River	Tree rings on the Qilian Mountains	Dates of Indian monsoon onset	Yearly rainfall in Calcutta	Hight of the yearly nile flood at Cairo	Wetness/dryness index of Germany (50°N, 12°E)	Wetness/dryness index of Russia (50°N, 35°E)	Tree rings on the Mountains in California (37°30′N, 113°12′W)
Drought/rainfall indices of the acient Chang-an region	0.420	− 0.136	− 0.571	0.155	− 0.266	− 0.209	0.13	0.094
Samples	1500	935	64	132	128	83	83	80
Limit values of eonfidence r(α=0.05)	0.062	0.098	0.250	0.174	0.174	0.217	0.217	0.217

62

Acknowledgement. We are grateful to Zeng Zhaomei, Liu Anlin, and Li Li for their help in our work.

References

Atlas of the wet/dry distribution in China for the last 500 years (1981), Cartography Press.
Zhang De-er (1983) *Collections of science and technology in meteorology, Climate and wet dry,* Academy of Meteorological Sciences, State Meteorological Administration (ed). Chinese Meteorological Press pp 17–26

Climate of Japan during 1781—90 in Comparison with that of China

Takehiko Mikami[1]

Abstract — The climate of Japan in the 1780's is reconstructed using weather records of lod diaries. Dryness and wetness patterns in Japan and China during 1781–90 have close relations, being extremely variable year by year. The year 1783 is particularly worth noting in terms of remarkable wetness in Japan and middle China. These relationships are examined statistically using principal component analysis of the July precipitation data set. The results indicate that the first principal component represents simultaneous rainfall in Japan and middle China with opposite indications in north and south China. Finally, the connection between the general atmospheric circulation and unusual weather patterns in recent years are considered synoptically.

Introduction

The climate of Japan in the late 18th century is estimated to have been cool and wet, because sever famines due to crop failure occurred frequently (Yamamoto, 1970). In particular, from 1782 to 1787, Japan experienced the most severe famine in its history, which brought a decrease in population of approximately one million. This period also corresponds to the Little Ice Age in Europe (Lamb 1977; Maejima and Tagami 1983).

However, no instrumental records are available for this period, which indicate spatial and temporal climate variation in Japan. The purpose of this paper is to reconstruct the climate of Japan in the 1780's on the basis of historical proxy data, and to compare them with that of China, which has already been reconstructed.

Reconstruction of Weather Patterns

A weather record described in old diaries is one of the most useful and reliable data sources for estimating climates in historical ages (Maejima and Koike 1976; Mikami 1983). In this investigation, we used several old diaries as proxy data sources. Such diaries can be obtained in many districts in Japan. Figure 1 shows locations of weather records, most of which are derived from official diaries of feudal clans since the 17th century.

In order to examine regional differences in climatic variations, daily weather distribution maps were completed using daily weather records from 1781 to 1790. Descriptions of weather are somewhat different from diary to diary, therefore they were standardized into four symbols; fair (open circle), cloudy (double circle) showery (circle with dot) and rainy (solid dircle). Following this procedure, these

1. Ochanomizu University, Tokyo, Japan

Fig. 1. Locations of historical data.

weather maps were classified into several types according to their rainfall patterns (Fig. 2). For example, type G shows that there are no rainfall areas in the whole country, whereas in the case of type A, every district has rainfall. Type E is characterized by the rainfall in northern and eastern Japan. In the same manner, type C (rainfall in central Japan), type W (rainfall in western Japan), and type S (rainfall in southern Japan) can be explained.

Figure 3 displays a graph of percentage frequency for main weather type in summer (July and August). For the purpose of comparing them with those for recent years, the same kind of weather maps were made from 1976 to 1980. From an examination of the frequencies of various weather types in 1781–1990 and in 1976–1980, it has been found that the percentage of type G had a large year-to-year valiability in the 1780s, as in the late 1970s. Kington (1980) pointed out that the climate of England in the 1780s was also extremely variable with year-to-year variations being much more marked than in the 20th century. Particularly, the frequency of type G exceeded 50 per cent in 1781, 1785, and 1790, which was comparable to the very hot and dry summer in 1978. On the other hand, type A

65

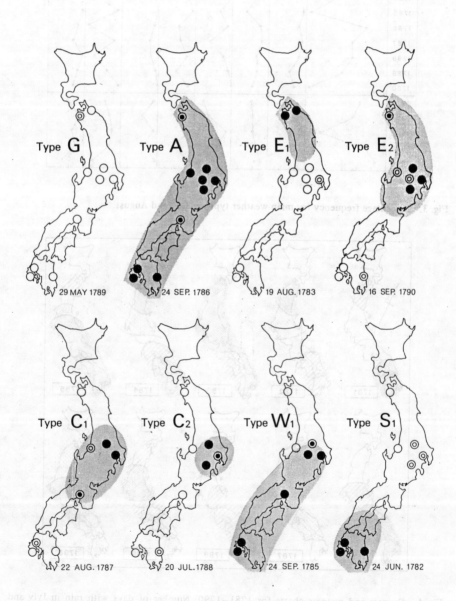

Fig. 2. Examples of weather distribution maps for main types. *Stippling* indicates rainfall area

66

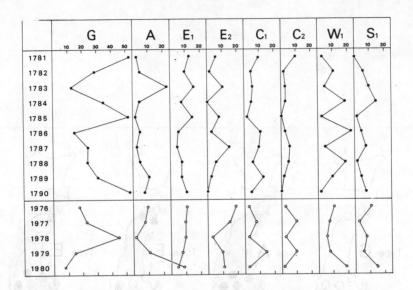

Fig. 3. Percentage frequency for main weather types in July and August

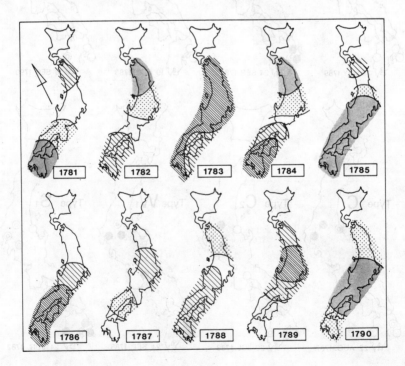

Fig. 4. Dryness and wetness charts for 1781–1790. Number of days with rain in Jyly and August for each district is divided into five grades; very wet (*dense hatching*), wet (*hatching*), normal, dry (*stippling*), and very dry (*dense stippling*)

appears more often in the cool and wet summer, such as in the remarkably cool and rainy summer of 1980. In 1783, when the heaviest famine occurred, the frequency of type A reached the highest percentage in the 1780s.

Dryness and Wetness Patterns

In China, the yearly charts of dryness and wetness for the last 500 - year period were already reconstructed on the basis of a great number of local annals and many historical writings (Central Meteorological Institute et al. 1981). In these charts, the dryness and wetness for each year is classified into five grades; very wet, wet, normal, dry, and very dry, denoted by the numbers 1, 2, 3, 4, and 5 respectively. Droughts and floods in China for the period 1470–1979 were examined by Wang and Zhao Zongci (1981) using these yearly charts.

The same kind of dryness and wetness charts could be reconstructed in Japan by overlapping the daily weather distribution maps as shown in Fig. 2. The number of days with rain in July and August for each district was divided into five grades on the basis of quinteil values for the period 1781–1790 (Fig. 4). It becomes clear that the dryness and wetness patterns varied markedly from year to year in this period. For instance, the north wet and south dry pattern in 1781 was reversed in the next year 1782; the reversal of climate in the north and in the south also appeared in 1784 and 1785. Changes of patterns from 1783 to 1786 were especially noticeable. Therefore, composite maps of dryness and wetness for China and Japan were completed in this period (Fig. 5).

From Fig. 5, the characteristic weather pattern in each year has been clarified: In 1783, when a severe historic famine was brought about, an extremely wet area in Japan extended far into southern China, whereas the northern part of China was relatively dry. In 1784, the north dry and south wet pattern appeared both in Japan and China. The wet area in southwestern Japan seems to have been linked with that in south China. In 1785, Japan had a notably dry climate, as dry as in China. There existed a very large dry area over northern and middle China, which expanded to Japan. In 1786, it was exceedingly wet in southwestern Japan and was also wet along the Changjiang River region in central China.

Statistical Analysis of Rainfall Patterns

In order to explain the spatial relationships in climate conditions statistically, a principal component analysis was applied to the July precipitation data set during 1951–1980 at 29 meteorological stations in China, Korea, and Japan. As a result of computation, it becomes clear that 82.1% of the cumulative variance can be explained by the first ten components. Figure 6 shows spatial eigenvector patterns for the first three components, which explain 43.6% of the total variance.

The first principal component explains 18.1% of the total variance. Positive eigenvectors are distributed from Japan to the central China Changjiang River region, in contrast with negative eigenvectors in the north and in the south. This appears more often in the cool and wet summer, such as in the remarkably cool and rainy summer of 1980. In 1783, when the heaviest famine occurred, the frequency of type A reached the highest percentage in the 1780s.

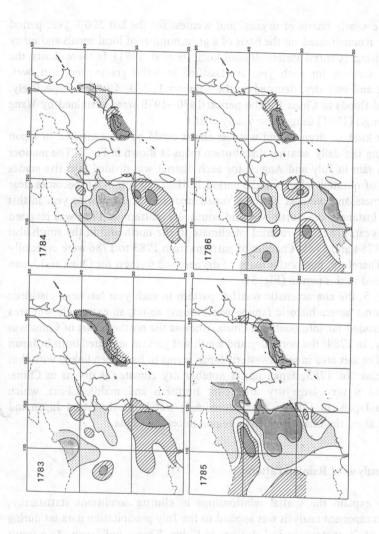

Fig. 5. Composite maps of dryness and wetness for China and Japan. Dense hatching, hatching, no marking, stippling, and dense stippling indicate very wet, wet, normal, dry, and very dry areas respectively

The first principal component explains 18.1% of the total variance. Positive eigenvectors are distributed from Japan to the central China Changjiang River, in contrast with negative eigenvectors in the north and in the south. This means that if the first component coefficients (component scores) are highly positive, for example in 1980, both Japan and middle China have very wet conditions simultaneously, but it becomes dry in the north and south of China. If the coefficients of the first component are highly negative, for instance in 1978, it means that remarkably dry areas extend from Japan to middle China, in contrast with wet areas in the north and in the south of China.

The second eigenvector explains 15.3% of the total variance. Positive values are distributed from northern China to northeastern Japan. Negative values can be seen from the middle and southern China to southwestern Japan. When this component contributes in highly negative, northern parts of both Japan and China have markedly dry climates.

The third eigenvector patterns are somewhat different from the first and the second components. This component indicates opposite signs between Japan and China except for the southern part, which explains 10.2% of the total variance.

Wang and Zhao Zongci (1981) applied an empirical orthogonal analysis to the flood and drought data at 25 stations in China for 1470−1977. The results obtained have a close resemblance to our results in the present paper. Eigenvector patterns for the first, second, and third components displayed in Fig. 6 are similar to those for the third, second, and first eigenvectors by Wang and Zhao Zongci (1981) respectively. It is noteworthy that eigenvector patterns using historical data in China and those using observational data in China, Korea, and Japan show strong similarities.

As mentioned before, the summer climate in 1978 was extremely dry and hot in Japan. It might be comparable to the very dry climate in 1785. On the other hand, it was exceedingly wet and cool in the summer of 1980, which seems to have been similar to the summer climate conditions in 1783. To compare these opposite climate conditions, distribution maps of monthly precipitation in July for 1978 and 1980 have been examined (Figs. 7 and 8). In 1978, unusually dry area extended from middle and southern part of China to Japan except for the Pacific coast. A rainfall distribution map in July 1980 indicates a remarkable feature of zonal patterns. It resembles the weather pattern in 1783 which is well known as a severe famine year. Heavy rainfall belts hatched with more than 200 mm extended from west to east. A zonal rainfall pattern will be explained by the activity of cyclones and polar fronts along these rainfall belts. Figure 9 shows an overlapped map of daily polar fronts for 31 days in July 1980. It is evident that polar fronts are concentrated along the heavy rainfall area from middle China to Japan.

Rainfall Patterns and Hemispheric Pressure Fields

Summer climates in Japan and China, which are characterized by the simultaneous variations of dry and wet areas, are considered to be linked with global circulation patterns. Figure 10 indicates sea surface pressure fields in July 1978 and 1980. In 1978, when it was extremely dry and hot in Japan and middle China, the axis of subtropical high (North Pacific High) was located northward, while in 1980, when it was exceedingly wet and cool in Japan and middle China, the axis of subtropical

70

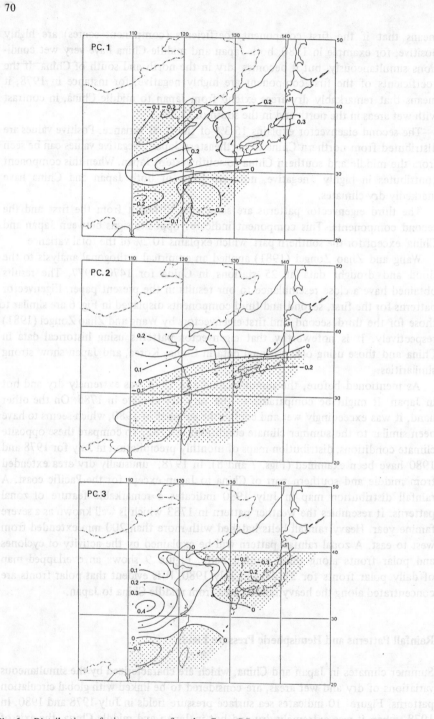

Fig. 6. Distribution of eigenvectors for the first (PC.1), second (PC.2) and third (PC. 3) Principal components

71

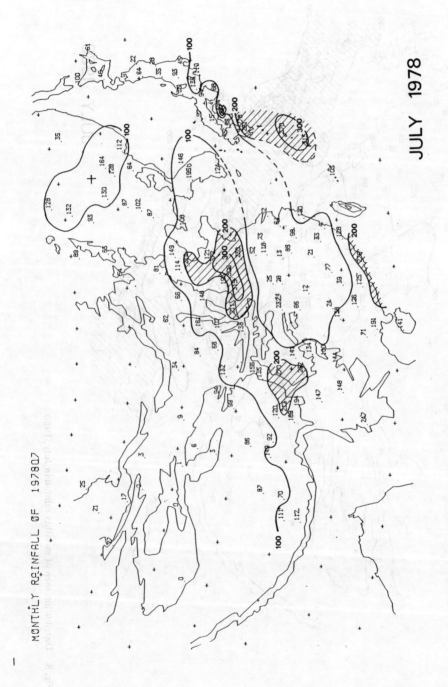

Fig. 7. Distribution map of monthly rainfall in July, 1978

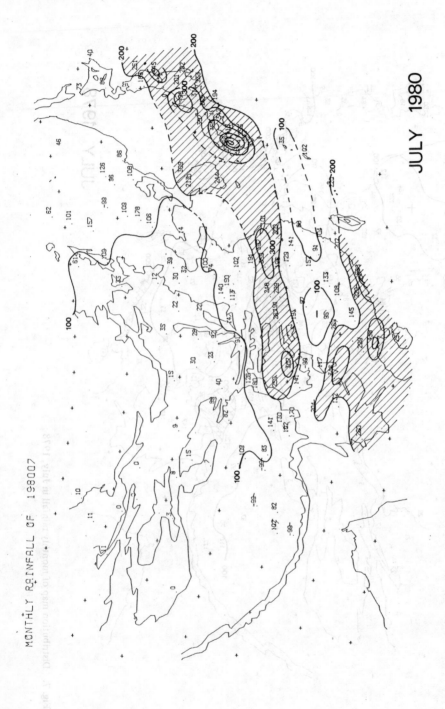

Fig. 8. Distribution map of monthly rainfall in July, 1980

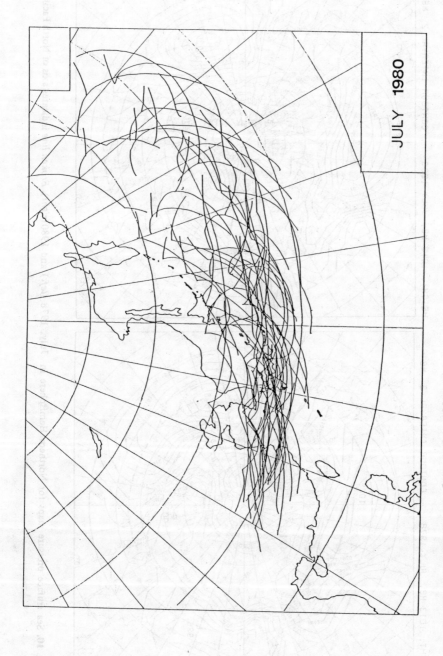

JULY 1980

Fig. 9. Overlapped map of daily polar fronts in July, 1980

74

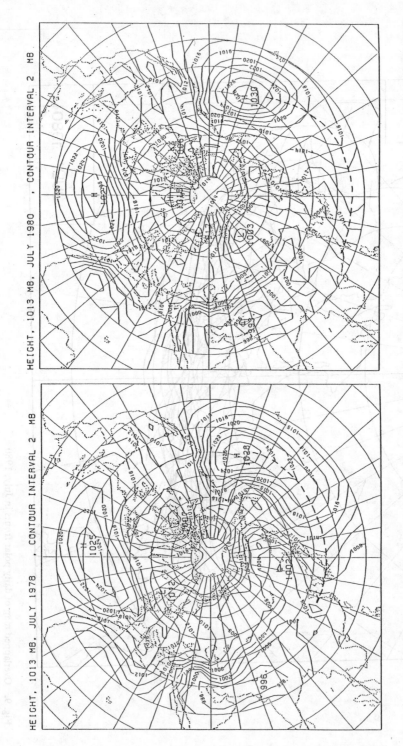

Fig. 10. Sea sruface pressure maps for Northern Hemisphere in July, 1978 (*left*) and 1980(*right*). *Broken line* indicates axis of North Pacific High

high was shifted far to the south, and polar frontal zone extended from middle China to Japan. Yoshino (1963) pointed out that the rainfall area along the frontal zone in June and July often runs from central China to south-western Japan. These results suggest that the circulation pattern in 1783, when it was unusually wet and cool, is estimated to have been similar to that in 1980, and that the circulation pattern of an extremely hot and dry year 1785 might have had resemblance to that in 1978.

Acknowledgement. The author is grateful to prof. Masatoshi M. Yoshino and the Institute of Geography, Academia Sinica, Beijing for providing the meteorological data of China.

References

Central Meteorological Institute et al. (1981) *Yearly charts of dryness/wetness in China for the last 500-year period.* Beijing, 332p

Kington J A (1980) Daily weather mapping from 1781. *Climatic Change* 3: 7–36

Lamb H H (1977) *Climate, present, past and future* Vol 2.Methuen & Co Ltd, London, 835p

Maejima I, Koike Y (1976) *Geogr Rep* Vol 11. Tokyo Metropol Univ, pp 1–12

Maejima I,Tagami Y (1983) *Geogr Rep* Vol 18. Tokyo Metropol Univ, pp 91–111

Mikami T (1983) *Geogr* (Tokyo Geogr,Soc)92: 105–115

Wang Shaowu and Zhao Zongci (1981) Droughts and floods in China, 1470–1979. In: Wigley T M L, Ingram M J, Farmer G (eds) *Climate and history,* pp 271–288. Cambridge, 530p

Yamamoto T (1970) *Meteorol Res Note* (Meteorol Soc Jpn) 105:333–343

Yoshino M M (1963) Bonner Met Abh 3: 1–127

Secular Fluctuations of Temperature over Northern Hemisphere Land Areas and Mainland China since the Mid-19th Century

R. S. Bradley[1], H. F. Diaz[2], P. D. Jones[3] and P. M. Kelly[3]

Abstract — A comprehensive set of long-term temperature station records from throughout the Northern Hemisphere has been used to produce a gridded data set for studies of long-term climatic variability. Areally weighted and normalised seasonal average temperatures for Northern Hemisphere land areas and for mainland China have been computed for 1851–1980 and 1881–1980, respectively. Both records show similar trends; temperature increased from the late 19th century to a maximum in the 1940s followed by a cooling trend which has reversed over the last 10–15 years. In the Northern Hemisphere record, temperatures appear to have decreased from the 1850s to the 1880s but the spatial coverage is poor during this period. Data from China are highly correlated with Northern Hemisphere land area data, suggesting that long-term proxy records from China will provide valuable indications of climatic fluctuations over a much larger area. Extremely sharp drops in temperature, particularly in Fall months, occurred after several major volcanic eruptions. High temperatures are sometimes associated with major El Niño years. On occasions, when an El Niño event has followed a large explosive eruption, the two opposing effects have tended to minimize the climatic fluctuation which would otherwise have resulted.

Introduction

The concentration of carbon dioxide in the atmosphere has increased ~26% since the middle of the last century (Stuiver 1978; Keeling et al. 1982). According to recent projections (Edmonds ° s al. 1984) by about the year 2060 the concentration may reach ~600 ppmv, more than double the level prior to the industrial revolution. In view of the important role that CO^2 plays in the global energy balance, concern has been expressed about the potential climatic effects this build-up may produce (MacCracken and Luther 1985). Two strategies have generally been followed to evaluate this question: (a) modelling and theoretical studies of future atmospheric conditions with enhanced CO^2 concentrations (e.g. Manabe and Wetherald 1980; Hansen et al. 1981) and (b) empirical studies of past climatic conditions to evluate if a "CO^2 signal" can be detected in the climatic record of the last 100–150 years (e.g. Wigley and Jones, 1981; Wigley, et al. 1985) and to assess the role of other factors which may have contributed to climatic variability over this period, and which may be important in the future (e.g. Mitchell 1983; Kelly and Sear 1984).

Here we describe the results of a study of long-term temperature records from land areas of the northern hemisphere. Analysis of long-term data may facilitate the detection of CO_2 effects by providing a better estimate of the range of

1. University of Massachusetts, Amherst, Ma., U.S.A.
2. NOAA/ERL, Boulder, Co., U.S.A.
3. Climatic Research Unit, University of East Anglia, Norwich, U.K.

tate the detection of CO^2 effects by providing a better estimate of the range of natural climatic variability. Particular attention is paid to the instrumental record from China to assess whether long, proxy climatic records from this region may provide a surrogate measure of temperature fluctuations over a much larger area for the period prior to the start of instrumental records (cf. Jones and Kelly 1983).

Data and Methodology

After a careful search of data archives and early meteorological publications, a high quality set of long-term temperature records has been assembled and digitized (Bradley *et al.* 1985). In order to make these records comparable, and in order to avoid combining data which varies widely in absolute terms, all values were expressed as departures from a reference period to enable the maximum number of station records to be compared over the longest interval of time. The optimum period for this purpose was 1946—1960. Accordingly, each set of station data was converted thus:

$$x_{ij} = x_{ij} - x_i$$

were x_{ij} is the data value for month i, year j and x_i is the 1946—60 average for month i, at the same station. The resulting station anomaly data were then used to interpolate values at grid points spaced at $5°$ latitude and $10°$ longitude intervals. By gridding the data, those areas which have dense concentrations of stations are not over-represented in areal averages and the effects of any minor inhomogeneities and locally unrepresentative data are minimized. The interpolation procedure involved fitting an inverse distance least-squares, best fit plane to anomaly values at the six stations nearest the grid point (see Jones *et al.* 1982 for details). This procedure was repeated for all grid points, for all months from January 1851 to December 1980.

Grid coverage is limited to land areas and adjacent oceanic regions, and is not evenly distributed through time (Fig. 1). The maximum coverage occurred in the 1950s, for which period ~57% of the surface area of the Northern Hemisphere can be gridded. Only ~8% of the area of the hemisphere can be gridded for the 1850s, when data collection was restricted mainly to Europe and eastern North America. This change in spatial coverage through time is an important characteristic of the Northern Hemisphere long-term temperature data set and must be considered in any study of hemispheric or regional averages over time.

Temperature Fluctuations: Northern Hemisphere Land Areas

To examine trends in temperature over the last 130 years, gridded anomaly data were averaged by zonal bands ($5°$–$25°$N, $30°$–$55°$N and $60°$–$85°$N) and for the hemisphere as a whole ($5°$–$85°$N). Grid values were weighted by the cosine of latitude to produce spatially representative averages. Figure 2 shows the resulting values. Each value plotted is a seasonal average (Winter = December, January,

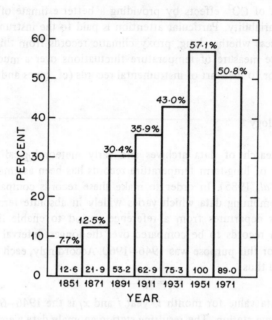

Fig. 1. Percentage of surface area of Northern Hemisphere "represented" by gridded land-based temperature data set, at 20 year intervals. Optimum coverage is in the 1950s when data are available for ~57% of the hemisphere. One hundred years carleer the area which can be gridded was only 7.7% of the surface area of the hemisphere (12.6% of the 1950s coverage)

February; Spring = March, April, May; Summer = June, July, August; Fall = September, October, November). Clearly seen are the generally lower temperatures of the 19th century (particularly the 1880s) and the marked warming episode from ~1915–1925 leading to maximum anomalies in the 1930s and 1940s, followed by declining temperature to the mid 1960s. Since 1965 a warming trend is evident. Of particular note is the pronounced increase in interannual variability from low to high latitudes (cf. Kelly *et al.* 1982). To take this into account, the data were next normalized by dividing each monthly grid-point anomaly value by the standard deviation of the grid point for the respective month in the 1946–60 reference period, viz:

$$x'_{ij} = \frac{x_{ij}}{\sigma_i}$$

where x_{ij} is the anomaly value for month i, year j at each grid point and σ_i is the standard deviation of month i in 1946–60 at the same location.

The resulting values (again weighted by the cosine of latitude) were then averaged to produce normalized, areally weighted hemispheric averages of seasonal temperature anomalies from the 1946–1960 reference period (Fig. 3). Seasonal data were also combined to produce a record of mean annual anomalies (Fig. 4).

WEIGHTED ZONAL RESIDUAL TEMPS. FOR 5N-25N, 30N-55N, 60N-85N, 5N-85N.   84/08/17.

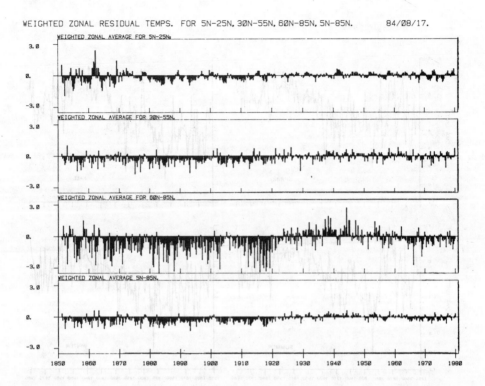

Fig. 2. Temperature anomalies (°C) from the reference period (1946–60) for three latitude bands and for the hemisphere as a whole (bottom). The difference in variability between different latitude bands is clearly seen. Values, plotted are 3-month (one season) averages, beginning with Spring (March, April, May) 1851

All seasons show the same basic low frequency characteristics; **temperatures** were steady, or increased from mid-1850s to the early 1870s, then fell, to reach minimum values in the 1880s. Temperatures increased to a maximum around 1940, mainly in two stages, from ~1880 to 1900 and from 1915 to 1930. After the 1940s peak, temperatures decreased to the early 1970s, followed by a slight increase in recent years. More recent data (not shown) indicate that this recent warming trend has continued into the early 1980s. In fact, the 5–year average (1980–1984) is the highest in the 130 year record (Kelly *et al.* 1985). Among the more important differences between the seasonal records, the relatively high temperature in summer months (and to some extent in Spring) during the late 1860s and 1870s is of interest. In summer months, temperatures at this time generally exceeded that of the reference period (1946–1960). However, it must be recognized that data coverage in the 1870s was only ~25% of that in the 1946–60 period (Fig. 1). Furthermore, differences in methods used to compute daily and monthly mean temperatures have changed over time (Bradley et al. 1985) so the two periods may not be directly comparable.

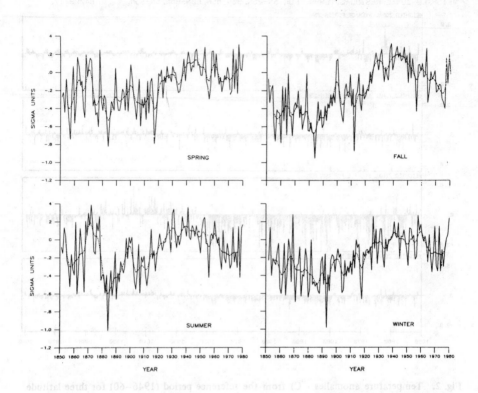

Fig. 3. Normalized, areally weighted seasonal temperature anomalies (from the 1946–1960 reference period) for Northern Hemisphere land areas. Spring = March, April, May; Summer = June, July, August; Fall = September, October, Norvember; Winter = December, January, February (value plotted for the year in which the December occurs, i.e. 1851/2 plotted as 1851). Dark line is a binomially weighted filter which smooths out high frequencies with a period of <10 years

Temperature Fluctuations in Cnina

To investigate whether temperature fluctuations in China over the last 100 years were similar to those over Northern Hemisphere land areas as a whole, a subset of the hemispheric gridded data set was extracted for the area of mainland China (Fig. 5). The same computational procedure already described was carried out on the smaller "China grid" data set. However, unlike the Northern Hemisphere, the best data coverage for China is in the period 1961–1970. Consequently, all anomaly values in the China data set were normalized with respect to 1961–1970. The resulting areally weighted, normalized temperature anomaly graphs are shown in Figs. 6 and 7. Before 1881, there were insufficient data to permit a meaningful regional average to be computed for Mainland China.

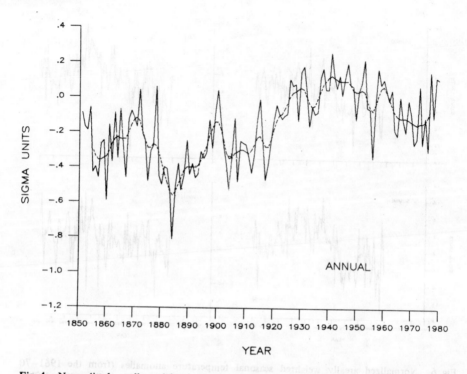

Fig. 4. Normalized areally weighted annual temperature anomalies (from 1946–60 reference period) for Northern Hemisphere land areas. Annual values produced from averaging seasonal values shown in Fig. 3

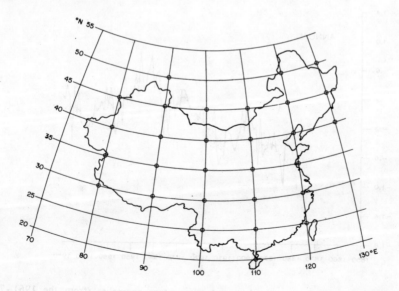

Fig. 5. Grid points (circled) selected for computation of temperature anomalies of China

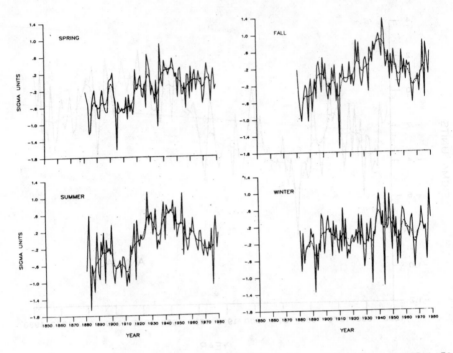

Fig. 6. Normalized areally weighted seasonal temperature anomalies (from the 1961–70 reference period) for China grid (shown in Fig. 5). Dark line is a binomially weighted filter which smooths out high frequencies with a period of <10 years

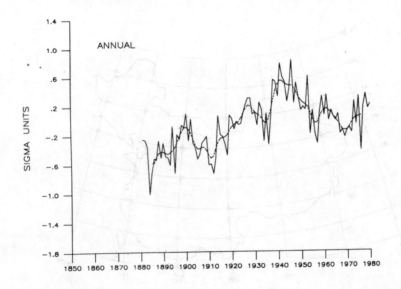

Fig. 7 Normalized, areally weighted annual temperature anomalies (from the 1961–70 reference period) for China grid (shown in Fig. 5)

The general trends (low frequency variations) were similar in China compared to the hemisphere land areas as a whole. However, the Chinese subset shows a more pronounced warming period in the late 1930s, particularly in Fall and Winter months (September–February). This is seen most clearly in Fig. 8 where the low frequency trends have been plotted for direct comparison with those of the Northern Hemisphere (Fig. 9). Both data sets indicate the warmest period of

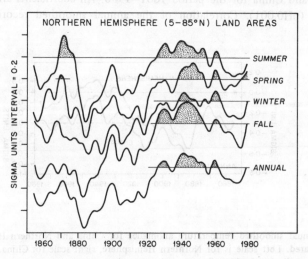

Fig. 8. Seasonal and annual low frequency temperature fluctuations of Northern Hemisphere land areas (from Figs. 3 and 4). The zero reference point has been displaced arbitrarily to enable trends to be compared. The same ordinate scale is used for each plot

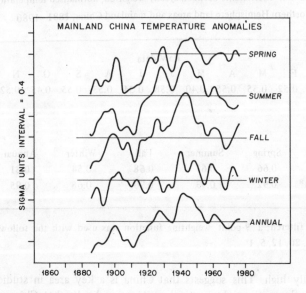

Fig. 9. Seasonal and annual low frequency temperature fluctuations of China (from Figs. 6 and 7). The zero reference point has been displaced arbitrarily to enable trends to be compared. The same ordinate scale has been used for each plot

the 20th century was the 1940s, followed by a cooling trend which seems to have reversed in the last decade or so. Although the Chinese data comprise a subset of the hemispheric land area data, they comprise a small fraction of the total number of data points, so the correlations are quite remarkable. Table 1 shows correlation coefficients between temperature anomalies for Northern Hemisphere land areas and China for the period 1881–1980. All coefficients are statistically significant (pH0.001). Correlation of the annual smoothed records (Fig. 10)

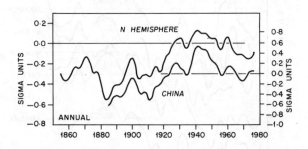

Fig. 10. Annual smoothed temperature anomalies for China and Northern Hemisphere land areas compared. Left scale is for Norhtern Hemisphere, right scale for China. Zero reference point has been displaced arbitrarily to permit comparison

Table 1. Correlation coefficients between areally weighted, normalised temperature anomalies of Northern Hemisphere land areas and mainland China, 1881–1980

						Raw data					
J	F	M	A	M	J	J	A	S	O	N	D
0.46	0.52	0.55	0.55	0.40	0.58	0.54	0.53	0.55	0.45	0.52	0.41

	Spring	Summer	Fall	Winter	Annual
Raw Data	0.66	0.67	0.68	0.54	0.81
Smoothed Data [a]	0.92	0.93	0.90	0.68	0.95

a Binomially filtered; a 9-point weighting function was used with the following weights: 1, 5, 12, 20, 24, 20, 12, 5, 1

is exceptionally high. This suggests that China is a key area in studies of long-term climatic fluctuations. In particular, it seems likely that Chinese historical records of temperature (e.g. Zhang and Gong 1979; Zhang 1980) could be used as a proxy record of temperature fluctuations over Northern Hemisphere land area as a whole, particularly for studies of low frequency trends.

High Frequency Climatic Fluctuations

Of particular interest in the Northern Hemisphere and China area seasonal temperature records are the "spikes" of unusually low and (to a lesser extent) high temperatures which can be seen in the inter-annual data (Figs. 3 and 6). It is significant that the extremely low temperatures which occur in Fall months generally coincide with major explosive volcanic eruptions (Simkin *et al.* 1981), the most important of which appear to have been Krakatau (1883), Katmai (1912) and Bezymianny in 1956 (Table 2). The major effect is short-lived and only occurs

Table 2. Major explosive volcanic eruptions (VEI ⩾ 5), 1851 − 1980 (after Simkin, *et al.*1981)

Name of Volcano	Location	Date of eruption	Volcanic explosivity index (VEI)
Sheveluch	56.78°N, 161.58°E	02.18.1854	5
Askja	65.03°N, 16.75°W	03.29.1875	5
Krakatau	06.10°S, 105.42°E	08.26.1883	6
Tarawera	38.23°S, 176.51°E	06.10.1886	5
Santa Maria	14.76°N, 91.55°W	10.24.1902	6
Ksudach	51.83°N, 157.52°E	03.28.1907	5
Novarupta (Katmai)	58.27°N, 155.16°W	06.06.1912	6
Quizapu (Azul)	35.67°S, 70.77°W	04.10.1932	5
Bezymianny	56.07°N, 160.72°E	03.30.1956	5
Mount St. Helens[a]	46.20°N, 122.18°W	05.18.1980	5

[a] Main volcanic blast directed laterally, not vertically

in the same year as the eruption if the eruption takes place in the first half of the year (perhaps before the stratospheric vortex breaks down). Not all seasons were affected by these eruptions; low winter season temperatures show no apparent relationship to major eruptions, even following Krakatau, the effect of which is clearly visible in all the other seasons.

Extremely warm years are also apparent in the record. Some of these are clearly related to major El Nino events, particularly those years in which El Nino events recurred or persisted from the preceding year [e.g. 1878, 1900, 1926, 1941, 1958, 1973; see Quinn *et al.* (1978) for a chronology of major El Nino events] Interestingly, several of the largest explosive eruptions of the last 100 years have been followed by El Ninos which have tended to both minimise the (cooling) effect of the eruption and the (warming) effect of the El Nino. The remarkable decline in temperatures after the eruption of Katmai in 1912 is perhaps related to the absence of any major El Nino at the time, or in the immediate period thereafter. Further study of circulation anomalies following these events is currently underway.

Summary

Long-term temperature data from stations throughout the Northern Hemisphere indicate that most of the temperature increase of the last 130years took place between 1880 and 1930, particularly in the periods 1880–1900 and 1915–1930. Prior to 1880, temperatures appear to have increased from the 1850s to the 1870s, then declined. The lowest temperatures of the last 130 years during the early 1880s were probably accentuated by effects of the explosive eruption of Krakatau in 1883. After the 1930s, average temperatures decreased to levels similar to those of the 1920s. A slight upturn in temperatures has occurred over the last 10–15 years. The effects of major explosive volcanic eruptions and of exceptional El Nino events are seen as pronounced short-term cooling and warming events, respectively, superimposed on the low frequency temperature trends of the last 130 years. Data from China and the Northern Hemisphere land areas as a whole are highly correlated, particularly on time scales of the order of decades. This suggests that long-term proxy temperature data, from Chinese historical records, may be a useful indicator of temperature fluctuations over the Northern Hemisphere.

Acknowledgement. This work was supported by the U.S. Department of Energy, Contract DE–AC02–81– EV10739.

References

Bradley R S, Kelly P M, Jones P D, Diaz H F, Goodess C, M (1985) A climatic data bank for Northern Hemisphere land areas, 1851–1980. *Tech Rep TR* 017, Carbon Dioxide Research Division U S Dept Energy, Washington D C (in press)

Edmonds J A, Reilly J, Trabalka J R, Reichle D E (1984) An analysis of possible future atmospheric retention of fossil fuel CO_2. *Tech Rep TR* 013, Carbon Dioxide Research Division. U S Dept Energy, Washington, DC

Hensen J, John Son D, Lee P, Rind D, Russell G.(1981) *Science* 213; 957 – 966

Jones P D, Wigley T M L, Kelly P M (1982) *Mon Wea Rev* 110: 59 –70

Jones P D, Kelly P M (1983) *J Climato* 3: 243 –252

Keeling C D, Bacastow R B, Whorf T B (1982) in: Clark W. C. (ed) *Carbon Dioxide Rev* 1982. Oxford Univ Press, New York, pp 377 –385

Kelly P M, Sear C B (1984) *Nature* 311: 740 –743

Kelly P M, Jones P D, Sear C B, Cherry B S G, Tavakol R K (1982) *Mon Wea Rev* 110: 71 –82

Kelly P M, Jones P D, Wigley T M L, Bradley R S, Diaz H F, Goodess C M (1985) The extended Northern Hemisphere surface air temperature record: 1851 – 1984. In: *Extended Sum-*

maries 3rd Conf Climate Varia Symp Contemp Climate: 1850–2100. Am Meteorol Soc Boston, Mass, pp 35 – 36

MacCracken M C, Luther F M (eds) (1985) *The Potential Climatic Effects of Increasing Carbon Dioxide.* U S Dept Energy, Washington D C (in press)

Manabe S, Wetherald R T (1980) *J Atmos Sci* 37: 99 –118

Mitchell J M Jr (1983) Empirical modeling of effects of solar variability, volcanic events and carbon dioxide on global-scale average temperature since A.D. 1880. in: McCormac B M (ed) *Weather and Climate Responses to Solar Variations.* Colorado Associated Univ Press, Boulder, pp 265 – 272

Quinn W H, Zopf D O, Short K S, Kuo Yang R T W (1978) *Fish Bull* 76: 663 – 678

Simkin T, Siebert L, McClelland L, Bridge D, Newhall C, Latter J H (1981) *Volcanoes of the world: a regional Directory, Gazetteer and Chronology of Volcanism, During the Last 10,000 years Hutchinson Ross,* stroudsburg, 232 pp

Stuiver M (1978) *Sicence* 199: 253 –258

Wigley T M L, Jones PD (1981) *Nature* 292: 205–298

Wigley T M L, Angell J, Jones P D (1985) Analyses of the atmospheric temperature record, Chap 4 in: MacCracken MC, Luther F M (eds) *Detecting the Climatic Effects of Increasing Carbon Dioxide.* US Dept Energy, Washington D C (in press)

Zhang Deer (1980) *Kexue Tongbao* 25: 497 – 500

Zhang Peiyuan, Gao-Fa Gong (1979) *Acta Meteoro Sin* 34: 238 – 247

Climatic Variation in Xinjiang Province for the Past 3000 Years

Li Jiangfeng[1]

Abstract

It is difficult to study climatic variation of Xinjiang due to lack of sufficient sources of data. By using limited indicator of climate variation, historical records, tree-ring chronology and glacier age, we have drawn the temperature variation curve for the 3000 or more years. Thus, we can compare the climate of Xinjiang Province with that of other areas in the world.

There is a maximum of precipitation formed in this dry-continental climatic region, surrounded by high mountains and influenced by the westerly jetstream. This precipitation supplies not only the glacier and snow covering the mountain areas, but is also the source of river waters. The richness or insufficiency of river waters, as well as the shortening or lengthening of rivers should be a distinguishing mark of a dry-wet climate. Besides, in the continental region, snow is not only a precipitation, but also indicates a degree of severe cold. It must be cold in years having more snow, and warm in years with less snow. A dry-warm period corresponds to the climate of a cold-wet period. The specific features of this regularity were also shown for the modern climate. Hence, the variation of precipitation and temperature can be inferred by means of water conditions such as abundance of waters and evolution of rivers, and the climate variation can also be inferred from the variation in snow line and glaciers.

On the Ancient Silk Road, the ancient fortresses alternately disappeared, were restored, or were covered by desert. The lakes and rivers became dry or shifted. Some of these are also a result of climatic variation. Even the abandonment of the Ancient Silk Road and climate are also closely linked. In addition, the climate for that time has been inferred by means of a plant climatic indicator. To sum up, we widely searched the data and made comparison as well as inference.

We have analyzed the climatic variation of the last 500 years and this century by more than 10 chronologies (one hundred or more tree-ring series) in Tian Mount and glacier data. The general conditions and developmental trend of climatic variation have been established.

---

1. Xinjiang Meteorology Research Institute, Xinjiang, China

Relationships Between the Climates of China and North America over the Past Four Centuries: a Comparison of Proxy Records

J. M. Lough[1], H. C. Fritts[1], Wu Xiangding[2]

Abstract – Interannual precipitation variations in China as represented by a documentary record of drought and flood are shown to be significantly correlated with the patterns of large-scale sea-level pressure over the North Pacific. These sea-level pressure anomalies appear to be correlated with the occurrence of widespread drought or flood in both east Asia and North America. The statistical associations are most pronounced when Chinese precipitation lags by one year behind the North Pacific sea-level pressure and North American precipitation. The same relationships are also found between Chinese documentary records and dendroclimatic reconstructions of annual sea-level pressure over the North Pacific and annual precipitation in central North America obtained form 65 North American tree-ring series. The Chinese documentary record was developed independently from the dendroclimatic reconstructions yet the time series exhibit statistically significant linkages over the past 360 years, even though they are from opposite sides of the North Pacific. The strength and character of these linkages can, however, vary from one decade to the next and appear to be dominated by particular lowfrequency variations representing time periods of 20 or more years. These teleconnected patterns are weak in that there are years and decades where the relationship in indistinguishable from random variations. However, there is some stability of the relationships among the records for each of the past four centuries and thus the identified teleconnections are likely to continue in the future. The stability of the relationships also serves as one type of verification of the dendroclimatic reconstructions because the patterns of variation identified in the 20th-century instrumental period are also present in the reconstructions obtained for earlier years.

Introduction

Prior to the widespread introduction of instrumental meteorological records in the mid-19th century, the spatial and temporal variations of past climate have to be inferred from a variety of proxy climate indicators (Kutzbach, 1975). No single proxy series can provide an exact estimate of past climate, as each is likely to contain bias and error terms unrelated to climatic factors (NAS, 1975). Comparison, cross-checking and integration of independently derived proxy series are, therefore, necessary to reach the most reliable spatially and temporally detailed description of past climate variations (Gordon et al., in press; Williams and Wigley, 1983). Such comparisons can be hampered by a) different time resolution and frequency characteristics of the proxy series, b) differences in climatic parameters they represent and c) proxy data from different localities.

In the present study we avoid problem (a) by considering two accurately

1. Laboratory of Tree-Ring Research, University of Arizona Tucson, Arizona 85719
2. Institute of Geography, Chinese Academy of Sciences, Beijing, China

dated, annual proxy climate series. We overcome problems (b) and (c) by first establishing whether significant correlations can be found between climate in different regions using spatial grids from the instrumental record. If there are also statistically significant correlations between the same regions and time period when proxy data are used in place of the instrumental record, one can conclude that the proxy data appear to be responding, directly or indirectly, to the same climatic factors. However, if the proxy series were calibrated or derived from analyses of the instrumental record over the modern time period, then the two proxy series cannot be considered truly independent for that time period. It is necessary to obtain correlations from independent data, in this case from intervals before the 20th century. If the correlations are significant and show similar patterns in the proxy data from the earlier time period, then it can be concluded that proxy data are consistent and hence appear to be valid records of past variations in climate.

Dendroclimatology can provide accurately dated, yearly estimates of spatial arrays of past climate (Fritts, 1976; Hughes et al., 1982). Documentary records can also provide accurately dated information about interannual climate variations (Ingram et al., 1981). The Central Meteorological Bureau (1981) has assembled and analyzed a variety of historical information from China to produce charts of yearly wetness and dryness indices at 120 sites covering the period 1470 to 1979. Large interannual variations in the Chinese monsoon cause widespread floods and drought. Xu and Zhu (1983) demonstrate that these are linked with global sea-level pressure patterns, especially those of the Pacific. Such variations can have a large social and economic impact upon China (Lau and Li, 1984).

In this study we examine how variations of precipitation in China, as repre-sented by the long documentary records of drought and flood, compare with a) the instrumental record of sea-level pressure and surface climate of the North Pacific and North America for the 20th century, b) the dendroclimatic reconstructions for the 20th century and c) the dendroclimatic reconstructions spanning the period 1602–1900. The following questions were considered:

a) Are Chinese precipitation variations correlated with the instrumental sea-level pressure variations over the North Pacific Ocean?
b) If a significant statistical relationship exists, are surface climate conditions of the United States and southwest Canada also correlated with Chinese preci-pitation?
c) Are the reconstructions sufficiently accurate to show similar relationships for the same time period as the instrumental record?
d) Are relationships identified from the instrumental record and the dendrocli-matic reconstructions during the 20th century also evident in the reconstruc-tions prior to the 20th century when the dendroclimatic data are independent of the calibration period?

Once it is established that the same relationships are found in the independent period, it is possible to examine the stability of these relationships by considering different periods of time. If the recurrence of these relationships is consistent over time, this fact itself would serve as verification of the reconstruc-tions and demonstrate that the reconstructions and the proxy precipitation data derived from different sides of the Pacific Ocean provide valid information about climate at least back to 1602. If stability is low, then it may be inferred that either the teleconnections or linkages in climate causing the correlations, or the statistical associations between the proxy data, vary over time.

Data

Dendroclimatic Reconstructions

Dendroclimatic reconstructions of sea-level pressure, temperature and precipitation have been developed following the multivariate-transfer-function techniques described by Fritts et al. (1979). A network of 65 well-dated tree-ring chronologies, each of which was developed from ten or more trees at a particular site (from western North America) (Fritts and Shatz, 1975), was used to reconstruct seasonal sea-level pressure at 96 grid points between 100°E to 80° W and 20° N to 70° N. (Grid points are located at 10-degree latitude intervals and at 10-degree longitude intervals between 20° N and 50° N and at 20-degree longitude intervals at 60° N and 70° N.). The seasonal temperatures came from 77 stations and the precipitation records, from 96 stations in the United States and southwest Canada.

The seasons that were used for the reconstructions are: winter, December to February; spring, March to June; summer, July to August; and autumn, September to November. They were chosen to allow for differences in the tree-ring response during the course of the year. Transfer-function models (Fritts 1976; Blasing 1978; Lofgren and Hunt 1982) were calibrated with seasonal instrumental data over the period 1901 to 1963 (1899 to 1963 for sea-level pressure). Verification of the seasonal temperature and precipitation estimates was performed using independent data prior to 1901. For sea-level pressure, where insufficient data prior to 1899 were available for statistical verification, subsampling during the calibration period was used for verification. This verifies the general form of the models but not their final estimates (Gordon 1982). Thus, comparisons with the Chinese precipitation data could provide additional verification of the estimates prior to 1899 (Gordon et al., in press).

Several models involving different lags and numbers of predictor and predict variables were developed for each season and were variable. The estimates from the two or three models with the best calibration and verification statistics were averaged, thus increasing the reliability of the final reconstructed seasonal climate. Annual estimates (December to November) were obtained by averaging the estimates from the four seasons for temperature and pressure and by totaling the seasonal precipitation estimates. Details concerning the sea-level pressure and temperature series are given in Fritts et al. (1979) and Fritts and Lough (in press). Generally, the annual estimates calibrate more variance than the seasonal estimates. An average of 47 percent of the annual sea-level pressure variance, 49 percent of the annual temperature variance and 26 percent of the annual precipitation variance was calibrated. The reconstructions have the largest errors and are least reliable in regions furthest removed from the tree-ring sites: over the northern Asian continent for sea-level pressure, and in the eastern United States for temperature and precipitation.

Chinese Precipitation

Numerous historical writings have been processed to produce yearly wetness and dryness indices for 120 sites in China during the period 1470 to 1979 (Central Meteorological Bureau, 1981). The dryness or wetness of each year is assigned to one of five grades: 1 = very wet, 2 = wet, 3 = normal, 4 = dry and 5 = very dry.

There are, however, considerable variations in the number of sites with assigned index values for a given year. For the period after 1600, 45 percent of the site-years (number of sites multiplied by the number of years) are missing in the 17th century, 40 percent in the 18th century, 32 percent in the 19th century and 15 percent in the 20th century. To overcome these temporal variations in spatial coverage we selected 44 out of the total of 120 sites with 10 percent or fewer missing values during the period 1600 to 1979 (see Fig. 1). The 44 selected precipitation sites span the area from the south China coast (20° N) to 40° N and from the eastern coast (120° E) inland to 105° E. Relaxation of the selection criteria to 20 percent or fewer missing data values did not substantially increase the spatial coverage into northeast China or west of 105° E, so we limited the analysis to the 44 sites. All missing values within this set were replaced by class 3 (normal) precipitation conditions.

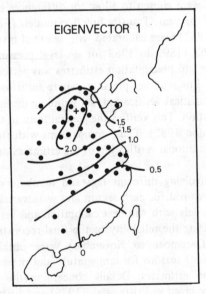

Fig. 1. Eigenvector 1 of 44 Chinese precipitation series 1600 – 1979. *Heavy dots* represent focation of sites

Principal component analysis using the correlation matrix was performed on the series from the 44 sites over the period 1600 to 1979. The results are very similar to those presented by Wang and Zhao (1981) based on 25 sites over the period 1470 to 1977. The first eigenvector which explains 20.1 percent of the total variance, describes rainfall variations of the same sign throughout most of the region (Fig 1). The time series of the amplitudes of eigenvector 1 (referred to as the principal component) indicate some low-frequency variations (Fig. 2, top). An extended dry period (high values of the plot) which occurs in the early 17th century is particularly marked.

The second eigenvector (not shown) accounts for 10.9 percent of the total variance. It describes precipitation variations of opposite sign north and south of about 35° N. The third eigenvector (not shown) accounts for 6.6 percent of the total variance, and describes three zones of precipitation variations. The region centered about 35° N experiences variations of opposite sign to the regions to the north and south. These lower order eigenvectors are again similar to those presented by Wang and Zhao (1981).

Comparison with Instrumental Data

Annual precipitation in east and north China is mainly dominated by the summer southwesterly monsoon from June to August with only a small amount of precipitation associated with the northeasterly winter monsoon. Wang and Zhao (1981) suggest that the information contained in the documentary precipitation indices relates primarily to the summer monsoon season. To test this hypothesis we compared instrumental precipitation data with the documentary record over the period 1951 to 1979.

Table 1. Correlation coefficients between the amplitudes of the first principal components of the 44-site precipitation data and various seasonal combinations from the 17-station instrumental data. Period = 1951 to 1979 (** indicates significance at the 99% confidence level).

17-station season	r
Annual	0.86**
March to May	0.61**
March to September	0.91**
April to September	0.89**
June to August	0.56**
May to September	0.87**
September to November	0.29
December to February	0.21

Monthly instrumental precipitation totals were available for 17 of the 44 stations used (see Appendix). Principal component analysis was performed (using the correlation matrix) on these data for annual totals and various seasonal combinations. The first principal component of the 44-site analysis was then correlated with the first principal components from the instrumental data analyses. Only the first principal component which describes large-scale precipitation variations, was considered here. The results (Table 1) support Wang and Zhou's hypothesis that the documentary records are primarily descriptive of the March to September precipitation totals. There is no apparent linkage with autumn or winter precipitation.

Methods

To examine the linkages between Chinese precipitation variations and the climate of the North Pacific and North America, we analyzed the data in the following ways. First, we calculated the correlations of the Chinese precipitation principal component with the instrumental records of North Pacific and North American climate over the period 1901 to 1961. Second, we examined the correlations of the principal component with the reconstructed data over the same period. Finally, we examined the correlations of the principal component with the reconstructed data from 1602 to 1900. In each case the first principal component was correlated with the time series of sea-level pressure, temperature, and precipitation, and maps were produced showing the areas of positive and negative correlation of each variable with the Chinese precipitation series. One set of correlations between Chinese precipitation and the three climatic variables was produced using the same year for each; then the data were offset to create lags of +1 and −1 years and the correlations were mapped again. In this paper all correlation coefficients are tested for significance at the 95 percent confidence level.

The reconstructions for the 20th century are from the calibration period and, therefore, are forced by the calibration procedure to look like the instrumental record. As a result, the 20th century reconstructions contain the same general teleconnections as the instrumental data. Reconstructions prior to the 20th century are independently derived and thus will consistently show similar teleconnections only if the calibration models reproduce the same relationships. Both the seasonal and annual series were examined, but the annual data gave the most consistent results. This was to be expected, as we had averaged the reconstructions from two to three models and over the four seasons to obtain the annual results. The error in these reconstructions has been shown to be much smaller than the errors of the original seasonal estimates (Fritts and Lough, in press). Thus we present only the analysis of the more reliable annual results.

To assess the collective significance of a group of significance tests based on finite and intercorrelated data sets, the procedures recommended by Livezey and Chen (1983) were followed. First, the critical percentage (significant at the 95 percent confidence level) of grid points or stations of the finite set of values was determined from the binomial distribution. For sea-level pressure and precipitation (with 96 grid points or stations) this percentage was 9.4 and for temperature (with 77 stations) 10.1. Next, 500 random simulations of the correlation procedures were performed for each instrumental and reconstructed annual and seasonal data set. From these simulations we determined the threshold percentage of grid points of stations necessary to produce a collective set of significant statistics while allowing for the effects of spatial correlation on the degrees of freedom of the data set.

The results for the annual data (Table 2) show that the reconstructions contain generally higher spatial correlation with fewer spatial degrees of freedom than the instrumental records. The greater spatial correlations of the reconstructions occur because only the largest principal components of climate were selected for calibration and reconstruction.

Table 2. Critical percentages of grid points or stations significant at the 95% confidence level for instrumental and reconstructed annual sea-level pressure, temperature and precipitation based on 500 random simulations.

	Sea-level Pressure	Temperature	Precipitation
Instrumental 1901–1960	14.5	16.5	9.5
Reconstructed 1901–1960	22.5	18.5	16.5
Reconstructed 1602–1900	22.5	20.5	14.5

Results

Instrumental Data

The correlations between the annual instrumental sea-level pressure gridpoint values and the first principal component of Chinese precipitation are shown in Fig. 3 for data matched to the same year and for the sea-level pressure preceding the Chinese precipitation series by one year. The maps of correlations with the precipitation preceding the sea-level pressure were also calculated and exhibited no distinctive or significant correlations, so are not shown.

We also tested the correlations between the instrumental sea-level pressure distribution and the second and third principal components of Chinese precipitation variation. The latter principal components describe small-scale precipitation variations, predominantly in zonal bands, over China. No widespread, significant linkages were found between these components and the larger scale sea-level pressure distribution over the North Pacific. This result can be attributed to the small scale of the lower order precipitation eigenvectors and the relative coarseness of the sea-level pressure grid used in this study. We, therefore, confined our analyses to the first principal component of Chinese precipitation variations, which, as we show below, is singificantly correlated with sea-level pressure over the North Pacific on a spatial scale that is also likely to be identified in the dendroclimatic reconstructions.

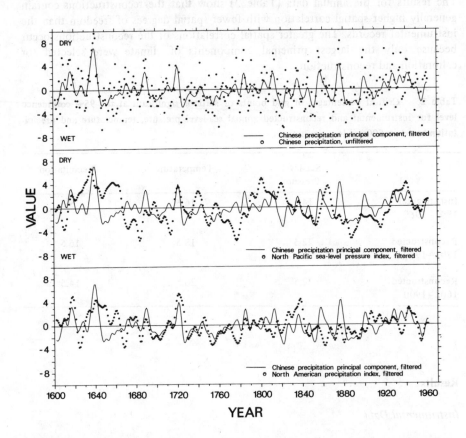

Fig. 2. Unfiltered (dots) and filtered (line) amplitudes of Chinese precipitation principal component 1 (top); filtered amplitudes of Chinese precipitation principal component 1 (line) and the North Pacific sea-level pressure index (dots) (middle); and filtered amplitudes of Chinese precipitation principal component 1 (line) and the North American precipitation index (dots) (bottom). The data are smoothed with a digital filter passing 50 percent of the variance at 8 years

When the data are matched to the same year, 13.5 percent of the sea-level pressure grid points are significantly correlated with the first principal component of Chinese precipitation. An even higher percentage of grid points, 35.4, is significantly correlated when the sea-level pressure leads the precipitation by one year. This percentage of significant correlations greatly exceeds the threshold value of 14.5 percent expected by chance (Table 2). For both cases, high positive values of the first precipitation principal component, representing drier conditions throughout China, are statistically linked with lower sea-level pressure in the central North Pacific. Conversely, wetter conditions are linked to higher sea-level pressure.

We also examined the correlations with seasonal sea-level pressure and noted that the annual values appear to resemble the spring (March–June) and winter

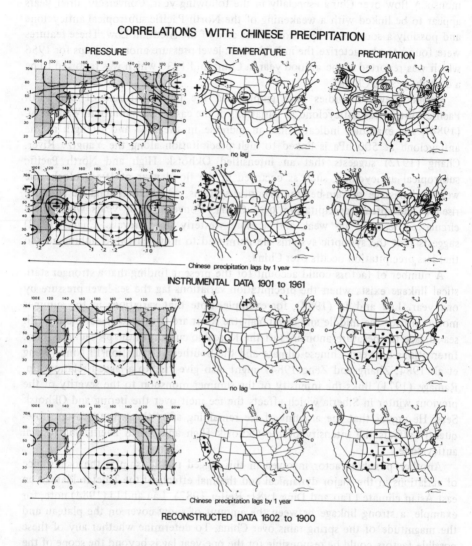

CORRELATIONS WITH CHINESE PRECIPITATION

Fig. 3. Correlations between Chinese precipitation principal component 1 and instrumental and reconstructed annual sea-level pressure, temperature and precipitation. *Heavy dots* represent grid points or stations with correlation coefficients significant at the 95% confidence level

(December—February) sea-level pressure fields where 25.0 and 21.9 percent, respectively, of the correlations were significant at a lag of one year. No clear association was found for summer (July—August) and autumn (September—November).

Wet years in China, as measured by the 44-site principal component, therefore, appear to be linked with an intensified and expanded North Pacific subtropical anticyclone and, to a lesser extent, an increase in pressure of the Siberian High.

These conditions presumably lead to an intensification of the south/southwesterly monsoon flow over China especially in the following year. Conversely, drier years appear to be linked with a weakening of the North Pacific subtropical anticyclone and possibly a southerly expansion of the area of the Aleutian Low. These features were found to characterize the individual sea-level pressure anomaly maps for 1956 which was reported to be a flood year in China and 1936 which was reported to be a drought year in China.

Several previous studies have emphasized the important role of the North Pacific subtropical anticyclone to variations of the east Asian monsoon. Wang et al. (1984), for example, indicate that an increase in the area and strength of the anticyclone at 500 hPa is linked to high precipitation along the Yangtze River. Chang (1972) suggests that an intensified Okhotsk High and North Pacific subtropical anticyclone, such as occurred in the flood year of 1954, blocks the westward passage of rain-bearing synoptic systems, which then stagnate and give rise to widespread precipitation over China. Conversely, in years when these circulation features are weak and a more westerly flow prevails, Chang (1972) suggests that the synoptic systems are permitted to move eastward unhindered so that less precipitation occurs over China.

A number of factors could account for the present finding that a stronger statistical linkage exists when the precipitation variations lag the sea-level pressure by one year. Lau and Li (1984), for example, note that variations in the summer monsoon over the middle and lower Yangtze River are significantly correlated with sea-surface temperature anomalies of the Kuroshio current in the preceding winter. Interactions between Chinese climate and the Southern Oscillation/El Nino (Wang et al. 1984; Zhang and Zeng 1984) might also give rise to a lagged relationship. Ramage (1971) links the intensity of the summer monsoon to the severity of the previous winter in Siberia which affects the ice melt over the Bering and Okhotsk Seas. He argues that the amount of ice remaining after the winter season subsequently affects the intensity of the Okhotsk High and North Pacific subtropical anticyclone.

Another possible factor involved in this lagged relationship is the occurrence of variations in the major dynamical and thermal effects of the Tibetan Plateau on east Asian climate (Tao and Ding 1981; Reiter 1982). Lau and Li (1984) note, for example, a strong linkage between the amount of snow cover on the plateau and the magnitude of the spring rains over China. To determine whether any of these possible factors could be responsible for the one-year lag is beyond the scope of the present study, but these possiblities are certainly questions which require more attention.

There is no obvious pattern in the correlation at zero lag between Chinese precipitation and North American temperatures (Fig. 3). When the precipitation variations are lagged by one year following the temperature data (Fig. 3), however, 23.4 percent of the stations show significant correlations. This figure exceeds the critical percentage for significance (Table 2). Dry conditions in China are linked to warm temperatures in the western United States. Such a relationship can be attributed to enhanced southerly air flow into the western states due to the weakening of the North Pacific subtropical anticyclone. Conversely, wet years in China are associated with enhanced northerly air flow and cooler temperatures in the West.

Relationships between Chinese and North American precipitation variations are less clear (Fig. 3) possibly due to the greater spatial variability of this parameter. The percentages of stations significant at zero (13.5) and one-year lag (10.4) both exceed the critical percentage (Table 2). At both zero and one-year lags, dry conditions in China appear to be statistically correlated with wet years in the central United States and Gulf states, especially in the lower Mississippi drainage area.

Reconstructed Data

The·correlation maps for the period 1901 to 1960 of the reconstructed climate data are not shown, as they are very similar to the instrumental maps shown in Fig. 3. However, the magnitudes of the correlation coefficients are lower than for the instrumental data because there is always less information in the reconstructions of climate than in the original instrumental record from which they were derived.

The results for the independent period of the dendroclimatic reconstructions, 1602 to 1900, are presented in the lower part of Fig. 3. There is a strong resemblance in the pattern of correlation coefficients between the reconstructions for 1602–1900 and instrumental data during the 20th century (Fig. 3, top), both when the data are matched to the same year and when Chinese precipitation lags behind the sea-level pressure field by one year. The pattern correlations between the instrumental and reconstructed sea-level pressure data maps are 0.36 and 0.34 for zero and one-year lags, respectively. Both values are significant. In addition, the percentages of grid points with significant correlations (24.0 and 26.0 for zero and one-year lags, respectively) exceed the critical percentages given in Table 2. An area of significant correlations occurs in the vicinity of the North Pacific subtropical anticyclone. The analysis of the instrumental data shows that this circulation feature is most strongly associated with the Chinese precipitation variations during the 20th century. As one might expect, linkages over the east Asian Continent are poorly correlated because of the low calibrated sea-level pressure variance in this region.

The sea-level pressure correlations calculated for subperiods 1602 to 1700, 1701 to 1800 and 1801 to 1900 are not reproduced here as they show the same reconstructed patterns as those in Fig. 3. This similarity in patterns suggests that the average sea-level pressure linkages are relatively stable from one century to the next.

The resemblances between the results for the reconstructed and instrumental temperature patterns are less striking. The pattern correlations between the instrumental and reconstructed data maps are 0.33 and 0.34 for zero and one-year lags, respectively. Both are significant. The percentages of stations with significant correlations, 5.2 and 3.9 for zero and one-year lags, respectively, do not, however, exceed the critical values shown in Table 2.

A much-expanded area of significant correlations is apparent in the maps of correlation with precipitation (Fig. 3), when compared to those for the instrumental data. Dry conditions in China during the period 1602 to 1900 appear to be significantly related to wetter reconstructed conditions over much of the central United States. The pattern correlations between the instrumental and reconstructed data maps are significant with coefficients of 0.56 and 0.51 for zero and one-year lags, respectively. The percentages of stations with significant correlations, 33.3 and 28.1, respectively, also exceed the critical values (Table 2).

Extreme Dry Minus Wet 10-year Periods

To further examine the apparent linkages between the two proxy records, running 10-year means were calculated for the amplitudes of the first principal component of the 44-site precipitation data. The driest 10-year periods in each century prior to 1900 were found to be 1634–1643, 1714–1723 and 1856–1865 and the three wettest 10-year periods were 1645–1654, 1724–1733 and 1889–1898. The dendroclimatic reconstructions were averaged over these periods and the mean of the wet 10 years subtracted from the mean of the dry 10 years. Each century was examined separately, partly to test for stability of the patterns and partly to prevent longer-term trends, if they were present in the dendroclimatic reconstructions, from distorting the results.

The difference between the dry and wet extremes in the 17th century (Fig. 4, top) shows a region of anomalously low pressure centered south of the Aleutians in the dry years when compared to wet years. This distribution suggests a weakening of the North Pacific subtropical anticyclone and an associated southward expansion and intensification of the Aleutian Low. The most marked temperature change at this time is a general cooling in the northern central states, probably associated with increased air flow from the North American Arctic. Over most of the United States, generally wetter conditions were reconstructed, possibly resulting from a southerly displaced storm track.

In the 18th century (Fig. 4, middle), a low-pressure anomaly is centered in the vicinity of the North Pacific subtropical anticyclone, indicating that this circulation feature was less intense in the dry years as compared to the wet years in China during this century. This circulation anomaly would give rise to increased southerly air flow and warming in western North America. Cooling, which could arise from enhanced northerly air flow, is confined to the East. Drier years in China are also associated with wetter conditions over the south and central United States and drier conditions in the North. The wetter conditions reconstructed in the Southwest are consistent with lower pressures over the eastern North Pacific and possible southerly displacement of storm tracks. The reconstructed characteristics of the dry- and wet-year periods in the 17th and 18th centuries appear to be broadly similar to each other and to the teleconnection patterns described previously.

The average differences in dry-minus-wet periods estimated for the 19th century extremes (Fig. 4, bottom) show, however, little similarity to those in the two prior centuries. A marked weakening of the Arctic High, with a possible strengthened and displaced Aleutian Low, is associated with dry years in China. This pattern is, in turn, associated with warming in the north and central states. Drier conditions were generally reconstructed over most of the United States. These

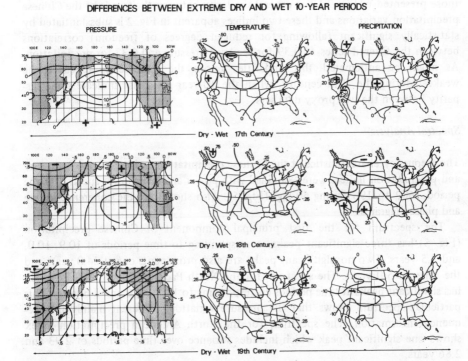

Fig. 4. Differences of reconstructed sea-level pressure, temperature and precipitation between average dry 10-year periods and average wet 10-year periods in China in the 17th, 18th and 19th centuries. *Heavy dots* represent grid points or stations with values significantly different at the 95% confidence level

conditions would presumably result from a northward shift of the major storm tracks, as suggested in the sea-level-pressure difference map. Although pressure is estimated to be lower in the region of the North Pacific subtropical anticyclone, the anomaly is less marked than that of the 17th and 18th centuries. These differences in the 19th century 10-year data could be due to different factors dominating the various climatic variables or they could simply reflect inaccuracies in the tree-ring derived data. Whatever their cause, such differences are to be expected given the low level of statistical correlation between the Chinese precipitation series and the dendroclimatic reconstructions.

To examine these relationships further, two indices were developed of the North Pacific sea-level pressure and North American precipitation variations most closely associated with the Chinese precipitation variations. These indices (shown in the lower part of Fig. 2) were computed by multiplying the statistically significant correlation coefficients from the instrumental analyses at one year's lag by the reconstructed sea-level pressure or precipitation values and forming the sum of these values for each year from 1602 to 1960. For sea-level pressure 33 grid points entered the index and for precipitation 15 stations. (Indices were also computed from regression rather than correlation coefficients but were essentially the same as those presented in Fig. 2.) The general similarity between the trends of the Chinese precipitation variations and these two indices apparent in Fig. 2 is subsstantiated by statistically significant (allowing for reduced degrees of freedom) correlations between the filtered series of 0.39 for sea-level pressure and 0.37 for precipitation. As might be anticipated, the similarities among the three series appear to be weakest during the 19th century, when the 10-year maps also indicate some disparity between the two proxy data sets.

Spectral Analysis

The frequency characteristics of the Chinese precipitation and the sea-level pressure and precipitation indices were examined with power and crosspower spectra for the period 1602 to 1961 using 60 lags (Fig. 5). Also shown are the coherence-square and phase diagrams.

The spectrum for the first principal component of Chinese precipitation (Fig. 5) has three significant peaks, corresponding to time periods of 10.9, 10.0, and 2.5 years. Two nonsignificant peaks are also prominent, one at 24.0 years and the other at 5.0 years. The spectrum for the North Pacific sea-level pressure index has significant peaks at time periods corresponding to 120.0, 4.0 and 2.6 years. This particular spectrum shows that the pressure variations represent mainly low-frequency information. The spectrum for the North American precipitation index shows one significant peak which includes variance over time periods of 2.93 and 2.86 years.

The coherence-square values indicate the frequencies at which significant agreement occurs. Significant coherence-square values between Chinese precipitation and the North Pacific sea-level pressure index can be noted for time periods of infinity to 120.0 years and for 60.0 years. The coherence-square values between the Chinese precipitation and the North American precipitation index are significant at infinity, 24.0 years and 10.0 years. For both the pressure and precipitation indices, the phase angles (see Fig. 5) for the frequencies with significant coherence-square values are near zero, indicating an in-phase relationship. Neither the sea-level pressure nor the precipitation indices show significant coherence with the Chinese precipitation series at high frequencies with periods shorter than 10 years. This suggests that the relationship between the dendroclimatic data and Chinese precipitation series involves primarily low-frequency variations.

In the light of this conclusion, it is interesting to note that the most extreme dry and wet 10-year averages that were used to calculate the 17th-and 18th-century maps in Fig. 4 represented two adjacent 10-year periods approximating one wavelength per 20 or more years. This is close to the 24-year period for North American precipitation that had a significant coherence-square value. Unlike the other 10-year comparisons, the difference between the extreme dry and wet periods in the 19th century involves a time span of 30 to 40 years. As noted earlier, the pattern of the difference in these two 10-year periods was different from the patterns in the previous two centuries. The coherence-square results are consistent with these maps in that there are no significant peaks around 30–40 years. These results suggest that

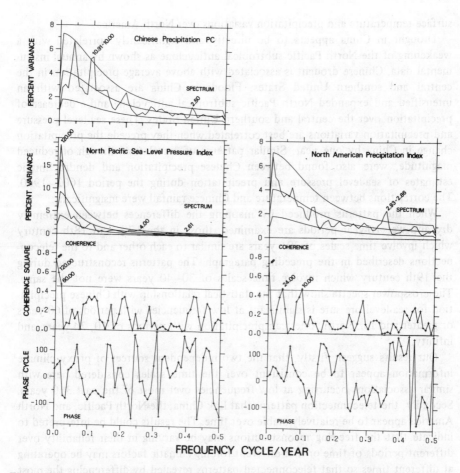

Fig. 5. Power spectra of Chinese precipitation principal component 1, North Pacific sea-level pressure index and North American precipitation index. Coherence-square and phase for cospectra of Chinese precipitation principal component and sea-level pressure and precipitation indices. Period = 1602 to 1961 with 60 lags. (Significant coherence-square values indicated by *dotted line*)

the difference maps would have been in more agreement if they had considered primarily time periods at 10, 24 and 60 or more years.

Discussion and Conclusions

In this study we have attempted to compare two different types of proxy climate records developed from data located on opposite sides of the North Pacific Ocean. To do this, we first established from the instrumental record that drought and flood over China are linked to distinctive sea-level pressure distributions over the North Pacific. The anomalous sea-level pressure distributions were in turn linked to surface temperature and precipitation variations over North America.

Drought in China appears to be directly and significantly correlated with a weakening of the North Pacific subtropical anticyclone as shown by annual instrumental data. Chinese drought is associated with above average precipitation in the central and southern United States. Floods in China are associated with an intensified and expanded North Pacific subtropical anticyclone and a decrease of precipitation over the central and southern United States. These sea-level pressure and precipitation variations are best correlated when they precede the precipitation change in China by one year. Similar patterns of correlation, though of reduced magnitude, were also found between Chinese precipitation and dendroclimatic estimates of sea-level pressure and precipitation during the period 1602–1900. The correlations between temperature and Chinese rainfall were insignificant.

When the patterns produced by mapping the differences between extremely dry and wet 10-year periods are examined, those in the 17th and 18th century which involve time scales of 20+ years are similar to each other and to the teleconnections described in the preceding paragraph. The patterns reconstructed during the 19th century which involve time scales of 30–40 years were not the same. The crosspower spectra show that the statistical relationship with Chinese precipitation for sea-level pressure is significant at low frequencies with periods of 60 years or greater and for North American precipitation with periods of 10, 24 years, and infinity.

Our results suggest, firstly, that the two independent sources of proxy climate information appear to be consistent over the time scales considered here with similar associations occurring at low frequencies over most of the past 360 years. Secondly, the teleconnection patterns that link China, the North Pacific, and North America appear to be relatively stable over time. The results could be interpreted to indicate that the tree-ring reconstructions may be varying in their reliability over different periods of time or that different forcing climatic factors may be operating at different times so that teleconnected patterns revealed by differencing the most extreme 10 years are not always consistent from one time period to the next. Finally, the study as a whole demonstrates how different types of proxy climate series from different geographical locations can be compared by making use of teleconnections established from the instrumental records of climate. Such comparisons are essential for the development of the most reliable description of past climate. Similarities can help to verify that the reconstructed climate patterns are correct, and dissimilarities suggest where errors in the record may exist.

Acknowledgement. The important contributions of T. J. Blasing, G. R. Lofgren and G. A. Gordon to the development of the dendroclimatic reconstructions are gratefully acknowledged. This research was supported by research grants GA-26581, ATM-7517034, ATM-7522378, ATM-7719216 and ATM-8115754 from the Climate Dynamics Program of the National Science Foundation. The assistance of K. Hillier is gratefully acknowledged.

References

Blasing TJ (1978) Time series and multivariate analysis in paleoclimatology. In: Shugart HH

Jr (ed) Time series and ecological processes. SIAMSIMS Conf Ser No 5. Soc Indust Appl Mathemat, Philadelphia, pp 212 = 226

Central Meteorological Bureau (1981) *Yearly charts of dryness/wetness in China for the last 500–year period.* Cartographic Publ House, Beijing, 332 pp

Chang J (1972) *Atmospheric circulation systems and climate.* The Oriental Publishing Company, Honolulu, Hawaii, 328 pp

Fritts H C (1976) *Tree rings and climate.* Academic Press, London, 567 pp

Fritts H G Shatz DJ. (1975) *Tree-Ring Bull* 35: 31–40

Fritts H C Lough JM (in press) *An estimate of average annual temperature variations for North America,* 1602 to 1961. *Climatic Change*

Fritts HG, Lofgren GR Gordon GA (1979) *Quat Res* 12:18–46

Gordon GA (1982) Verification of dendroclimatic reconstructions. *In:* Hughes M K, Kelly P M, Pilcher J R, LaMarche V C, Jr (eds), *Climate from tree rings.* Cambridge Univ Press, Cambridge, pp 58 – 61

Gordon G A, Lough J M, Fritts H C, Kelly P M (in press) Comparison of sea-level pressure reconstructions from western North American tree rings with a proxy record of winter severity in Japan. *J Climate App Meteorol*

Hughes M K, Kelly P M, Pilcher J R, LaMarche V C, Jr (eds) (1982) *Climate from tree rings.* Cambridge Univ Press, Cambridge, 223 pp

Ingram M J Underhill D J, Farmer G (1981) The use of documentary sources for the study of past climates. In:Wigley T ML, Ingram M J Farmer G (eds) *Climate and history.* Cambridge Univ Press, Cambridge, pp 180 – 213

Kutzbach J E (1975) Diagnostic studies of past climates. In: *The physical basis of climate and climate modelling.* GARP Publ Ser No 16. ICSU/WMO, WMO, Geneva, pp 119–126

Lau K, Li M (1984) *Bull Am Meteorol* Soc 65: 114 – 125

Livezey R E, Chen W Y (1983) *Mon Wea Rev* 111: 46 –59

Lofgren G R, Hunt JH (1982) Transfer functions. In: Hughes M K, Kelly P M, Pilcher J R, LaMarche V C (eds) *Climate from tree rings.* Cambridge Univ Press, Cambridge pp 52 – 56

Natl Acad Sci (1975) *Understanding climatic change.* Natl Acad Sci Wash DC

Ramage C S (1971) *Monsoon meteorology.* Academic Press, London 296 pp

Reiter ER (1982) *Bull Am Meteorol Soc.* 63: 1114 – 1122

Tao S, Ding Y (1981) *Bull Am Meteorol Soc* 62:23 – 30

Wang S Zhao Z (1981) Drought and floods in China, 1670–1979. In: Wigley T M L, Ingram M J Farmer G (eds) *Climate and history* Cambridge Univ Press, Cambridge pp 271 – 288

Wang S, Zhao Z, Feng M (1984) *Trop Ocean-Atmos* Newsle 23:13 –14

Williams L D, Wigley T M L (1983 *Quat Res* 20:286 – 307

Xu R, Zhu S (1983) Global sea-level pressure patterns and the drought/floods in eastern China. In: *Collected papers on meteorology science and technology,* No 4. Acad Meteorol Sci. Meteorology Press, pp 27–39

Zhang M, Zeng Z (1984) *Trop Ocean-Atmo Newslett* 23:5 – 7

Astronomical Frequencies in Paleoclimatic Data

A. Berger[1] and P. Pestiaux[1]

Abstract – Among the longest astrophysical and astronomical cycles which might influence climate (and even among all forcing mechanisms external to the climatic system itself; Berger, 1981) only those involving variations in the elements of the Earth's orbit have been found significantly related to the long-term series available in the geological record. Spectral analyses and a paleoclimatic data bank have allowed us to perceive the overall physical meaning of the isotopic spectra collected from deep-sea cores. A typical shape of such power spectra of paleoclimatic variations clearly shows significant peaks of astronomical origin (100 kyr, 41 kyr, 23 kyr, 19 kyr), superimposed on a continuous red noise spectrum. Moreover, the slope in a log-log diagram of the low frequency part of these spectra is definitely in −1, although the high frequencies (periods smaller than 10 kyr) display a more stochastic slope in −2. The stability of these frequencies in deep-sea cores, tested by an evolutive maximum entropy spectral analysis, shows that the dominant 100-kyr cycle is only present during the last 800 kyr roughly.

Relevance of the Astronomical Theory for Research on Present Climatic Variations

Although the astronomical theory of paleoclimates (Imbrie and Imbrie 1979; Berger *et al.* 1984) is usually thought of as being only related to long-term climatic variations at the geological time scale, there are at least four reasons for which its relevance to research dealing with our actual and next future climate can be assessed:

1. Through its direct results: The combination of knowing the exact form of the astronomical input and of observing the geo-climatological output gives a splendid opportunity for *validating* climate models and studying the internal dynamics of the system (Berger 1980a), namely its seasonal cycle, the results of which contribute to an improved theory of climate useful for further applications as the study of man's impact on climate (Berger 1985).

2. Through the hypothetical transitivity of a super-interglacial stage: The effect of increasing the atmospheric CO_2 concentration on the global mean temperature (Berger 1984a) must be viewed in the frame of the glacial-interglacial cycle of the past 150 kyr (Mitchell 1977) and of our near future. Consumption of the bulk of the world's known fossil fuel reserves would plunge our planet into a super-interglacial era. This man-made perturbation of our present warm climate would be totally different if exerted against a different climatic background. As we are still in a interglacial phase, global mean temperature induced by CO_2 and other trace gases during the next century (ies) (Tricot *et al.* 1984) could reach levels several degrees higher than these experienced at any time during the last

1. Univercite Catholique de Louvain, Institut d' Astronomie et de Geophysique G. Lemaitre, 2 Chemin du Cyclotron, B – 1348, Louvain-la-Neuve, Belgium

million years. Moreover, the slow long-term cooling towards the next "ice age" (Berger 1980b) would have to wait until this warming had run its course more than a thousand years from now, if ever, depending on whether the climate system is transitive or not (Lorenz 1968).

3. Through the natural CO_2–climate link: Following recent results by Broecker (1982), Lorius and Raynaud (1983), Pisias and Shackleton (1984), Broecker and Peng (1984), and Oeschger et al. (1984), it seems definite that ice volume, CO_2, and orbital changes are closely related. This relationship means that a natural long-term decrease in the atmospheric CO_2 concentration during the next 5 kyr is expected, but will most probably not exceed 50 ppm in volume. The rate at which this natural lowering of the CO_2 content in the atmosphere will occur is nevertheless very important and requires further investigation.

4. Finally, through transfer of techniques: Very valuable tools and mathematical techniques, shaped and tested in this overall validated astronomical frame (Pestiaux and Berger 1984), will be immediately transferable to the investigation of human-scale climate variations.

These reasons were sufficiently convincing for some of us, at least, to initiate a numerical data bank for long-term climatic change (PACDATA; Berger and Pestiaux 1982). This paper will in fact be limited to the presentation of some of the results that have been recently obtained in the frequency domain thanks to PACDATA.

Astronomical Frequencies in Geological Records

As the climate system is thermally driven by solar insolation, there is a real interest in analyzing the relationships between insolation parameters and climate on the global scale. The difficulty which then first arises is to determine which are the most realistic latitudes and during which periods of the year (Berger 1980a). The adjustment may be tested in the time and/or the frequency domains. This is why the different kinds of insolation which are supposed to be used for modeling the climate or for simulating the climatic variations have been first carefully analyzed as far as their computation, accuracy, and spectrum are concerned (Berger 1984b; Berger and Pestiaux 1984).

Over the last 700 kyr record, $^{18}O/^{16}O$ isotope ratios and other climatic proxy data revealed clear periodicities of 100 kyr, 40 kyr, 23, and 19 kyr (Hays et al. 1976; Berger 1977). According to the now-respected Milankovitch theory (Berger 1980b; Imbrie 1982; Berger et al. 1984) ice ages are driven at these periodicities by changes, respectively, in the eccentricity of the Earth's orbit, the tilt and the precession of the Earth's rotation axis.

Not only these fundamental frequencies are alike, but the climatic series are also phase-locked and strongly coherent with orbital variations (Imbrie et al. 1984). Moreover, most recent models outputs compare favorably with data of the past 400 kyr (Imbrie and Imbrie 1980; Kukla et al. 1981; Berger et al. 1981).

To objectively support the correlations found between the orbital parameters and the past climate, spectral multivariate analyses have been performed between

δ^{18}O deep-sea cores data (Hays *et al.* 1976) and the zonal monthly mean insolations (Berger 1978a, 1978b).

The close agreement between the simulated and the geo-ecological paleoclimlates (87% of the climatic variation is explained; Berger 1980b) authorizes the prediction of the future natural climate. This, however, will only materialize if man's impact on land and the atmosphere has not yet modified the mechanism of climate change and does not do so in the future. The first cold peak will arrive 4 kyr AP, and the model foresees an improvement peaking at about 15 kyr AP, followed by a cold interval centered around 23 kyr AP. Major glaciation, comparable to the stage four of the last glacial cycle, is indicated at 60 kyr AP.

Frequency Analysis of Paleoclimatic Data

There are numerous justifications to a spectral approach for paleoclimate modeling (Imbrie et al. 1984). Its purpose is here to use a combination of different spectral analysis techniques (Pestiaux and Berger 1984) in order to extract an optimal information about the continuum and the peaks characterizing the paleoclimatic variability at periodicities ranging from 10^5 to 10^3 years.

Fourteen of the most accurate existing oxygen isotope records have been selected from the paleoclimatic data bank (PACDATA). They have been classified in three groups (Pestiaux 1984) covering respectively more than 700 kyr for L3, 400 kyr (the last 10 oxygen isotope stages) for L2 and less than 125 kyr (the last Interglacial-Glacial cycle) for L1. The chronologies of these records have been built either by absolute dating, or/and by isotopic stratigraphy without making any a-priori assumption of a relationship between a master curve and the analyzed isotopic record.

Blackman-Tukey Spectral Analysis has been first applied for its confidence in the amplitude estimation and its good statistical properties. Then, Maximum Entropy Spectral Analysis by Least Squares has been used in order to increase the frequency resolution of the statistically significant spectral peaks. An experimental bandwidth has also been added to the statistical one, in order to take into account the uncertainty in the chronology. These upper and lower limits associated with each selected spectral peak have been used to define different frequency bands of increased paleoclimatic variability.

For the 6 cores of L3, three such distinct frequency bands are apparent over the last 800 kyr: B6(200–80 kyr), B5(70–30 kyr), and B4(30–18 kyr), whose mean and standard deviation are respectively 103– 24, 42–8, and 23–4 kyr. All the spectra thus present peaks that can be associated to the 100, 41, and 23 kyr astronomical quasi-periodicities (Fig. 1a illustrates this for core V28238, whose geological data are from Shackleton and Opdyke 1973). Evolutive Maximum Entropy Spectral Analysis applied to the whole records (data window of 400 kyr; for example Fig. 1b) exhibits a satisfying stationarity up to 600 kyr; further back in time, the location of the peaks and their amplitude change strongly, probably in relation to climatic nonstationarities affecting also the sedimentation rate and therefore the chronology.

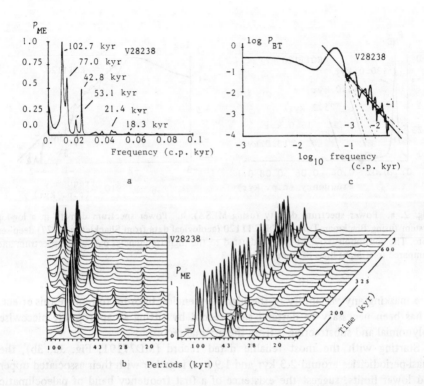

Fig. 1. a. Power spectrum density (using Maximum Entropy Spectrum Analysis-MESA). **b.** Evolutive Maximum Entropy Spectral Analysis. **c.** Power spectrum density in a log-log system (using Blackman-Tukey) of deep-sea core V28 −238 (geological data from Shackleton and Opdyke 1973). The continuous straight line is adjusted to the high frequencies part of the spectrum and compared to straight lines of slopes −1, −2, −3

The Maximum Entropy Power Spectra of the 4 L2 deep-sea cores (Fig. 2a, data from Shackleton 1977) present only a poor frequency resolution characterized by a few number of peaks in the range of the astronomical frequencies. All the peaks are not significantly different from the mean periodicities which are calculated from the four cores: 120 ± 18 kyr, 41 ± 14 kyr and 23 ± 8 kyr.

High sedimentation rate and refined sampling are necessary to identify high frequency paleoclimatic variability. Following these two criteria, four deep sea cores (L1 group) have been selected in the Indian Ocean (data from Duplessy 1982). For each of these cores two foraminifera have been analyzed, one of them being a surface living species. All these cores are strongly influenced by monsoon circulation and therefore by the hydrological cycle. In all the records, a high frequency variability is superimposed to the glacial-interglacial oscillation, implying a strong nonstationarity in the mean. If spectral analysis was misused on these original nonstationary records, the shape of the power spectrum would have been reddened artificially, leading to an amplification of the low frequencies and there-

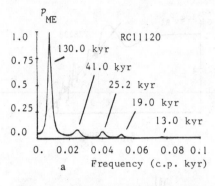 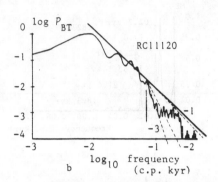

Fig. 2. a. Power spectrum density (using MESA). **b.** Power spectrum density in a log-log system (using Blackman-Tukey) of RC 11120 (geological data from Shackleton, 1977) deep-sea core. The continuous straight line is adjusted to the high frequencies part of the spectrum and compared to straight lines of slope $-1, -2, -3$.

fore masking any variability in the high frequencies. In order to remove this effect, it has been necessary to detrend each record by fitting an appropriate piecewise polynomial and substracting it from the original record.

Starting with the most reliable dated record (MD77191, Fig. 3a, 3b), the quasi-periodicities around 2.3 kyr and 1.9 kyr, together with their associated upper and lower limits, suggest the existence of a first frequency band of paleoclimatic variablity. This (B1) will include the quasi-periodicities ranging from 1.3 kyr to 3.2 kyr. The spectral peaks of the other records, which have a large part of their extreme possible range in this first frequency band, are considered as belonging to it.

Two other frequency bands have been defined in a similar way, one including quasi-periodicities ranging from 3.5 kyr to 6.5 kyr (B2) and the other, those from 7 kyr to 14 kyr (B3). All these quasi-periodicities, refined in frequency resolution and belonging to the same frequency band, have been averaged to give mean quasi-periodicities of 10.2 kyr, 4.6 kyr and 2.3 kyr with corresponding standard deviations of 1.2 kyr, 0.3 kyr and 0.2 kyr.

The Log-Log Shape of the Paleoclimatic Spectra

The only known deterministic inputs to the climate system acting at the scale of the Quaternary paleoclimatic variations are the insolation parameters with quasi-periodicities of 41 kyr, 23 kyr, 19 kyr, and a very weak of 100 kyr. To which extent can the typical power spectrum of paleoclimatic variables be explained by such a deterministic input and/or a stochastic one corresponding to a white noise?

The shape of the continuum and its logarithmic slope can provide indications about the formulation of the basic equations governing the underlying dynamical climatic system. For example, the Power Spectral Density of the output of some

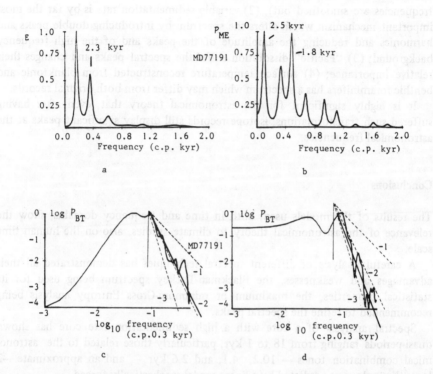

Fig. 3. Power spectrum density of deep-sea core MD77191 (geological data from Duplessy 1982) using MESA for planktonic (**a**) and benthic (**b**) species. The same in a log-log system (**c** and **d**).

simple linear differential systems have a logarithmic slope in -2. Hasselman (1976) has studied the integration of the meteorological fluctuations by such a linear system with two different characteristic times. This model was also used by Kominz and Pisias in 1979 to discuss the spectral slope of paleoclimatic data. In this respect, it is worthwhile to point out that the logarithmic spectra of the oxygen isotope data may differ from core to core. Moreover, the long and medium records have slopes of the order of -1 (Figs. 1c and 2b), whereas the Blackman-Tukey spectra transformed in a logarithmic scale have a -1.8 slope for the planktonic species, while it is even steeper for the deeper living species (Fig. 3c and 3d).

In order to illustrate how the observed spectral shape might be explained by a purely stochastic autoregressive model (a particular case of the linear models just mentioned) or an autoregressive model forced by both insolation and white noise, a set of simulations have been performed (Pestiaux 1984).

To model these spectra, one must first simulate the effects on the spectrum of the physical transformations that the climatic variables have been subject to, before giving their shape to the isotopic variations recorded in deep-sea cores. It was possible to show that (1) bioturbation reduces the amplitude of the peaks (high frequencies are smoothed out); (2) variable sedimentation rate is by far the most important mechanism which alters the spectrum by introducing double peaks .and harmonics and reducing the amplitude of the peaks and of the high frequency background; (3) "Pacific" dissolution shifts the spectral peaks and changes their relative importance; (4) surface temperature reconstructed from planktonic and benthic foraminifera has a spectrum which may differ from both original records.

It is highly significant for the astronomical theory that, even after having suffered such transformations, isotope records still display significant peaks at the astronomical frequencies.

Conclusions

The results of the models used, both in time and frequency domains, show the relevance of the astronomical theory to climate studies, also on the human time scale.

A careful analysis of different spectral techniques has demonstrated all their advantages and weaknesses, the Blackman-Tuckey spectrum being used for its statistical properties, the maximum or minimum-Cross Entropy analysis being recommended to refine the spectral peaks.

Spectral analysis of a core with a high sedimentation rate core has shown quasi-periods ranging from 18 to 1 kyr, particularly those related to the astronomical combination tones − 10.2, 4.1, and 2.6 kyr −, and an approximate −2 logarithmic slope as predicted by a linear model stochastically forced.

Spectral analysis of deep-sea cores covering the last 2 million years and even only the last 10 oxygen isotope stages show significant spectral peaks associated to the astronomical frequencies, with a −1 logarithmic slope. This increase from a −1 to a −2 slope when going from the low to the high frequency parts of the spectra is supposed to be related to a variable sedimentation rate and to the time scale inaccuracies. Evolutive Maximum Entropy power spectrum analysis allows to detect nonstationarities before 600 kyr, most probably linked to the modification of the internal characteristics of the climatic system.

Finally, it is expected that the paleoclimatic data bank will be enlarged to other reliable cores (land, ice, and deep-sea) (any contribution is welcome) and that all the techniques applied here will be used to confirm the results found in this study. In order to help resolve the nonstationarity of paleoclimatic data and the transitivity of the climatic system, Walsh analysis and a boolean approach are also underway. A more complete presentation of the results can be found in Pestiaux (1984) and in Berger et al.(1984).

Acknowledgement. This work was supported in part of the Climate Programme of the Commission of the European Communities under Contract CLI-026-B. The authors would like to express their appreciation to J. Backman, J. Cl. Duplessy, C. Emiliani, W. Prell, M. Sarnthein, N. Shackleton and H. Zimmerman for providing the deep-sea cores data. Thanks are due to Mrs N. Materne and F. Mercier for their technical assistance.

References

Berger A (1977) *Nature* 268: 44—45

Berger A (1978a) *Contribution nr 36, Institut d'Astronomie et de Geophysique G. Lemaitre.* Univ Catholique de Louvain-la-Neuve

Berger A (1978b) *J. Atmos Sci* 35 (12): 2362—2367

Berger A (1980a) *Sun and Climate*,CNRS, CNES, DGRST, Toulouse, 30 Sept. — 3 Oct. 1980, pp 325—326

Berger A (1980b) *Vistas Astronomy* 24(2): 103—122

Berger A (1980) In: Berger A (ed) *Climatic Variations and Variability: Facts and Theories.* Reidel, Dordrecht, Holland pp 411 — 432

Berger A (1984a) Flohn H, Fantechi R. (Eds) *The climate of Europe: past, present and future. natural and man induced climatic changes: an Europenan perspecitve.* Dordrecht, Holland, pp 134—197

Berger A (1984b) Berger A, Imbrie J, Hays J, Kukla G, Saltzman B (Eds), *Milankovitch and Climate* Reidel Dordrecht, Holland, pp 3—39.

Berger A (1985) The astronomical theory of paleoclimates. *World Climate Programme Newslett* (in press).

Berger A, Pestiaux P, (1982) Contribution nr 28, *Institut d'Astronomie et de Geophysique G. Lemaitre,* Univ Catholique de Louvain-la-Neuve

Berger A, Pestiaux P (1984) Berger, A. Imbrie, J. Hays, J. Kukla, G. Saltzman B. (Eds), *Milankovitch and Climate* Reidel Dordrecht, Holland, pp 83—112

Berger A, Guiot J, Kukla G, Pestiaux P, (1981) *Geol Rundsch,* 70: 748—758

Berger A, Imbrie J, Hays J, Kukla G, Saltzman B (Eds), (1984a) *NATO ASI Ser C* vol 126. Reidel Dordrecht, Holland, 895pp

Berger A, Pestiaux P, Gallee H, Gaspar P, Tricot C, Van der Mersch, I., (1984b). *Sci Rep* 1984/6, Institut d'Astronomie et de Geophysique G. Lemaitre, Univ Catholique de Louvain-la-Neuve

Broecker W S (1982) *Prog. Oceanogr* 11: 151—197

Broecker W S, Peng T H (1984) Proc Ewing symp Lamont-Doherty Geol Observatory

Duplessy J C (1982) *Nature* 295: 494—499

Hasselman K (1976) I *Tellus* 23 (6): 473—484

Hays J, Imbrie J, Shackleton N J (1976) *Science* 194: 1121—1132

Imbrie J (1982) *Icarus* 50: 408—422

Imbrie J, Imbrie K P, (1979) *Oce Ages. Solving the Mystery.* Macmillan, London, 224pp

Imbrie J, Imbrie J L, (1980) *Science* 207: 943—953

Imbrie J, Hays J, Martinson D G, McIntyre A, Mix' A C, Morley J J, Pisias N G, Prell W L, Shackleton N J, (1984) Berger A, Imbrie J, Hays J, Kukla G, Saltzman B (Eds) *Milankovitch and Climate* Reidel Dordrecht, Holland, pp 269—306

Kominz M A, Pisias N G, (1979) *Science* 204: 171—173

Kukla G, Berger A, Lotti R, Brown J, (1981) *Nature* 290: 195—300

Lorenz E N, (1968) *Meteorol Monogr* 8(30): 1—3

Lorius C, Raynaud D, (1983) Bach W, Crane A, Berger A, Longhetto A (Eds) CO_2, *Energy and Climate.* Reidel Dordrecht, Holland, pp 145—178

Mitchell J M Jr (1977) *Carbone Dioxide and Future Climate,* EDS, Environmental Data Service (NOAA), March

Oeschger H, Beer J, Siegenthaler U, Stauffer B, Dansgaard W, Langway C C, (1984) In: Hansen J E Takahashi T (Eds) *Climate Processes and Climate Sensitivity* Am Geophys Un pp 299—306

Pestiaux P, (1984) *Approche Spectrale en Modelisation Climatoqie* Dissertation Doctorale, Institut d'Astronomie et de Geophysique G. Lemaitre Univ Catholique de Louvain-la-Neuve

Pestiaux P, Berger A (1984) In: Berger A, Imbrie J, Hays J, Kukla G, Saltzman B (Eds) *Milankovitch and Climate* Reidel Dordrecht, Holland, pp 417—446

114

Pisias N G, Shackleton N J, (1984) *Nature* 310: 757–759

Shackleton N J, Opdyke N D, (1973) *Quat Res* 3(1): 39–55

Shackleton N J, (1977) *Phil as Trans R Soc Lond* 280: 169 – 182

Tricot C R, Berger A, Brasseur C, (1984) *Progress Rep* 1984/4, Institute of Astronomy and Geophysics G. Lemaitre. Catholic Univ of Louvain-la-Neuve

Evidence of Holocene-Late Glacial Climatic Fluctuations in Sediment cores from Tibetan Lakes

K. Kelts[1], J. Francheteau[2], C. Jaupart[2], J. P. Herbin[3],
M. Delibras[4], Shen Xianjie[5]

Abstract

The basinal sediments of modern and ancient lakes commonly contain high resolution, sensitive records of environmental conditions. The problem is to interpret these records in terms of direction and rate of climate change. This paper first reviews the basic principles and methods. It then attempts to relate the sediment record of two 5-meter piston cores obtained during the Sino-French Heat Flow Project (Francheteau et al., Nature 1984) in the lakes Yamzho-yumco and Puma-Yumco, south of Lhasa at altitudes of 4.5 and 5.0 km.

The cores were subjected to analysis for: gamma-ray density, carbon-carbonate, organic carbon, mineralogy, microfossils, stratigraphy and sedimentology. Detailed radiocarbon dating by accelerator mass spectroscopy and pollen analysis are in progress.

Puma Yumco is a clastic lake, receiving meltwater from distant glaciers. It thus has a sedimentation rate about double of the Yamzho Yumco which is more isolated in a drowned drainage. Both lakes have suprisingly low sedimentation rates and provide evidence for wetter and drier alternations rather than a unidirectional trend on the Tibetan Plateau during the interval 23 000 years B.P. to present. Indicators of increased salinity and thus evaporation include the presence of magnesian carbonates, aragonite and traces of gypsum. Lowered lake levels are shown by beds of in-situ grass mats, littoral carbonates or soil horizons. Higher levels are indicated by diatom-rich, laminated oozes which also indicate bottom water anoxia during more nutrient-rich episodes.

These lakes do not seem to show evidence of an ice cap in Tibet during the Late Quaternary, which is consistent with the general model of aridity. The initial results from these cores already show the potential held in a transect of Tibetan lakes which may eventually provide details of the Glacial and Postglacial fluctuations of monsoonal influences across the Himalayan chain.

1. Geol. Inst. ETH-Zurich, CH8092 Switzerland
2. Inst. Physique du Globe, Pl. Jussien, 75230 Paris, France
3. Inst. Francais du Petrole, Rœil-Malmaison, France
4. CNRS, Gif-sur-Yvette, France
5. Inst. Geology, Academica Sinica, Beijing, China

Some Features on Climatic Fluctuation over the Qinghai-Xizang Plateau

Lin Zhenyao[1] and Wu Xiangding[1]

Abstract – Based on an analysis of the atmospheric circulation types and thermal corrections, this paper discusses the climate of the Xizang Plateau before and after the uplift. During late Tertiary, the annual mean isotherms ran latitudinally and the south-north thermal gradient was not as great as present. Since the rapid uplift, it has undergone a great change. During the late Mid-Pleistocene, the thermal distribution types were quite similar to the modern condition.

Introduction

As a huge highland area, the Qinghai-Xizang plateau (Tibet) plays an important role in hemispherical atmospheric circulation, especially in the circulation over East Asia. The climatic fluctuation over the Tibet plateau has seldom been analyzed over a long period because of the severe climate and poor data.

With a comprehensive scientific survey, some data which might indicate local climate variation have been collected. It is necessary and possible, using the data, to describe the climatic change on the plateau, even though the description is preliminary and incomplete.

The Climates During Two Critical Periods

The uplifting of the Tibet plateau has changed the atmospheric circulation and induced the southern Asiatic monsoon since the late Tertiary, when the plateau average level was only 1,000 m. We call the critical period typical for before the uplift. The other is the late Middle Pleistocene when the average level of the plateau was about 3,000 m rather than the 4,500 m today.

Figure 1 a,b,c indicates the annual average isotherms respectively over the plateau during the different periods.

Based on three kinds of correction: i.e., altitude, circulation, and large scale climatic fluctuation, we have discussed the climates of the plateau before and after the uplift (Lin Zhenyao and Wu Xiangding 1981). It is obvious that the annual average isotherms ran latitudinally during the late Tertiary, and the thermal gradient from south to north was not as steep as it is today. Generally speaking, the annual average temperature then was 10–20°C higher than it is now (average 1950–1970) over the Tibet plateau.

After the rapid uplift, a great change occurred. During the late Middle Pleisto-

---

1. Institute of Geography, Academia Sinica, Beijing, China

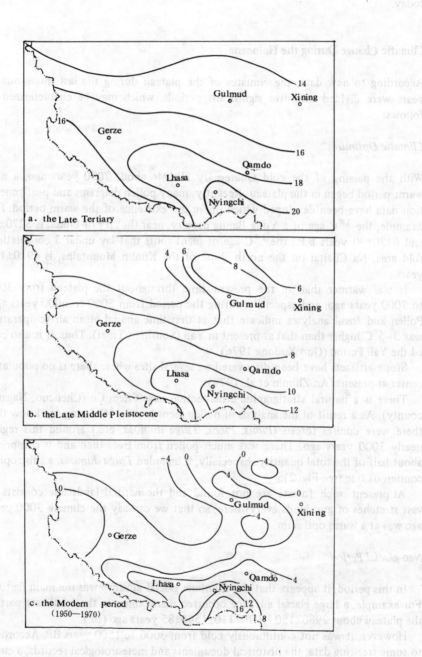

Fig. 1. Distribution of the annual average temperature over the Tibet plateau

cene, the thermal distribution types were already quite similar to the modern condition. The average temperature then was only 4–8°C higher than that of today.

Climatic Change During the Holocene

According to new data, the climates of the plateau during the last thousands of years were divided into five significant periods, which may be characterized as follows:

Climatic Optimum

With the passing of the cold Quaternary climate about 7000 years ago, a new warm period began in the plateau. Recently, many pollen diagrams and peat-generation data have been developed which prove the existence of the warm period. For example, the ^{14}C age of a Yung Bajing peatery, near the city of Lhasa, is 3270±70 and 6130±90 years B.P.; the ^{14}C age of plant roots that lay under a congelation-fold area, Na Chaitai on the north slope of the Kunlin Mountains, is 4910±100 years.

It was warmer than in the present time throughout the plateau from 3000 to 7000 years ago, and especially during the period from 5000 to 6000 years ago. Pollen and fossil analyses indicate that at that time annual mean air temperature was 3–5°C higher than that at present in Yali (Southern Tibet). Thus, it is also called the Yali Period (Guo Xudong 1976).

Stone artifacts have been discovered on several sites where there is no populated center at present (An Zhimin et al. 1979).

There is a natural stratigraphic cross-section near Chem Co (Chencuo, Nagarze county). As a result of the analysis of pollen records and ^{14}C dating, we know that there were conifer forests (*Pinus, Picea, Tsuga dumosa*, etc.) around this region nearly 3000 years ago. There was much pollen from trees then and it comprised about half of the total quantity. Especially, it included *Tsuga dumosa*, a subtropical coniferous tree (see Fig. 2).

At present such forests are not found and the natural landscape consists of vast stretches of grasslands everywhere, so that we can say the climate 3000 years ago was at a warm optimum.

Neo-glacial Period

In this period, it appears that the montane glacial advance was the main feature. For example, a large glacial advance occurred many times in the southeast part of the plateau about 2980±150 1920±110, 1540±85 years ago (BP).

However, it was not continuously cold from 3000 to 1500 years BP. According to some tree-ring data, the historical documents and meteorological records, a curve showing grades of variation of temperature for the last 2000 years in the plateau is shown in Fig. 3. A grade of 3.0 represents a normal condition; a grade greater than 3.0 represents warmer conditions less than 3.0, colder than normal condition (Wu Xiangding and Lin Zhenyao 1981a).

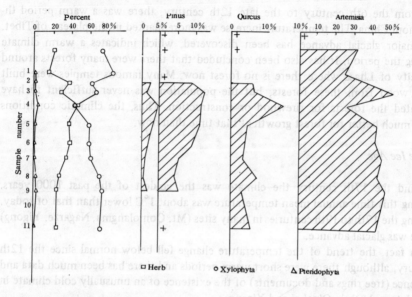

Fig. 2. Pollen diagram near Chem Co

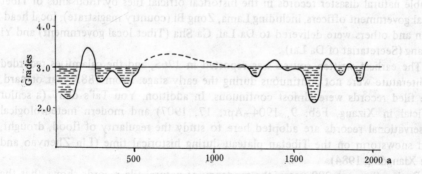

Fig. 3. Grades of temperature for the last 2000 years in the plateau

It is obvious that there was a short warm period toward the end of the second century AD and during the first half of the third century, although for the rest of the time the climate was colder.

Warmer Period

From the 6th century to the late 12th century, there was a warm period throughout China. The temperature increase was more marked in the plateau of Tibet. No major glacial advance has been discovered, which indicates a warm climate during the period. It has also been concluded that there were many forests around the city of Lhasa where there is no forest now. Many famous temples were built with wood from these forests, but the population was never sufficient to have depleted the forests for firewood or construction. Thus, the climatic conditions were much better for forest growth at that time than now.

Little Ice Age

Around the 17th century the climate was the coldest of the past 1000 years. During this time annual mean temperature was about 1°C lower than that of today. During the 17th to 19th centuries in many sites (Mt. Qomolangma, Nagarze, Yigong) there was glacial advance.

In fact, the trend of the temperature change fell below normal since the 12th century, although there were short warm periods and there has been much data and evidence (tree rings and documents) of the existence of an unusually cold climate in the plateau, both in Qinghai and Xizang.

Last Warm Period

It has been getting warmer since the mid-19th century over the plateau. All available natural disaster records in the historical official files by thousands of Tibet local government officers, including Lama, Zong Bi (country magistrate), local headman and others were delivered to Da Lai, Ga Sha (Tibet local government) and Yi Chang (Secretariat of Da Lai) .

The earliest written source was reported in 1765, and the calamities recorded in literature were not continuous during the early stage. From 1883 year onward, the filed records were almost continuous. In addition, You Tai's diary (a senior official in Xizang, Feb. 9, 1904—Apr. 17, 1907) and modern meteorological observational records are adopted here to study the regularity of flood, drought, and snowstorm on the Tibetan plateau during historical time (Lin Zhenyao and Wu Xiangding 1984).

During the past 200 years, the tendency of natural file records shows that the snowstorm was the main feature, the drought frequency was small, with almost no flood in the Qiangtang, but drought and flood occurred frequently in southern Xizang.

The main features of the flood, drought and snowstorm in Tibet resulted as follows (also see Fig. 4):
A. During the past 100 years, flood and drought appeared alternately, and can be divided into 3 rainy periods (1883—1906, 1916—1934, 1947—1962). An extraordinary year was 1905, when the Lhasa urban area was submerged, the de-

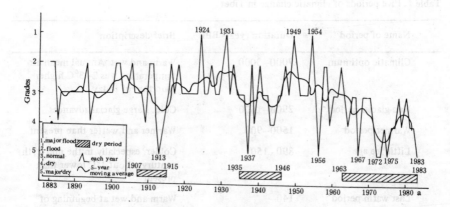

Fig. 4. Grades of floods and drought in Tibet (1883-1980)

pth of water being 0.6–0.8 m. There are 3 dry periods (1970–1915, 1935–1946 1963–1980), the last dry period seems to be extending longer into the future.

B. By analysing this series, we found a 2.6–3.6 year periodicity of flood and drought over the plateau.

C. The total annual precipitation in Yadong was heavy at the beginning of the 20th century. The average value was over 1000 mm and the maximum was above 1500 mm. From the end of the 1940s to the beginning of the 1950s, annual rainfall was only about 700 mm and in most years the annual rainfall amount did not exceed 1000 mm. Al though rainfall of southern Xizang was double the normal rainfall, still no large-region flood disaster in this area. If the rainfall is smaller than normal rainfall by 100 mm, severe dry damage occurs in some local areas.

D. It is expected that the climate will tend to become dry and warm over the plateau in the future. In the past 20 years, irrigated area and cultivated land were expanded, particularly in the plant area of winter wheat. As a result, the drought damage may increase.

In order to understand these five periods clearly, the climatic change of the Plateau of Tibet during historical times is summarized in Table 1.

Some Characteristics of Climatic Change

The climatic change of the plateau depends both upon factors of climatic control in the world and of the geographic environment over the Plateau. Thus the characteristics of climatic change on the Plateau are not only similar to those in other areas, but also have distinct features.

Table 1. Five periods of climatic change in Tibet

Name of period	Duration (years BP)	Brief description
Climatic optimum	7000–3000	Warm and wet. Annual mean air temperature was $2-3°C$ higher than that of today
Neo-glacial period	2900–1600	Cold. Large glacial advances
Warmer period	1500–900	Warmer and wetter than present
Little ice age	890–150	Colder, especially in the mid-17th century, about 1°C lower than present
Last warm period	140–	Warm and wet at beginning of 20th century, becoming drier

The Relationship of Climatic Change

In accordance with the general trends of climatic change in the Northern Hemisphere, the climatic periods in Tibet are similar to the temperature changes in other places If we compare several periods of climatic change of the Plateau for the last thousand years with the conclusions of well-known scientists such as Chu Kochin (China), H.H. Lamb (England), M.M. Yoshino (Japan) and others, we find that the gross features of long-term climatic fluctuation in the Plateau are similar to those in East China, Europe, and East Asia (Wu Xiangding and Lin Zhenyao 1981b).

Amplitudes of Climatic Change

Usually, the variability of mean air temperature is related to the local amplitude of temperature change. We found that the plateau has high values of amplitude, the average being much larger than that of East China.

Within the plateau the amplitudes of variation in the north and west are larger, and the higher the altitude, the greater the amplitude .

Sensitivity to Climatic Change

Recently, many climatologists have pointed out that the polar regions are the areas most sensitive to climatic change, because the ice cap is huge and the stability of the near-ground stratification is large. Both facts cause rising temperatures to intensify in the beginning of a warm period.

The latent stability, similarly, shows that the near ground stability on the plateau is stronger than that on the plain. In fact, the process of rising temperature of the Plateau was greater than that of the plain.

In a word, the Plateau may be considered as an area more sensitive to climatic change than other areas of the same latitude.

Periodicity of Climatic Change

By carrying out analysis on various series, we found some periodicities in the climatic change of the plateau.

Using the Power Spectrum analyses and filtering techniques, many cycles can be calculated from the series which indicate climatic change. For example, the "quasi-biennial pulse" appears in both the proxy data and the meteorological data, and is quite widespread over the plateau. Apart from this cycle it is also in harmony with two principal cycles of relative sunspots which occur at intervals of about 11 or 22 years.

In addition, a proportion of the tree-ring series also have cycles of ca. 30 years, which is in agreement with the well-known Bruckner Cycle.

References

An Zhimin et al, (1979) *Archaeology* 681–491

Guo Xudong, (1976) Reports of science investigation in the Qimolangma Area-Geology during the Quaternary, Science Press Beijing, pp 63–78

Lin Zhenyao, Wu Xiangding (1981) Proc Symp Ages Amplitudes Qinghai-Xizang Upheaval. Science Press, Beijing, pp 159–166

Lin Zhenyao, Wu Xiangding (1984) *Kexue Tongbao* 29: 768

Wu Xiangding, Lin Zhenyao (1981a) Proc Symposium Climatic Change (1978). Science Press, Beijing, pp 18–25

Wu Xiangding, Lin Zhenyao (1981b) *Acta Meteorol* Sin 39: 90–96

Climatic Fluctuations in the Central Sanara during the Late Pleistocene and Holocene

Dieter Jakel[1]

Abstract – Geyh and Jakel (1974) discussed climatic development in the central Sahara on the basis of histograms of ^{14}C dating of material from the Sahara and sediment analyses of middle terrace profiles in Wadi Tabi near Bardai (Tibesti). These results were complemented by field studies in December 1974 and by new dates, resulting in a Tibesti climate curve of the Late Pleistocene and Holocene based entirely on geomorphic evidence and sediment analysis (JAKEL 1979).

The great number of ^{14}C dates gained from Tibesti (Chad) and Fezzan (Libya) (cf. Fig. 1) enabled new histograms to be calculated (Jakel and Geyh, 1982) for a fairly uniform climatic area, bringing confirmation and only slight modification of the 1974 histogram of ^{14}C dates. This indicates that first interpretations were correct and that with the more recent histograms a more differentiated picture of smaller areas is possible.

The results are discussed here once more, since the informative value and reliability of the method has been repeatedly questioned. Compared with the climatic curve published in 1979, a modification is shown in Fig. 4 in that the dates from the period 16 200–13 000 BP may have been contaminated and, therefore, the climate curve may have to be corrected for this period. As regards the histograms of ^{14}C dates, it should be emphasized again that they are a valuable aid in tracing trends for comparative purposes. They are not direct climatic evidence, however. Together with palaeobotanic, prehistoric, hydrologic and geomorphic evidence of climate they may be used to extend knowledge of climatic conditions. However, care should be taken in interpreting such histograms, since a lack of concrete knowledge of the sampling sites can easily lead to erroneous conclusions. There is no doubt that the method, properly used, represents an extremely helpful research aid.

Introduction

Climatic change in North Africa has long been the subject of research and controversy in the various disciplines of earth science. First attempts to clarify the problem concentrated on the high mountain areas such as Hoggar and Tibesti; later research also included the mountain forelands and plains. As a result a vast quantity of data is available for the Quaternary period. An important and productive contributor to this collection was the Bardai Research Station, maintained by the Free University of Berlin in the Tibesti Mountains (central Sahara) form 1964 to 1974.

As a result of interdisciplinary cooperation at the Station, enough data were obtained to construct a climatic curve of the Late Pleistocene and Holocene. The curve is based on geomorphic climatic indicators dated by radiocarbon analysis. In order to verify the curve by other methods, it was compared with chronohistograms based on geomorphic, prehistoric, and hydrologic ^{14}C data (Jakel and Geyh

---

1. Institute of Physical Geography, Free University of Berlin, Grunewaldstrasse 35, D-1000 Berlin 41

1982). The frequency distributions of these data are particularly suitable for establishing periods of transition since, as experience shows, more data are available from periods of pronounced morphologic change. In order to identify such periods as accurately as possible, researchers concentrate on obtaining datable material to uocument the start or end of accumulation, whereas there is generally a lack of data covering periods of erosion. Thus. distribution patterns may be established and compared according to study area and subject. Hence similar trends within a particular area may be identified and results may be compared with those from distant areas. Therefore, in the context of large-scale atmospheric circulation, chronohistograms are important evidence and may contribute to the understanding of general climatic events. For this reason, some histograms and their interpretation will be discussed again below with a view to comparing results of Tibesti research with other regions of the earth.

A Climatic Curve of the Tibesti Mountains Based on Geomorphic Factors

In the course of extensive studies of Tibesti valleys, stream terrace sediments were observed at different altitudes. They were formed during alternating phases of accumulation and erosion (Fig. 2). The Zoumri-Bardague-Araye wadi system, which flows north towards Serir Tibesti, was surveyed in long and cross profile over a stretch of 150 km from Zoui up to the mountain fringe (Fig. 3). Results showed that the terraces were not caused by tectonic movements but by climatic fluctuations alone (Jakel 1971).

Unconformities in the areas surrounding the oasis of Bardai showed a sixfold alternation of erosion and accumulation processes in the valleys. According to the principle of actualism, we may assume that in arid phases like the present, with a mean annual precipitation of about 10 mm at Bardai, sand, pebbles and cobbles were deposited in the middle reaches of the valleys. In wetter phases erosion occurs in mountainous areas, and sedimentation in the forelands. Thus, with increasing humidity, fluvial deposition progresses further and further from the edge of the mountains (where corrosion prevails during arid phases) and, depending on its intensity, may exceed the length of the river by over 100 km. During periods of decreased rainfall the river becomes correspondingly shorter. The discharge of the Bardague river was observed at Bardai in 1965 and 1966. Only during the rare storm rainfalls over wide mountain areas did the waters of the Zoumri-Bardague-Araye network flow as far as the pronounced playa at the edge of the Tibesti Mountains. In most cases, discharge caused by less intensive rainfall dried up in the middle reaches, up to 100 km away from the present playa. This indicates that in the course of a long profile there are very different terrace cross-profiles and the terrace deposits in the middle river courses (which, with their sand, pebbles and cobbles, correspond to present-day deposits in the low, mean, and flood channels) belong to arid phases. This applies mainly to the pre-high terraces and the upper layer of fine-grained limnic sediments. The lower terrace deposits are also an exception, consisting mainly of coarse cobbles. The valley fill of the high terraces. very rich in tuff and up to 40 m thick, may be assigned to a phase of volcanic eruption, mainly in western Tibesit, where climatic conditions were slightly wetter than today.

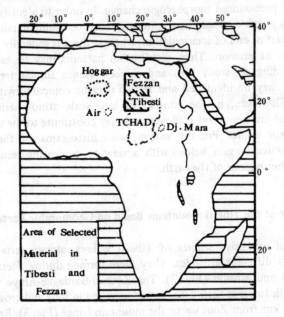

Fig. 1. Location map of study area: Tibesti (Chad) and Fezzan (Libya)

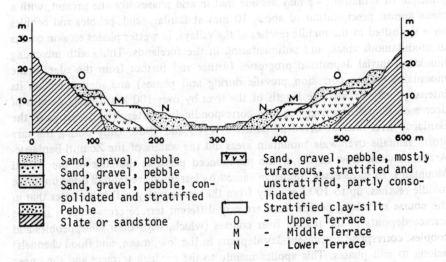

Sand, gravel, pebble

Sand, gravel, pebble

Sand, gravel, pebble, consolidated and stratified

Pebble

Slate or sandstone

Sand, gravel, pebble, mostly tufaceous, stratified and unstratified, partly consolidated

Stratified clay-silt

O   Upper Terrace

M   Middle Terrace

N   Lower Terrace

Fig 2. Shematic cross-profile of the Enneri Bardague near Bardai

127

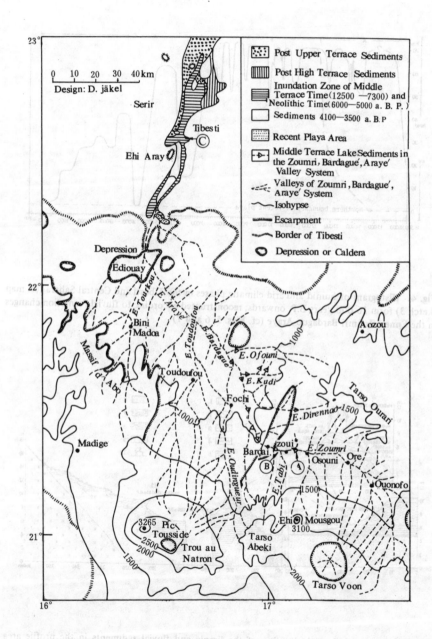

Fig. 3. Map of the Wadi System of Enneri Zoumri, Bardague, Araye in the Tibesti Mountains.
After Carte de l'Afrique 1:1 Mill., Djado, NF 33, IGNO. Jakel (1971): Karte des Enneri Araye
(Tibesti) im Endpfannenbereich 1: 200 000 (Map of Wadi Araye, Tibesti, in the inumdation
zone)

128

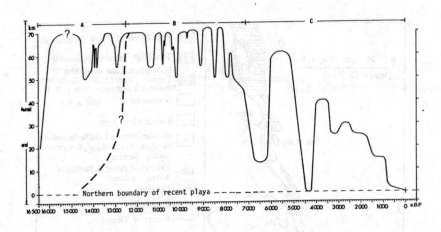

Fig. 4. Histogram of humid and arid climatic phases in the Tibesti and Central Sahara (cf. map sketch 3) form 16 000 years B.P. onwards; reconstructed according to fluvial formation changes in the Enneri Zoumri-Bardague-Araye (cf. Figs. 5, 6 and 7)

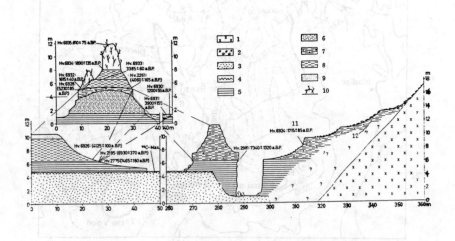

Fig. 5. Geomorphological situation of the limnic and fluvial sediments in the profile area 18 km north of the recent playa of the Bardague-Araye. 16°37'E, 22°36'N. 1 – Granite; 2 – slope debris; 3 – Stratified sand and gravel; 4 – Ditto with soil formation; 5 – Stratified clay; 6 – Strat. fine gravel, sand and silt; 7 – Strat. silt and clay; 8 – Strat. sand and silt; 9 – Strat. fine gravel and sand; 10 – Tamarisk mound; 11 – Roots of Tamarisk; 12 – Possible unconformity

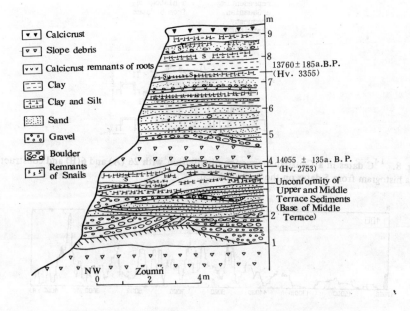

Fig. 6. Middle terrace profile of the Enneri Zoumri (Outcrop 43 after Molle 1969, 1971)

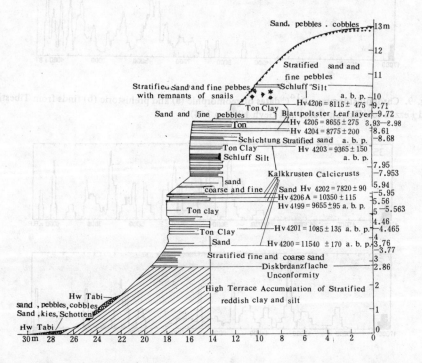

Fig. 7. Middle Terrace Profile 3 of Enneri Tabi, Tibesti. Situation on 17.4. 1970 (cf. Geyh and Jakel 1974) 21°20′N, 17°03′E 1030 m a.s.l.

Ways of representing a Gaussian frequency distribution

Constructing a histogram from 5 dates

Fig. 8. ^{14}C dates in a Gaussian frequency distribution with 26 (A) and 6(B) and construction of a histogram from 5 dates on the basis of A±26

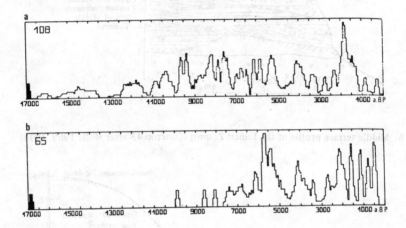

Fig. 9. Chronohistograms of ^{14}C dates of geomorphic (a) and prehistoric (b) finds from Tibesti and Fezzan (cf. Fig. 1). (Jakel and Geyh 1982, p 144)

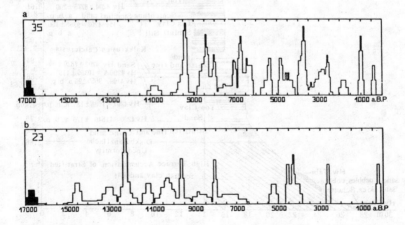

Fig. 10. Chronohistograms of ^{14}C dates of Tibesti groundwater samples at and near Bardai (a) and Fezzan (b) (cf. Fi.g 1) (Jakel and Geyh 1982, p 144)

The penultimate terrace deposits, belonging to the middle terrace, differ from all the other terrace sediments. Owing to their organic content they could be dated more accurately, permitting a more detailed picture of climatic deveopment from their accumulation onwards (Fig. 4). They were formed in some valleys behind calcsinter barriers and show a stepped gradient in long profile. The corresponding sediments in the forelands are the limnic playa sediments which radiocarbon dates have shown to be coeval (Jakel 1979). Since the terraces do not occur in all valleys and are only found above an altitude of 850 m a.s.l. in the Bardague basin (the mountain edge starts at about 750 m a.s.l.), it would seem that the mountain accumulations were deposited during a more humid phase of erosion, when calcsinter and the resulting barriers were formed only at rapids in narrow stretches of valley. Investigations of these sediment profiles and stratigraphic sequences, and analyses of sediment, pollen, fresh- and brackish water snail ecotopes indicate that clayey layers in the lakes behind the barriers correspond to wetter periods and sandy horizons to dryer ones (Geyh and Jakel, 1974). Radiocarbon analysis of organic substances gave dates between 12 500 and 8000 yrs BP. It was shown that the individual barriers and lakes were not coincident, but were formed earlier or later depending on the topographic situation. However, in all cases a change in sedimentation occurred around 8000 yrs BP, documented by an abrupt change to sand, pebble, and cobble accumulations which were deposited on top of the limnic sediments until about 7300 yrs BP, before dissection of the barriers and removal of the middle terrace sediments began.

The climate curve shown in Fig. 4 was based on three different sediment profiles. Part A shows the profile of a former lake in the upper valley of the study area. According to Molle (1969), it was radiocarbon dated at 14 000–13 500 yrs BP (Fig. 6). Samples taken from the base of calcsinter barriers were dated at 16 120 ± 215 BP (Hv5624) and 14 445 ± 180 BP (Hv 5625). The topographic situation and the nature of the material point to the start of a pluvial-phase sedimentation. FAURE et al. (1963) analyzed diatoms (dated at 12 400 ± 400, 14 970 ± 400 and 14 790 ± 400 BP (Gif 378–380) which document the existence of a 350-m-deep lake in the closed caldera of the Trou au Natron. This depth of water presupposes a lengthy humid period and, together with the base date of 16 120 ± 215 BP, led to the assumption that a pluvial period existed between 16 000 and 15 000 yrs BP. This interpretation was questioned by M.A. Geyh (pers. comm., 1982) on the grounds that the dates were possibly unreliable and ought not to be used as evidence of a pluvial period at this time. He drew attention to the fact that no corroboratory evidence had been found elsewhere in the Sahara for this period. In order to illustrate this problems a dashed line was drawn on the climate curve to indicate that a wetter climate possibly only prevailed in the Sahara around 13,000 BP after a lengthy arid period (after 20 000 BP).

In order to clarify this question it would be necessary to undertake further field studies and collect additional samples. This is impossible, however, owing to the current political situation in northern Chad.

Part B of the curve is based on sediment studies of the middle terrace profiles near Bardai (Geyh and Jakel 1974; Jakel and Schulz 1972) and in the middle reaches of the Zoumir-Bardague-Araye basin (Fig. 7). Part C represents processes of sedimentation, deflation, and corrosion in the foreland of this river network (Jakel 1979). The measure of humidity on the y-axis is the distance of the limnic and fluvial deposits north of the mountain edge.

Chronohistograms of ^{14}C Dates and Their Values in the Context of Climate Fluctuations and Oscillations

In order to increase the informative value of ^{14}C dates, Geyh (1971, p 40; 1980) developed chronohsitograms on the basis of Gaussian normal curves (Fig. 8), enabling individual results to be set in a wider context. Chronohistograms permit comparisons of differnt problems and thus indirect statements on questions of climate. For this reason chronohistograms were compiled on the basis of relvant geomorphic and prehistoric data (Fig. 9) and hydrologic data (Fig. 10) form the Tibesti mountains and the province of Fezzan (southern Libya) (Jakel and Geyh 1982).

A comparison with terrace profiles which have undergone radiocarbon dating and sediment analysis (Geyh and Jakel 1974) showed that such histograms may be correlated with actual changes in processes of surface formation. There are characteristic differnces between the histograms.

It should be empasized most strongly that the height of the peaks in the histograms is not a direct expression of climatic conditions. The relationship between peaks and gaps, which represent wetter and dryer climatic conditions respectively, only indicates the existence of periods with frequent occurrence of prehistoric artifacts, increased production of organogenic substances, or groundwater recharge. Naturally, the older the prehistoric or geomorphic material, the fewer samples are still available. In addition, for methodological reasons, older dates with higher sigma values produce lower peaks than younger dates with lower sigma values. Therefore, a climatological evaluation is only possible if comparison is made with other field results.

The chronohistogram of geomorphic data from Tibesti and Fezzan (Fig. 9a) was compiled from 108 radiocarbon dates (Jakel and Geyh 1982). The values at 16 000 yrs BP (2 dates) and between 15 500 and 13 500 yrs BP (5 dates) were gained from diatoms and calcretes. This material, taken from terrace remnants and lake deposits at Trou au Natron, points to wet climatic conditions in Tibesti. No dat are available for the period between 13 500 and 12 500 BP. Between 12 500 and 11 500, however, 6 dates are known. Geomorphically speaking, they belong to the start of a lake phase with calcsinter formation in the Tibesti valleys. There is a data gap between 11 500 and 11 000 BP, but dates become more frequent again between 11 000 and 10 000 yrs BP (10 dates). There is a further gap from 10 000 to 9800 BP, whilst the following periods from 9800 to 9200 (9 dates), 9000 to 8000 (14 dates) and 8000 to 7300 yrs BP (10 dates) are well documented. The samples were mainly taken from calcretes and lake deposits in Tibesti and indicate a continuous pluvial phase. Only 2 dates exist between 7300 and 7000 BP. Eight dates fall within the period between 7000 and 6300, and 14 between 6200 and 5000. On the histogram a clear differentiation and a data gap are visible at 6300 BP.

Geomorphic evidence points to a relatively dry phase with extreme climatic variations. This phase clearly differs from the period between 6200 and 5000 BP, since during this period there was a persistent desert lake in the Tibesti foreland (cf. Fig. 3). The period between 5000 and 4200 (4 dates) was very arid, as is confirmed by morphological evidence. An oscillation towards increased humidity occurred again between 4200 and 3700 BP (5 dates). From then onwards, the climate became more and more arid with an apparently rhythmic pattern of fluctuations. An exception to this pattern occurred between 2000 and 1200 BP (14 dates). The samples dated were all taken from the forelands, from well-preserved, groundwater-fed tamarisks. They are therefore evidence of an arid climate, even though conditions were slightly more favorable than at present. These examples in particular show how necessary it is to take into account the site conditions of the sample when interpreting a histogram and how important it is to recognize that such a histogram is not a direct expression of climatic development (Geyh and Jakel 1974, p 97).

A remarkable feature of the prehistory chronohistogram (Fig. 9b) is that there are no dates from Tibesti and Fezzan older than 10 000 BP. Only 3 values are available for the period between 10 000 and 8000 BP. This is astonishing since geomorphic evidence suggests that this period may have been a favorable time for settlement in the central Sahara. The only conclusion to be drawn is that Man first entered these areas in any great number at a later time. Data begin to amass in the period between 7500 and 7300 BP, when dissection of the middle terrace sediments began in the Tibesti valleys. The quantity of data remains at about the same level until 6200 BP, increases considerably up to 5000 BP, and decreases again up to 4700 BP. This reflects the favorable conditions for settlement here during the Neolithic pluvial phase, which is generally assigned to this period. There is a good correlation with the geomorphic chronohistogram (Fig. 9a), with the accumulation of data up to 4700 BP lagging behind (Fig. 9b) by about 300 to 400 years. This might be due to Neolithic Man's continuing to occupy the area despite deteriorating ecological conditions. The rapid increases in data around 4200 BP are probably also linked to the climate: the number of occupation sites and corresponding artifacts increased with improving climatic conditions. According to this analysis the population level remained high until 3000 BP. Here again there is a time lag compared to the geomorphic data, and there is no gap around 3500 BP.

The peaks and gaps from 3000 BP up to the present are difficult to interpret in climatic terms. Although they may express a fluctuation in the population, this may be due to improved techniques rather than to climatic oscillations. It is therefore possible that the population remained high despite increasing aridity and that the peaks could well express strong fluctuations within this high population range.

It proved difficult to correlate the chronohistograms of groundwater data of Tibesti and Fezzan (Fig. 10a and b) with the histograms of Fig. 9. Because of the lime content of the groundwater the conventional ^{14}C ages are not directly comparable with organic samples.

According to Klitzsch et al. (1976), errors of 1300 years are possible. Geyh and Obenauf (1974) found 1800 to 2800 years to be a better approximation (Jakel and Geyh 1982, p 145). Furthermore, 20 of the 35 Tibesti samples analyzed

134

contained tritium, which suggests that mixed samples were dated. For this reason no comparison was made with the histograms in Fig. 9.

Comparison of the groundwater data from Tibesti and Fezzan reveals a clear time shift. The majority of the Fezzan dates are older., those from Tibesti younger than 10,000 BP. The more recent age of the Tibesti samples may be due to older water having flowed into the forelands, whereas it remained in place in the Fezzan basins. A further possible explanation is that Fezzan is situated nearer to the Mediterranean than Tibesti. However, the histograms do not suggest a satisfactory answer to this problem: a further reason to emphasize caution when interpreting them.

References

Faure H, Manguin E & Nydal R (1963) *Bull BRGM* Paris 3: 41–63

Geyh M A (1971) Die Anwendung der ^{14}C-Methode. Die Entnahme, Auswahl und Behandlung von ^{14}C-Proben sowie Auswertung und Verwendung von ^{14}C-Ergebnissen. Calusthaler Tektonische Hefte Clausthal-Zellerfeld 11: 118 S

Geyh M A (1980) Einfuhrung in die Methoden der physikalischen und chemischen Altersbestimmung. Wiss Buchges Darmstadt, Darmstadt, 2365

Geyh M A, Jakel D (1974) *Z Geomorph NF* 18(1): 82 – 98

Geyh M A, Jakel D (1974): ^{14}C-Altersy bestimmungen im Raimen der Forschungsarbeiten der Auoenstelle Bardai/Tibesti der Freien Univ Berlin. Pressedienst Wissenschaft FU Berlin Berling 5/74: 106–117

Geyh M A Jakel D (1977) *Nat Res Dev,* Tubingen 6: 64–79

Geyh M A, Merkt J, Muller H (1971) *Arch Hydrobiol,* Stuttgat 69: 366–399

Geyh M·A, Obenauf K P (1974) Zur Frage der Neubildung von Grundwasser unter ariden Bedingungen. Ein Beitrag zur Hydrologie des Tibesti-Gebirges. Pressedienst Wissenschaft FU Berlin, Berlin 5/74: 70–91

Jakel D (1971) *Ber Geogr Abh* 10: 55 S

Jakel D (1977) *Geogr J* 143: 61–72

Jakel D (1979) Run-off and fluvial formation processes in the Tibesiti mountains as indica-tors of climatic history in the Central Sahara during the late Pleistocene and Holocene. Palaeoecol Afr Rotterdam 11: 13–44

Jakel D, Schulz E (1972) Spezielle Untersuchungen an der Mittelterrasse im Enneri Tabi, Tibesti-Gebirge. *Z Geomorph N F Suppl* 15: 129–143

Jakel D, Geyh M A (1982) *Berl Geogr Ab* 32: 143–165

Klitzsch E, Sonntag C, Weistroffer K, Geol E M (1976) *Rundsch* 65 (1): 264–287

Molle H G (1969) *Berl Geogr Abh* 8: 23–31

Molle H G (1971) *Berl Geogr Abh* 13: 53S

Climatic Evolution of the Sahara in Northern Mali During the Recent Quaternary

Nicole Petit-Maire[1]

LENGTHY ABSTRACT

The part of the Taoudenni Basin situated in northern Mali, between 18° to 24° N and 1° to 5° W, is geologically stable (the West African shield), out of reach of any surface waters from exterior areas, far from any oceanic or mountains influences, and undisturbed by anthropic action. Thus, it is an excellent field for investigating past climatic variations, allowing the interpretation of hydrological or environmental changes in terms of changes in climate parameters only.

Geology, paleolimnology, paleontology, paleobotanics and paleoanthropology provided evidence for a recent (Upper Pleistocene and Holocene) drastic evolution of landscapes.

Presently, the studied area, covering nearly 500 000 km², is hyperarid and one of the driest in the world. Vegetation, animal life, and human population are reduced nearly to naught, annual precipitation ranging from about 50 mm (to the South and East) to about 5 mm (at the tropical latitudes), the Intertropical Front seldom going beyond the 17th parallel.

The following sedimentary and climatic phases could be evidenced from the sum and comparison of the geological and biological data.

Humid Phase, from About 40 000 to About 20 000 B.P.

This is evidenced by lacustrine silts, travertines, and calcretes which contain rare freshwater molluscs (generally *Melania tuberculata*) and by paleosoils. The deposits have not yet been isotopically dated but they are quite regularly associated with surface Aterian prehistoric sites. Consequently, the phase is either contemporary or anterior to the Aterian; the first hypothesis seems more likely (40 000–20 000 B.P.). However we cannot totally exclude contemporaneity with the mid-Pleistocene lacustrine phase recognized in Libya in the Shati valley (Petit-Maire 1982).

Arid Phase, Upper Pleistocene

In stratigraphy, it is immediately posterior to the Aterian, artefacts having

---

1. Laboratoire de Geologie du Quaternaire-CNRS-Case 907--Luminy--13288 MARSEILLE Cedex 9-FRANCE

been found in situ at the base of this formation.

The phase is characterized by intense aeolian ablation and dune formation. Probably an equivalent to the Ogolian and Kanemian phases described in occidental and oriental Sahara.

Humid Phase, Holocene

10 000 B.P. Sand grains from the former phase show hydric reworking. Marshes, then fresh water lakes, form in the whole area.

9000 to 7000 B.P. An important lacustrine phase is evidenced by fossiliferous silt and clay deposits in the topographical depressions and in the interdune troughs of the consolidated former ergs.

Freshwater Foraminifera, Ostracoda and Mollusca are present in all the pools. Fish, crocodile, hippopotamus are associated with the large lakes South of 22° N only. The shores are edged with reeds and terrestrial Mollusca live in the dampest banks throughout the year. Between the lakes, brown steppe soils extend widely, even to the north of the Tropic, in the Tanezrouft; they confirm the existence of a Gramineae cover implied from the number of large bovid species, rhinoceros, and elephant bones found in the food remains of prehistoric men who lived throughout this whole landscape in a hunting, fishing, and gathering economy. A *minimal* pluviometry of 300 mm is implied from such evidence.

During this bioclimatic optimum, the Niger river flows into a large interior delta North of Timbuktu; during the major floods, its effluents may reach up to 300 km North of its present curve (Petit-Maire and Gayet, M. 1984, Petit-Maire and Riser in press).

About 6500 B.P. A drier episode is testified to by an evaporitic phase in the lacustrine history: according to size and location of the deposits, mud-cracks, sandy layers, or thick salt beds indicate a change in the precipitation/evaporation ratio and to an increase of aeolian processes.

About 5500 B.P. A new lacustrine episode is recorded sediments. However, the water level is lower than during the Early Holocene; the smallest pools remain dry or seasonal; to the north of the 23th parallel, the lack of fauna and of human settlements posterior to 6000 B.P. shows that the climatic optimum was then over at the highest latitudes. To the south, on the contrary, the lacustrine conditions are maintained.

4500 B.P. The first sheet-floods evidence for a change in the rain regime, which switches to a "wadi" torrential type, associated with arid areas.

By 3800 B.P. The lakes dry up and a sebkha system settles in the largest depressions. The last sahelian mammals have left. Man groups around the water holes, then the wells, in the largest depressions: we find settlements on the dry lakes deposits.

Arid Phase

At 3000 B.P., aeolization begins anew; new dunes build; the "saharization" of a former Sahelian area starts.

This very recent major change in the climate of the most severe desert in the world being evidenced, we now leave to meteorology the interpretation of these data in terms of general atmospheric processes. We would like to emphasize the importance of *desert research* for the understanding of the present climatic evolution towards aridity and suggest that comparative studies in the Takla-Makan desert (Western China), could lead to a better understanding of paleoclimatology.

References

Petit-Maire N (Ed.) (1982) *Le Shati. Lac Pleistocene du Fezzan (libye)*. CNR, Paris, 118p

Petit-Maire N, Gayet M (1984) Hydrographie du Niger (Mali) a l' Holocene ancien. *C R Acad Sci* Paris (II) 298: 21–23.

Petit–Maire N, Riser J. (in press) Did the course of the Niger River change in the Holocene?

Section II

Ocean-Atmospheric Interaction in

Short-Period Climatic Variation

El Nino/Southern Oscillation and the Asian Monsoons

Eugene M. Rasmusson[1] and Phillip A. Arkin[1]

Abstract— Sufficient high quality data are now available to permit the description of the longterm mean annual cycle in the planetary-scale upper and lower tropospheric circulation, sea surface temperature and precipitation, as inferred from satellite measurements of outgoing longwave radiation (OLR). These data clearly show that the large-scale Asian monsoon circulations possess an annual cycle which is related to changes in large-scale boundary conditions over the ocean as well as over the land. For example, the annual interhemispheric migration of the major western Pacific—eastern Indian Ocean monsoon convective region is closely linked to a similar migration of the region of highest sea surface temperature (SST).

The large interannual variability of the monsoon also appears to be in part related to fluctuations in oceanic boundary conditions associated with the El Nino/ Southern Oscillation (ENSO) phenomenon. Although the complex pattern of SST, rainfall, and circulation anomalies during an ENSO episode varies throughout the year, it is characteristically phase-locked with the seasonal cycle.

The broadscale annual cycles in and relationships among tropical SST, OLR, and circulation in the tropics are described along with their interannual variability associated with ENSO. The ENSO signal in the Northern Hemisphere extratropics is examined in the context of teleconnections associated with anomalous tropical heating.

Introduction

The El Nino/Southern Oscillation (ENSO) phenomenon (Rasmusson and Wallace 1983) owes its existence to longitudinal asymmetries in the large-scale direct thermal circulation systems that dominate the tropics on time scales of weeks or longer. The moist ascending branches of these circulations, associated with the monsoons and oceanic convergence zones, mark the major tropical regions of condensation and latent heat release. The associated atmospheric heating, which extends throughout a deep layer of the troposphere, is a primary forcing mechanism for the large-scale circulation of both the tropics and extratropics. The resultant low-level wind field drives the low-latitude ocean circulation, resulting in longitudinal variations in SST that in turn influence the location of the oceanic rainfall zones, thus completing feedback loop between ocean and atmosphere. It seems possible (Philander et al. 1984) that under certain circumstances these interactions in the Pacific may become unstable with a perturbation in one or more elements of the rainfall—atmospheric circulation—SST linkage resulting in an evolving ENSO anomaly pattern with a lifetime of 1—2 years.

The climatological field of large-scale vertical motion in the tropics can be deduced from Fig. 1, which shows the mean outgoing longwave radiation (OLR) for the two solstice seasons (DJF; JJA) derived from polar orbiter satellite observations

1. Climate Analysis Center, National Meteorological Center, NWS/NOAA, Washington, D.C. 20233, U.S.A.

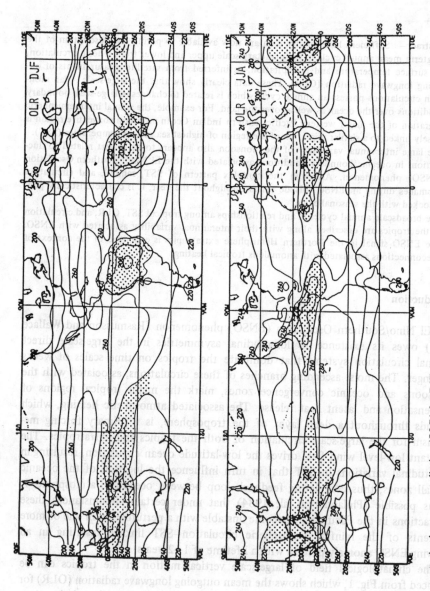

Fig. 1. Mean outgoing long wave radiation (OLR) for DJF (upper) and JJA (lower). Areas where OLR is less than 240 W m^{-2} are stippled. Contour interval 20 W m^{-2}

acquired during the period 1974–83. In the tropics, decreased OLR corresponds to increased coverage of high cold cloud tops, and is a good indicator of increased convective rainfall. Areas in the tropics with OLR values less than 240 W m^{-2} delineate regions of heavy convective rainfall and mean upward motion. Two ascending branches of the mean circulation are more or less anchored over summer hemisphere continental areas (Africa, South America/Central America and adjacent waters). The third and most extensive area of tropical convection forms the ascending branch of the Australasian monsoon system, with eastward extensions along the Pacific Intertropical Convergence Zone (ITCZ) north of the equator, and along the South Pacific Convergence Zone (SPCZ) toward the southwest Pacific.

The time mean circulation can be decomposed into its rotational and divergent components and their relationship studied. The stream function (ψ) defines the rotational part of the flow while the velocity potential (χ) field provides a picture of the divergent flow associated with the upper branches of the planetary scale thermal circulations. The fields are computed by relaxation of the vorticity and divergence fields derived from the U. S. National Meteorological Center (NMC) operational global optimum interpolation (OI) analyses. The relationship between these fields is given by:

$$\mathbf{V} = \nabla\chi + \mathbf{k} \times \nabla\psi = \mathbf{V_d} + \mathbf{V_r} \,.$$

Figure 2 shows the mean 200 hPa χ (5 years) and 200 hPa ψ (16 years) fields for DJF and JJA. The divergent flow is directed normal to the contours of from low values toward high, while the rotational circulation is parallel to the ψ contours, clockwise about maxima and counterclockwise about minima.

One should not confuse the planetary scale pattern of the χ field with the field of divergence $\nabla^2 \chi$, which has a much smaller spatial scale, and is more appropriately compared with the tropical OLR patterns. In terms of divergence, the χ field is an integral quantity, i.e. the average divergence over an area enclosed by a χ contour is the line integral of $\nabla \chi$ around the contour divided by the enclosed area. The divergent with field ($\mathbf{V_d}$) derived from the χ field is useful in describing the large-scale direct thermal circulations associated with the non-uniform surface boundary conditions in the tropics.

Both components of the flow show large longitudinal asymmetries which can be related to those of the OLR. In general, the χ charts show a planetary-scale 200 hPa cross-equatorial divergent flow from the convective regions of the tropics to the radiative sink regions of the subtropics, and flow towards the relatively cold water regions of the eastern tropical Pacific. Regions of tropical convection are characterized by low-level convergence and upper tropospheric divergence, with maximum vertical motion and latent heat release in the middle troposphere. Temperatures in these regions are relatively high throughout the troposphere. Thus, convective regions are capped by mean upper level anticyclonic flow. The strongest upper level anticyclones are part of the Australasian monsoon circulation system. During JJA, the 200 hPa anticyclone centered over southern Asia dominates the Northern Hemisphere circulation over approximately 180° of longitude (Fig. 2). A companion 200 hPa anticyclonic circulation is found in the same longitudes of the Southern Hemisphere. The tropical easterly jet (TEJ) (Koteswaram 1958; Flohn

144

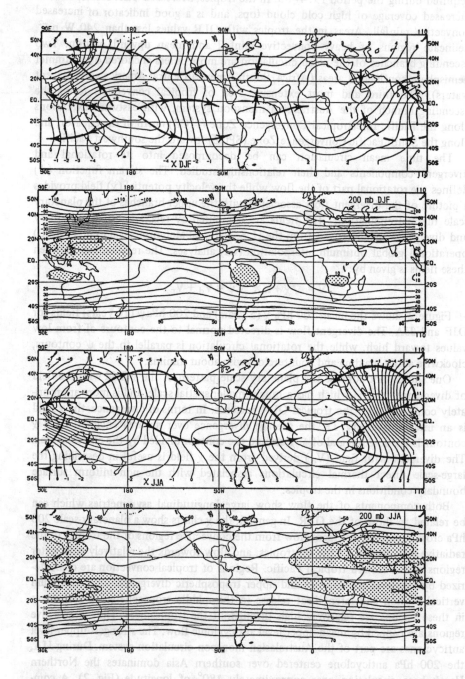

Fig. 2. Mean 200 mb velocity potential (χ) and stream function (ψ) for DJF (upper two panels) and JJA (lower two panels). χ contour interval is 10^6 m^2 sec^{-1}. ψ contour interval 10×10^6 m^2 sec^{-1}. χ mean is for 5 years; ψ mean for 16 years.

1964), with its core around 150 hPa lies between these anticyclones. The corresponding field (Fig. 2) shows that the region of upper level outflow over the west Pacific lies in the entrance region of the TEJ, while the Southern Hemisphere region of inflow centered near 20° S is in the longitudes of the TEJ exit region. The largest values of V_d are near the axis of the TEJ, which reaches maximum values near 10° N, 75° E.

During the Southern Hemisphere summer the monsoon anticyclones - shift eastward and are centered over the Pacific—Australian region, where the upper-level interhemispheric monsoon circulation is most vigorous. The major features of the DJF 200 hPa χ field are associated with the outflow from the convective region extending from Indonesia southeastward along the SPCZ, and the upper level inflow and subsidence over the cold land surfaces of southern and eastern Asia. Comparison of JJA and DJF χ fields shows a shift in direction of the major upper troposphere cross-equatorial divergent flow, which in both seasons is the reverse of the low-level cross-equatorial monsoon flow. This seasonal change is consistent with and reflected in the seasonal cycle of the zonally averaged Hadley circulations (Oort and Rasmusson 1971).

The upper level divergent flow from the monsoon convective region in the west Pacific to the subsidence region over the cold waters of the eastern tropical Pacific is of particular relevance to ENSO, since it relates to the much discussed Walker Circulation (Bjerknes 1969). As previously noted by Krishnamurti (1971), this east-west divergent flow is of broad latitudinal extent, rather than being narrowly confined to the equatorial zone, where the dynamics described by Bjerknes (1969) are valid. Furthermore, except for the western Pacific during DJF, the divergent component of the flow is typically a small fraction of the total flow.

The χ and ψ fields clearly show the important longitudinal asymmetries of the planetary scale mean flow and their relationship to the asymmetries in SST and precipitation. They also show pronounced differences between the hemispheres. It is clear from the OLR and χ fields that the summer hemisphere monsoon anticyclone is characterized by extensive cloudiness and upper level outflow associated with local anticyclonic vorticity generation. In contrast, relatively high values of OLR are observed over most of the area of the winter hemisphere monsoon anticylonic circulation, and the upper level divergent flow component is more difficult to characterize. Therefore, the maintenance of the upper tropospheric anticyclonic vorticity appears to involve somewhat different processes in the summer and winter hemispheres.

The distribution of tropical SST is illustrated in Fig. 3. The gradient between the eastern and western Pacific is striking. SST's in the west Pacific-eastern Indian Ocean are the highest in any of the large ocean basins, while in the upwelling region off Peru, SST's are the lowest observed at these latitudes. The location and seasonal migration of the west Pacific monsoon convection and the associated SST "warm pool", shown on Figs. 1 and 3; respectively, largely coincide. Throughout the year the vast tropical oceanic region where OLR is less than 240 W m^{-2} remains primarily within the area enclosed by the 28° C SST isotherm. Studies of Indian Ocean SST/cloudiness relationships (Gadgil et al. 1985) indicate that an SST of 28° C is a reliable threshold for widespread convective cloudiness. All these climatic features

146

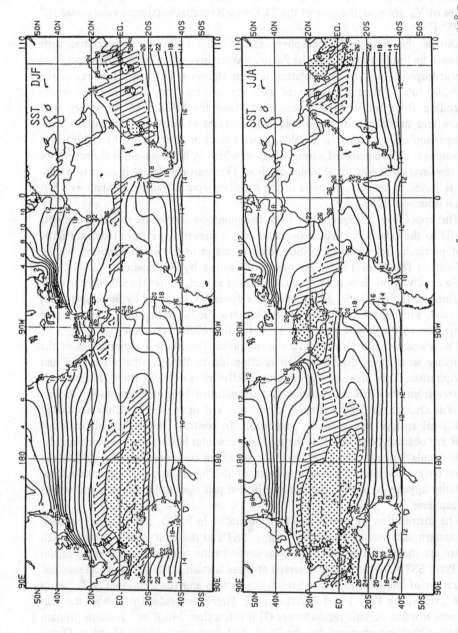

Fig. 3. Mean SST for DJF (upper) and JJA (lower). Areas where SST is between 27°C and 28°C are stippled; areas where SST is greater than 28°C are hatched. Contour interval (solid lines) 2°C

execute an interhemispheric migration associated with the seasonal cycle of the Australasian monsoon system which dominates the circulation over the Indian Ocean and the tropical Pacific west of the dateline.

The region east of the dateline is under the influence of the Pacific tradewind systems. The ITCZ marks the boundary between the Northeast and Southeast Trades. This convergence zone remains in the Northern Hemisphere throughout the year, with westward flow (easterly winds) prevailing over the equator. This is in contrast to the monsoon area, west of the dateline, where the east-west component of equatorial flow is small or toward the east.

ENOS and Interannual Variability

The pattern of interannual SST variability in the equatorial Pacific and its relationship to the annual cycle is illustrated in Fig. 4, which shows an SST time section along the equator spanning the period 1970–83. A pronounced seasonal cycle appears in the eastern Pacific, distorted but rarely overwhelmed by ENSO events. The annual cycle becomes progressively weaker toward the west (Horel 1982), and without multi-year averaging is almost indistinguishable west of $160°$ W. From there to Indonesia, in the region influenced by the monsoon circulation, the fluctuations are largely interannual in nature, reflecting changes in the location, extent and intensity of the west Pacific warm pool.

The major longitudinal excursions of the warm pool, as delineated by the shaded $27°$ $-28°$ C isotherm band on Fig. 4, are associated with major swings in the Southern Oscillation (Rasmusson and Wallace 1983). The pronounced eastward excursions of this isotherm band correspond to ENSO episodes. The strongest of these during the period shown occurred during 1972 and 1982/83.

It is apparent that both monsoon and trade wind systems are intimately involved in ENSO. While the largest warming occurs in the cold upwelling region of the eastern Pacific, the central equatorial Pacific, where the climatological SST is near the threshold value for widespread convection, appears to be the key region of ocean-atmosphere coupling. Because of the small east-west SST gradient in this region, a $1°$ C anomaly is sufficient to produce an eastward excursion of several tens of degrees of longitude in the $28°$ C isotherm. Such an eastward shift results in a comparable eastward extension of the West Pacific convection, and consequently the region of latent heat release. The configuration of the low-level wind field feeding moisture into this convective area adjusts in a manner that can best be described as an eastward extension of the monsoon circulation, i.e., the winds in the equatorial western and central Pacific become more westerly (Rasmusson and Carpenter 1982). The realistic response of relatively simple numerical and analytical circulation models to a prescribed equatorial heat source provides theoretical support for this scenerio (Webster 1972; Gill 1980).

The 1982/83 ENSO Episode

The sequence of events during the 1982/83 episode strongly support the view

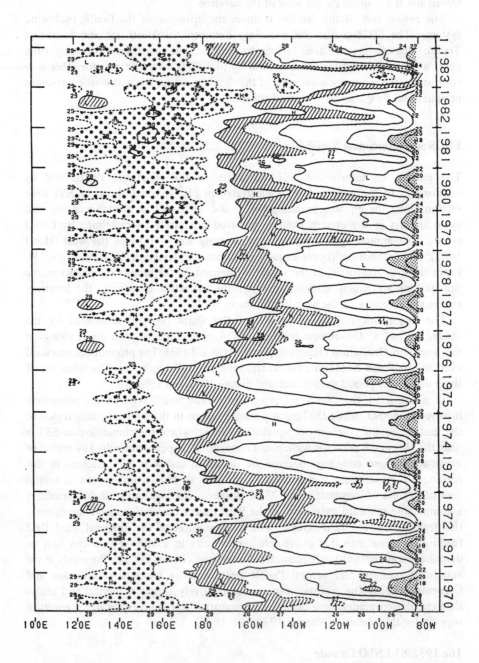

Fig. 4. SST time section. Section extends along the equator from 120° E to 95° W, then southeastward to the South American Coast at 8° S. Contour interval (sloid lines) 2° C

that large-scale tropical Pacific sea-air interaction plays a key role in the evolution
of the ENSO phenomenon. Perhaps the most striking evidence was the simultaneous eastward migration along the equator of the west Pacific warm water as
tracked by the 28° C SST isotherm, the region of enhanced rainfall, as tracked by
the negative OLR anomalies, and the low-level westerly and upper level easterly
wind anomalies (Quiroz 1983; Philander and Rasmusson 1984; Gill and Rasmusson 1983). This steady eastward migration, across more than 100 degrees of
longitude over a period of almost one year, is truly a remarkable example of the
relationship between ocean and atmosphere parameters, and provides compelling
evidence of the close coupling of the two media. Some features of the evolution
appear to be explainable in terms of ocean advection (Gill 1983; Gill and Rasmusson 1983), and the pattern of anomalies in the equatorial plane is broadly consistent with results from relatively simple linear models of the atmospheric response
to thermal forcing (Gill 1980).

The anomalies in the 200 hPa χ field were surprisingly large during the 1982/83
ENSO episode and also exhibited an eastward migration of the planetary scale
features (Figs. 5, 6). The anomalies exhibited a wave number one pattern over the
Indian Ocean-Pacific region which is reminiscent of the SO surface pressure seesaw
(Rasmusson and Arkin 1985). The eastward migration of this feature clearly
reflects the eastward shift of the west Pacific convection leading to heavy rainfall
over the central equatorial Pacific and drought conditions over Indonesia, eastern

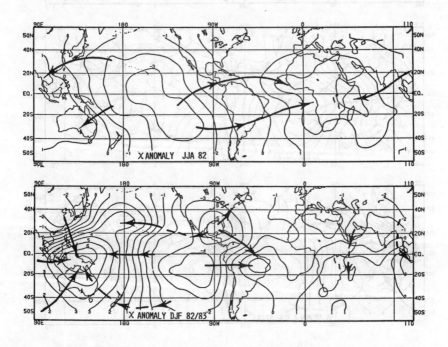

Fig. 5. 200 mb velocity potential (χ) anomaly for JJA 1982 and DJF 1982–1983. Contour
interval 10^6 m^2 sec^{-1}

150

Australia and the southwest Pacific. During the early stages of the episode (JJA 1982), an anomalous divergent flow extended from a 200 hPa outflow region over the central and south-central Pacific westward to Australia and Indionesia and eastward to North and South America and the Atlantic. This pattern migrated eastward, such that by DJF the major features over the Pacific were shifted far east of their normal climatological positions (Fig. 6). By this time, the anomaly χ field had intensified to such a degree that it was comparable in magnitude to the 5-year mean field itself. By JJA, the 200 hPa χ field had returned to a near normal pattern reflecting a rapid decay of the large-scale ENSO anomalies.

The DJF χ anomaly pattern appears to reflect the major regional rainfall anomalies over the world, which include dry conditions over the north Pacific subtropics, southern India and Southwest Africa, and heavy rainfall over the eastern

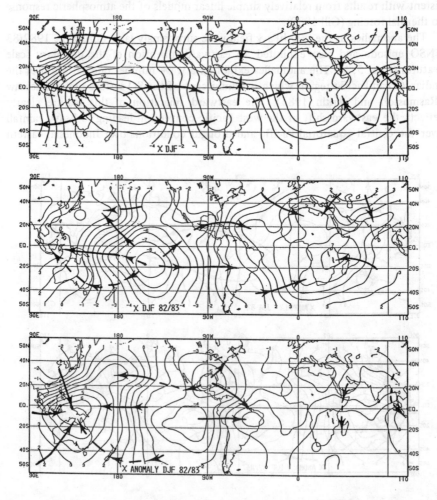

Fig. 6. 200 hPa velocity potential (χ). Upper: DJF five-year mean. Middle: DJF 1982/83. Lower; DJF 1982/83 anomaly. Contour interval 10^6 m^2 sec^{-1}

low-latitude Pacific and the southeastern United States, and Gulf of Mexico.

The pattern in the western North Pacific also indicates a subnormal northeast monsoon consistent with the composite ENSO surface wind anomalies derived by Rasmusson and Carpenter (1982). The high latitude extension of this and other features of the DJF 200 hPa anomaly χ field is illustrated in Fig. 7. Anomalous 200 hPa outflow is indicated over much of northern Eurasia and North America. The weakened wintertime subsidence over the two Northern Hemisphere continental areas was associated with major, long-lasting positive surface temperature anomalies over the two regions (Fig. 8). The relationship of these extratropical features to the

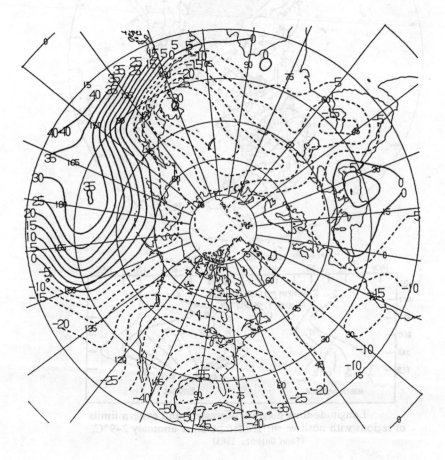

Fig. 7. 200 hPa Northern Hemisphere velocity potential (χ) anomaly for DJF 1982–1983. Contour interval 5×10^5 m^2 sec^{-1}

152

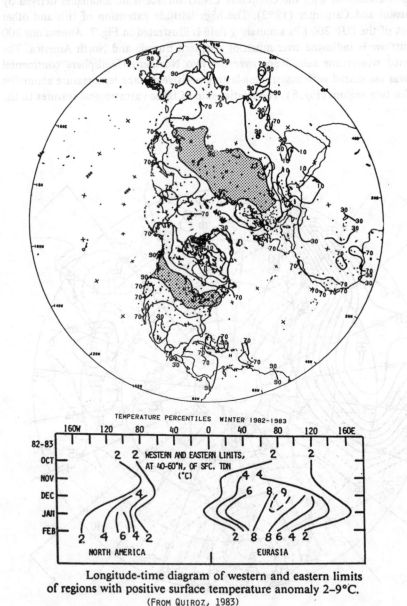

TEMPERATURE PERCENTILES   WINTER 1982–1983

Longitude-time diagram of western and eastern limits
of regions with positive surface temperature anomaly 2–9°C.
(FROM QUIROZ, 1983)

Fig. 8.   Percentiles of seasonal mean temperature for DJF 1982–1983 (upper panel) and longitude-time diagram of western and eastern limits of regions with positive surface temperatures
2° C (lower panel for Quiroz, 1983). In the upper panel, hatching denotes percentiles ⩾90 while
small dots indicate stations for which data were available

153

tropical anomalies deserves further investigation.

References

Bjerknes J (1969) *Mon Wea Rev* 97: 163–172
Flohn H (1964) *Bonner Meteorol. Abh 4*: 1–83
Gadgil S, Joseph P V, Joshi NV (1985) Ocean-atmosphere coupling over monsoon
  regions (Submitted to *Mon Wea Rev*)
Gill A E (1980) *Q J R Met Soc* 106: 447–462
Gill A E (1983) *J Phys Oceanogr* 13: 586–606
Gill A E, Rasmusson E M (1983) *Nature* 306: 229–234
Horel J D (1982) *Mon Wea Rev* 110: 1863–1878
Koteswaram P (1958) *Tellus* 10: 43–57
Krishnamurti T N, (1971) *J Atmos Sci* 28: 1342–1347
Oort A H, Rasmusson E M (1971) *Atmospheric Circulation Statistics.* NOAA Prof Pap
  No 5 US Government Printing Office (NTIS Com–72–50295)
Philander S G H, Rasmusson E M (1984) The Southern Oscillation and El Nino To be
  published in *Adv Geophy* (Academic Press
Philander S G H, Yamagata T Pacanowski R C (1984) *J Atmos Sci* 41: 604–613
Quiroz R S (1983) *Mon Wea Rev* 111: 1685–1706
Rasmusson E M, Carpenter T H (1982) *Mon Wea Rev* 110: 354–384
Rasmusson E M, Wallace J M (1983) *Science* 222: 1195–1202
Rasmusson E M, Arkin P A (1985) Interannual climate variability associated with the El
  Nino/Southern Oscillation. *Proc JSC/CCO Int Liege Colloqui "Coupled atmosphere-ocean
  models"* Elsevier Oceanography Series, Amslvdsm (In press)
Webster P A, (1972) *Mon Wea Rev* 100: 518–541

Some Aspects of Low-Frequency Phenomena of the Air-Sea Interaction A Review of the Recent Advances in the Institute of Atmospheric Physics (IAP)

Ye Duzheng (T. C. Yeh)[1]

Comprehensive Abstract

In recent years the air-sea interaction is one of the major study fields in IAP. This article reviews only the studies on the low frequency phenomena in the air-sea interactions. Based on the past 40 years's data (or more), it is observed that there are two dominate frequencies, namely 2-4 months and 3-4 years period, in many elements of the air-sea coupled system in the tropical Pacific, such as (1) the SST in eastern equatorial Pacific (Fu and Fletcher), (2) the intensity of North Equatorial Current (NEC), South Equatorial Current (SEC) and North Equatorial Countercurrent (NECC) (Fu), (3) the intensity of mean meridional circulation (M) and mean zonal circulation (Z) along the equator over the Pacific (Fu), (4) the intensity of the north Pacific subtropical high (SH) and the equatorial low (EL) (Chen, Fu), and so on. Of these two dominant frequencies, the lower frequency is much more important, The oscillations of these elements are coupled with each other. Fu and Su summarized the couplings in the following diagram (Fig. 1).

Chen and others further, analyzed the variation of the distributions of the time-lag correlations between the SST anomalies in equatorial eastern Pacific and the surface pressure anomalies at each grid point of the whole north Pacific Ocean. They indicate that the variation in intensity of the subtropical high in the eastern Pacific leads that of SST and in turn the variation of SST leads that of the intensity of subtropical high in western Pacific. The half cycle shown in the figure is of about 20 months. Therefore a complete cycle will be $3 - 4$ years.

Since the subtropical high in western Pacific is one of the most important pressure systems controlling the summer rainfall in China, the Chinese meteorologists have already used the anomalous SST in eastern equatorial Pacific as an important criterion for the long-range forecast of the subtropical Pacific high and then the summer rainfall in China. (Fu et al.).

For a period of 30 years (1951 — 80) SST in the eastern equatorial Pacific has further been correlated with the monthly 500 hPa height and surface air temperature over Northern Hemisphere. It was found that the correlation coefficient pattern in North America is quite different from that in Asia. In North America the correlation coefficients are positive to the north of $40° N$ and negative to the south of $40° N$, while in Asia the distribution is opposite (Zhang, Zeng and Pan). If attention is paid to the Pacific/N America area, the correlation pattern is similar to the PNA pattern. However, if attention is paid to Pacific/Eurasia area, there seems

---

1. Institute of Atmospheric Physics, Academia Sinica, Beijing, China

155

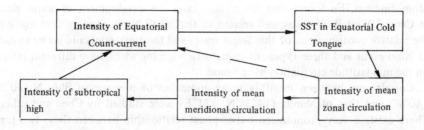

Fig. 1. Schematic diagram of low-frequency oscillation in the tropical Pacific air-sea system

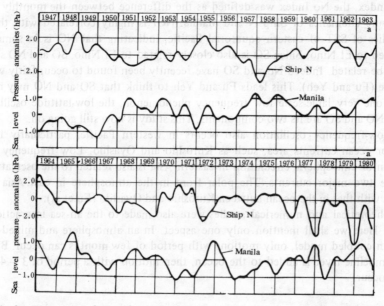

Fig. 2. Time series of monthly mean sea level pressure at stations of ship N and Manila

to be a PEA pattern, that is, the Asian trough develops deeply, and 500 hPa height in a broad area along the middle latitudes of Asia and western Pacific are below normal, while the 500 hPa height over the eastern part of Europe exhibit a positive anomaly center when SST of eastern equatorial Pacific is in a warm period (Zeng and Zhang). Dividing the 30 years into 3 decades, it was found that although the main features of the correlation fields for these three decades are similar, but there are differences among the decades. Further, a time lag correlation between the frequency of typhoon and SST was analyzed. When SST in warm (cold) the typhoon frequency is fewer (more) (Pan).

The attention to the influence of SST on the atmospheric circulation is paid not only to the Pacific, but also extended to the Atlantic. Comparisons of the responses of atmospheric circulation to SST of Pacific and Atlantic were studied (Zhu and Song).

The sequential process of evolution of El Nino was also a topic of study. Among other findings, Fu found that for certain cases the development of warm water in Central West Pacific was well related to that of El Nino. Studies were made to the relative contributions of this warm water and the equatorial cold water to each El Nino event and three types of equatorial warming which have different effects on the mid-latitude circulation were found.

Correlations between monthly mean anomalies of pressure of Ship N (30°N, 140°W) and that of Manila (14°31'N, 121°E) were studied by Chen and others. There exists a very pronounced out-of-phase relationship between these two time series, (Fig. 2), indicating a seesaw oscillation of pressure in east and west Pacific which was called the Northern Oscillation by Chen. Similar to the definition of the SO Index, the No Index was defined as the difference between the monthly mean pressure anomalies of Ship N and Manila. NOI is highly correlated with the anomalies of SST of eastern equatorial Pacific, indicating that NO is in some way related to EI Nino. Since SO is also closely related to EI Nino, SO and NO should also be related. Indeed No and SO have recently been found to occur fairly well in phase (Fu and Yeh). This leads Fu and Yeh to think that SO and NO may reflect just one very large-scale low frequency phenomenon, the low-latitude oscillation. The NO and SO are its two components. This study is now still going on.

Low-frequency oscillation also occurs in western Pacific, particularly in the major ocean currents area, such as Kuroshio and Oyashio. Low frequency oscillation of atmospheric circulation in eastern Asia is also related to the oscillation of these two major currents. The polar front in the atmosphere in east Asia oscillates with that in the ocean between Kuroshio and Oyashio (Fu et al).

Theoretical and numerical studies were also made to the air-sea interactions in IAP. Here we shall mention only one aspect. In an atmosphere and mixed-layer-ocean coupled model, only motions with period of few months can occur. But if a thermocline layer is added to the ocean, then motion with a period of 3 − 4 years can occur.

Typhoon, Monsoon in Asia and Sea Surface Temperature in the Equatorial Pacific Ocean

Pan Yi han[1]

Abstract — Statistical analyses from data of 35 years show that the monthly frequency of typhoons is reversely correlated with the sea surface temperature (SST) of key region for global climate in the east equatorial Pacific Ocean (EEP)−T_{130w}. In association with the general circulation pattern of atmosphere for different typhoon frequencies case, the summer monsoon over India and China are in teleconnection with SST in EEP. When the SST in EEP is cold (warm), typhoon formation is greater (fewer), the westerly over India is stronger (weaker), then the southeast (southwest) monsoon over China is prevailling. These relationships are mainly linked by the zonal vertical circulations in the equatorial atmosphere over the Pacific and Indian Oceans.

Typhoons and monsoon are important for the weather and climate in south and east Asia. They are ancient but also modern subjects and have always attract the meteorologists of China and many other countries. Therefore, although many findings have been already made, still many problmes remain unsolved (Tao Shiyan and Chen Longxun 1975; Chen Lianshou and Ding Yihui 1979).

Recently, accompanied by the spectacular El Nino episode of 1982–83, extremely abnormal weather and climate appeared all over the world once again. El Nino/Southern Oscillation (ENSO) has become one of the most attractive subjects nowadays. This paper will give some synoptic and statistical analysis about typhoon and monsoon formation from the view point of ENSO(Rasmusson and Wallace 1983).

Teleconnection Between Sea Surface Temperature (SST) in the Key Region with Typhoon Frequencies

As is well known, the oceanic thermal state SST$\geqslant$27°C and TSST$\leqslant$7°F/200ft had been given as the critical condition for the formation of Atlantic hurricanes and Pacific typhoons. In fact, a huge area of SST$\geqslant$28°C seasonally oscillates from arround the equator in the West Pacific, northward up to and beyond 20°N after April every year. But the interannual variability of typhoon is remarkable with the maximum of 40/yr and the minimum of 22/yr. Therefore, it is reasonable to search for the causes of typhoon formation not only in local oceanic thermal conditions, but also in the large scale circulation both in atmosphere and ocean; in the other words, from the view point of large-scale air-sea interaction.

Early in 1963, Xie et al. pointed out that more than 80% of typhoons are generated from the boundary belt where the equatorial westerly from the tropical

---

1. Institute of Atmospheric Physics, Academia Sinica, Beijing, China

158

Indian ocean meets the easterly from the northeast trade wind over tropical Pacific in the lower troposphere. Furthermore, it was found (Pan Yihong 1982) that this convergence belt (ITCZ) is mainly formed by two opposite zonal vertical atmospheric circulation over the tropical Pacific and Indian Oceans (so-called Walker cell over to Pacific and anti-Walker cell over the Indian Ocean). The comparision between a pair of more (August, 1967) and few (August, 1969) typhoon samples shows that typhoon appears frequently when the zonal vertical cells are stronger, see Fig. 1a, but fewer typhoons in weaker vertical cells (see Fig. 1b).

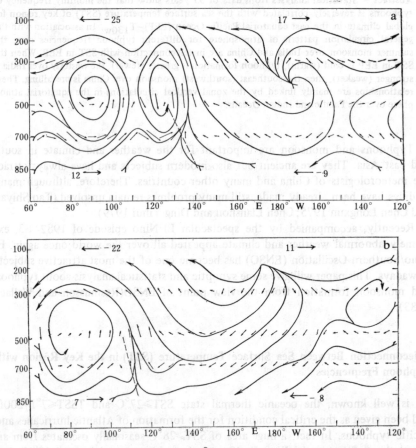

Fig. 1a. Zonal vertical circulation along $15°N$ over Pacific and Indian Ocean for August 1967.b. Same as Fig. 1a but for August 1969

The zonal vertical cell in the equatorial atmosphere over the Pacific is a thermally driven circulation with its intensity being basically decided by the zonal gradient of SST along the equatorial Pacific and then mainly by SST in EEP. The standard deviations of SST in EEP are much larger than in equatorial West Pacific, where the huge warm water mass is located permanently all the year. Therefore, to a certain extent, typhoon frequency is correlated with SST in EEP. Figrue 2 gives

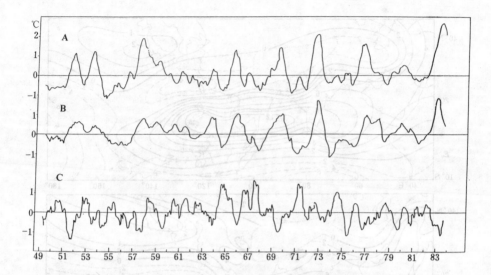

Fig. 2. Time series of seven-month running means of anomalies in 1949–1983 for A-SST at 85°W, 10°S–5°N; B-SST at 130°W, 10°S–5°N; C- Typhoon frequency

the sevenmonth running means for the anomalies of 35 years in 1949–1983 of typhoon frequency (curve C), SST off central and south America at 85°W (curve A) and SST at 130°W (curve B) over southern equatorial current in the belt of 10°S–5°N. However, from the view point of global change of atmospheric circulation for monthly and seasonal time scale, typhoon frequency, as one of its sensitive indexes naturally should be correlated with the SST in the key region $T_{130°W,10°S-5°N}$.

The Characteristics of Wind Field in Lower Troposphere for "Many" and "Few" Typhoons

August is the maximum month for typhoon frequency in the time interval averaged from 1949 to 1983. During this 35 years, 1967 with cold SST in EEP is the maximum typhoon year (40/yr) and 1969 with warm SST in EEP is one of the minimum typhoon year (22/yr). The number of typhoons formed in August is 9/month in 1967, but 4/month in 1969. August 1967 and August 1969 were thus chosen as a pair of more and few typhoon samples for comparison.

Since wind fields directly and clearly demonstrate the general circulation of atmosphere, we concentrate on analyzing the zonal and meridional wind fields. Figure 3a,b are zonal wind fields at 850 hPa for August 1967 and August 1969. They are quite different. The major differences are located in three areas. South of 20°N, one is over the tropical Indian ocean including the Peninsulas, the next is in the east of the Phillipines over the tropical north-west Pacific, and the third lo-

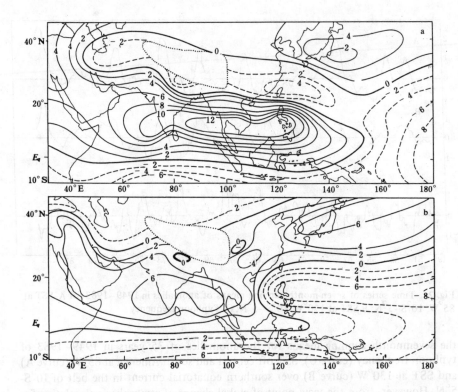

Fig. 3a. Zonal component of wind field at 850 hPa level of August 1967. **b.** Same as Fig. 3a for August 1969

cation of difference is over the continent of China and Japan north of 20°N.

In many typhoons with cold SST in EEP cases, the zonal current in tropics is much stronger. The strong westerlies from tropical India penetrate the intensive easterlies from the Pacific in a tongue shape and make the boundary just over the warmest sea surface of the tropical Pacific between 10°–20°N. The direction of the boundary at 850 hPa level is oriented from southeast to northwest with the isotaches concentrated adjacent it to make a more cyclonic convergent region in the lower atmosphere, which is favorable to typhoon formation.

Contrarily, in a few typhoon cases, the zonal currents in the tropics are both weaker, the easterlies from the Pacific penetrate the westerlies from the Indian ocean in tongue shape and make the boundary west of the Phillipines over South China Sea a relatively smaller area of sea surface than the tropical Pacific. Between 10°–20°N, the boundary orientation runs from southwest to northeast to make a more anticyclonic divergent atmosphere. As for the south part of boundary, although the direction is southeast to northwest, the latitude is too low and near the equator so that the Coriolis force is very small and approach to zero. All of these characters of wind fields U_{850hPa} are not favorable to typhoon generation.

Now, let us see the third different characters north of 20°N between the two cases when the westerlies over tropical Indian are weaker in the case of a few typhoons with warm SST in EEP and the westerlies over China and Japan of east Asia are strengthened. When the westerlies over tropical Indian are intensive in the case of many typhoons with cold SST in EEP, the westerlies in East Asia are weaker so that the easterlies penetrate into the continent of China. All the differences between August 1967 and August 1969 not only appear at U_{850hPa} but also at U_{700hPa}.

This interdependence of zonal circulation in atmosphere between tropics and subtropics over Asia is of interest in understanding the behavior of summer monsoon here.

The Interaction Between Summer Monsoon in India and China

The correlative differences of summer monsoon in India and China for the pair of opposite samples shown in the zonal wind fields mentioned above can be clearly revealed in the meridional wind fields. Figure 4a, b are meridional wind fields at 850 hPa—V_{850hPa} of August 1967 and 1969. Over tropical Indian, in stronger westerlies with cold SST in the EEP case, the southerlies over the sea surface and northerlies over the land are both stronger than those in the case of the warm SST in EEP with weaker westerlies. This means that there is stronger meridional exchange of air mass in the lower troposphere over South Asia in the case of cold SST in EEP. This is a favorable condition for monsoon precipitation. To a certain extent, it explains why the monsoon rains in India are heavier than when the SST in EEP is colder.

Over East Asia, the pattern of meridional isotaches in the two cases is also quite different. Corresponding to the zonal wind fields, when the SST in EEP is warm as August 1969, westerlies over East Asia are stronger so that isotaches of southerlies have an axis from southwest to northeast. This is a distinct pattern of southwest (SW) monsoon prevailing over East Asia. Then in the cold SST case, as in August 1967, easterlies blow around the continent of China so that the isotaches of southerlies give a clear pattern which indicates that southeast monsoon prevails over east Asia (SE). All the characters shown at V_{850hPa} also appear at V_{700hPa} fields. In other words, there is interaction in the atmospheric circulation, relative to SST in EEP, between tropical India and subtropical China.

Verification by More Samples

In order to understand whether the distinguishing characters in the general circulation of atmosphere in Asia with respect to SST in EEP between the given pair of samples are substantial or not, we will check it by more samples.

Firstly, the extremely warm SST in the EEP case of 1983 is given. The total amount of typhoons formed in 1983 is 22, one of the minimum years in the

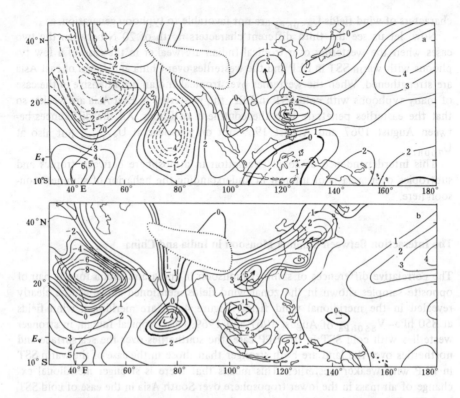

Fig. 4a. Meridional component of wind field at 850 hPa level for August 1967. **b.** Same as Fig. 4a but for August 1969

1949—1983 period, and only 11 typhoons appeared before September. Figure 5a,b shows U_{850hPa} and V_{850hPa} of July 1983. The patterns of both zonal and meridional wind fields are the same as in August 1969 with warm SST, only the absolute magnitudes of wind speed being much stronger everywhere in the three major aeras.

Then, the series of zonal wind speed averaged from a few individual rawinsonde stations, respectively representing tropical Indian and subtropical China are given in Fig 6a,b for comparison, in Fig 6a, curve A is the anomalies of U_{850hPa} averaged from Wuhan and Shanghai to represent China and curve B is that from Trivandrum over the Arabian sea and Port Blair over the Bay of Bengal to represent India for a period of 25 years in 1959 to 1983 of July, Curves A and B in Fig.6b are the same as in Fig 6a but for August. Both in July and August, A and B vary from year to year almost reversely. This means that in the southwest(SW) summer monsoon system over Asia, the westerly components over tropical India and subtropical China are connected and conditioned with each other, al least in the recent two decades, Usually, when the westerlies over tropical India are stronger (weaker), the westerlies over subtropical China are weaker (stronger). Since the summer monsoon rainfall between India and China also has an out of phase re-

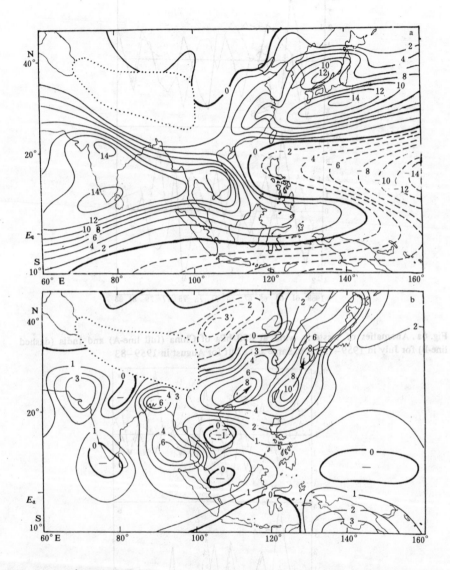

Fig. 5a. Zonal component of wind field at 850 hPa level for July 1983. **b.** Meridional component of wind field at 850 hPa level for July 1983

lationship for this period given by Fu and Fan (1983), the interdependence of summer westerlies in lower troposphere between India and China should be of great interest. In fact, early in 1934, based on the analysis of heavy floods and serious droughts over China from nearly 50 years data, Chu Ko–chen (1954) pointed out that the total amount and distribution of summer rainfalls in China depended on whether the prevailing monsoon is from the southeast (SE) or the southwest (SW).

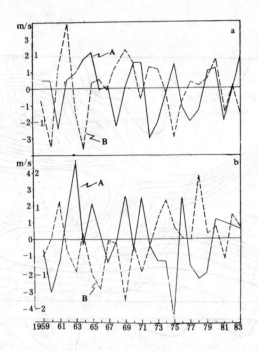

Fig. 6a. Anomalies of westerly wind at 850 hPa of China (full line-A) and India (dashed line-B) for July in 1959–1983. b. Same as 6a but for August in 1959–83

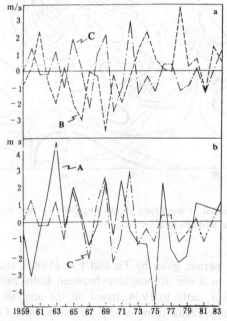

Fig. 7a. Anomalies of westerly over India (dashed line-B) and SST at 130°W, 10°S–5°N (dashed and dot line-C) for July in 1959–1983. b. Anomalies of westerly wind over China (full line-A) and SST at 130°W, 10°5°N (dashed and dot line-C) for July in 1959–1983.

As a result, the intensities of westerlies in South and East Asia should be correlated with SST in EEP. As given in Fig 7a,b curve C, which is the SST in the Key region $T_{130°W, 10°S-5°N}$ varies with the westerlies reversely over India and consistently over China in most years.

Remarks

Accompanied by the warmest area around the tropical West Pacific, the convergent rising branch between Walker and anti-Walker cells migrates near and over this area in the summer of the Northern Hemisphere. The westerlies in the lower troposphere over India and China may comprise the lower branch of the anti-Walker cell. Most of the typhoons form from the convergent rising belt — ITCZ between the two cells. In order to show the correlation between anti-Walker and SST in EEP, the zonal wind components at 200 hPa and 850 hPa of Singapore which is located near the equator over Indian ocean, are taken to represent the upper easterlies and lower westerlies of this equatorial zonal vertical circulation respectively. Figure 8 give the seven-month running mean for westerlies at 850 hPa of Singapore — A; easterlies at 200 hPa of Singapore — B; SST at 130°W, 10°S–5°N — C; and typhoon frequency of 20 years in 1964–83 — D. The correlation coefficients of different time lag between the atmospheric parameters and SST in EEP shown in Fig. 9 are considerably high. It seems that the teleconnection between typhoon frequency as well as summer monsoon in China and India with SST in EEP are linked by the zonal circulation in the equatorial atmosphere over the Pacific (Walker cell) and Indian Oceans (anti-Walker cell).

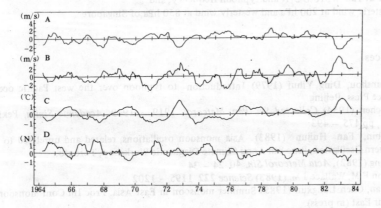

Fig. 8. Time series of seven-month running means of anomalies in 1964–1983 for A - Westerly wind at 850 hPa of Singapore; B - Easterly wind at 200 hPa of Singapore; C – SST at 130°W, 10°S–5°N; and D – Typhoon frequency

166

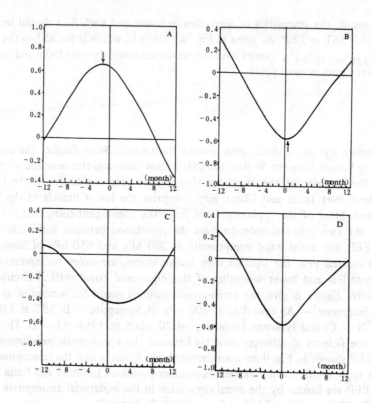

Fig. 9. Correlation coefficients for different lags of 20 years in 1964—1983 between
A – SST at 130°W, 10°S-5°N and easterly wind at 200 hPa of Singapore;
B – SST at 130°W, 10°S-5°N and westerly wind at 850 hPa of Singapore;
C – SST at 130°W, 10°S-5°N and typhoon frequency; and
D – Easterly wind at 200 hPa and westerly wind at 850 hPa of Singapore

References

Chen Lianshou, Ding Yihui (1979) Introduction to typhoon over the west Pacific ocean. Science Press, Beijing
Chu Ko-chen (1954) *Collected Sci. Pap. Meteorol.* 1919 – 1949. *Academia Sinica,* Peking, China, pp 475 – 493
Fu Congbin, Fan Huijun (1983) Asia monsoon oscillations, related and unrelated to the Southern Oscillation, Proc 8th Annu Climate Diagnostic Workshop, Toronto.
Pan Yihong (1982), *Acta Meteorol Sin.* 40 24 – 34
Rasmusson E M, Wallace J M (1983) *Science* 222 1195 – 1202
Tao Shiyan, Chen Longxun (1985) Summer monsoon in East Asia. Proc Int Conf Monsoon in the Far East (in press).
Xie Yibing, Chen Shoujun, Chang I liang, Huang Yinliang (1963) *Acta Meteorol Sin* 33 206–217

Seasonal Scale Ocean-Atmosphere Interaction over Tropical Areas of Indian Ocean and West Pacific

P. V. Joseph[1] and P. V. Pillai[1]

Abstract — During the Asian summer monsoon, the north Indian Ocean experiences cooling of its surface layer, when the other oceans of the Northern Hemisphere are warm. The monsoon cooling from May to August is largest in the western Arabian Sea and decreases to the east. The amplitude of the monsoon cooling is found to depend on the strength of the monsoon as represented by the area-weighted rainfall of India from 1 June to 30 September. Thus a poor (good) monsoon creates a warm (cold) SST anomaly. These anomalies which extend to the West Pacific cover 60 degrees in longitude and 30 degrees in latitude and persist from one monsoon to the next. Tropical Indian and East Pacific Oceans have warmed and cooled with only small phase lags during 1961–1973. Although the amplitude of the SST anomaly is much larger over the Pacific Ocean, the normal SST is much higher over the Indian Ocean. The poor (good) monsoon is able to produce a warm (cold) SST anomaly over the north Indian Ocean due to factors like (1) reduced (increased) wind mixing and reduced (increased) evaporation over the area and reduced (increased) upwelling over western Arabian Sea, and (2) reduced (increased) cloud cover over the area. The spatially large warm (cold) SST anomaly created by the poor (good) monsoon persisted during the following October to May. The large SST anomaly over tropical Indian Ocean was found to create anomalies in upper tropospheric temperature and winds over a large area in much the same way as in the Pacific as seen from observational and GCM sensitivity studies. The persistent circulation changes over South Asia in the months prior to the monsoon are related to the following monsoon rainfall. The SST anomalies over the tropical ocean produce large circulation changes in the atmosphere only in the winter hemisphere. Thus the SST anomalies of October to May of the Indian Ocean may have a role preconditioning the atmospheric circulation and weather particularly in the Northern Hemisphere, which should be important in the development of the Asian summer monsoon. It is important here to note that the tropical Pacific Ocean also warmed and cooled almost at the same time as the Indian Ocean.

Introduction

Sea Surface Temperature (SST) exhibits marked seasonal variations. Superposed on these are spatially large SST anomalies, which have persistence of months to seasons. In the 1960s, Bjerknes studied the SST anomalies over the equatorial Pacific Ocean and suggested physical mechanisms to show that the warm anomalies, in addition to increasing the rainfall in the region of the anomaly, increased the strength of the subtropical westerlies of the upper troposphere, consequently leading to the enhancement of the intensity of the Hadley circulation and in other ways also affected the atmospheric general circulation over large areas of the globe (e.g., Bjerknes 1969). Since then empirical evidence and studies using General Circulation Models (GCM) have shown that SST anomalies of tropical oceans are

---

1. India Meteorological Department, Pune 411005, India

168

important causes of the interannual variability of climate. SST anomalies in turn are caused by large scale (in time and space) anomalies in the atmospheric circulation. Thus there is seasonal-scale interactive coupling between tropical oceans and the global atmosphere.

SST Anomalies of the Tropical Indian Ocean

Figure 1 shows the annual variation of monthly mean SST obtained using SST data collected by ships of over 40 maritime nations (Voluntary Observing Fleet) during the period 1961–1970 for a few locations in tropical Indian Ocean and adjoining West Pacific. Locations 1,5, and 8 are mainly responding to the heating due to solar radiation. Location 4 on the equator has very small annual variation. Locations, 2, 3, 6, and 7 show the effect of monsoon colling of the ocean. Areas 2, 3, and 6 cool from May to August, while the other oceanic areas of the Northern Hemisphere warm. The amplitude of the monsoon cooling is largest at 2 and decreases toward the east. At location 7, in tropical West Pacific, the curve is flat from June to October. The amplitude of the monsoon cooling of the north Indian Ocean from May to August, is found to depend on the strength of the monsoon represented by the area-weighted monsoon rainfall of India as derived by Parthasarathy and Mooley (1978). This index of the monsoon rainfall of India is found to be significantly negatively correlated with the SST of the post-monsoon months over the north·Indian Ocean (Joseph and Pillai 1984). Thus a poor (good) monsoon creates a warm (cold) SST anomaly over the north Indian Ocean and the adjoining Pacific Ocean.

A study of the ocean and atmosphere data of 1972 (a drought monsoon year in India) and 1973 (a good monsoon year) by Joseph and Pillai (1982) has shown that a poor monsoon is able to produce a warm SST anomaly over the tropical Indian Ocean due to reduced monsoon cooling (May to August) of the surface layer of the ocean on account of (i) reduced wind mixing, reduced evaporation and reduced upwelling over the western Arabian Sea due to weaker surface winds and (ii) reduced cloud cover. The monsoon cooling of the ocean surface during 1972 and 1973 may be seen from Fig. 2. The reduced monsoon cooling in 1972 is that it transformed the cold SST anomaly of pre-monsoon of 1972 (e.g., April 1972) to a warm anomaly soon after this monsoon and the warm·anomaly thus created persisted till the following monsoon (e.g., April 1973), as may be seen from Fig. 3. The intensity and the spatial extent of the SST anomaly may also be seen from the figure. These anomalies over the tropical Indian Ocean and the adjoining West Pacific Ocean cover an area over 60 degrees in longitude and about 30 degrees in latitude, comparable in size to the El-Nino over the equatorial East Pacific Ocean (Rasmusson and Carpenter 1982), and persist from the end of one monsoon season to the beginning of the next (October to May). Although the amplitude of the SST anomaly is much larger over the East Pacific Ocean, the normal SST is much higher over the Indian Ocean (see Fig. 4).

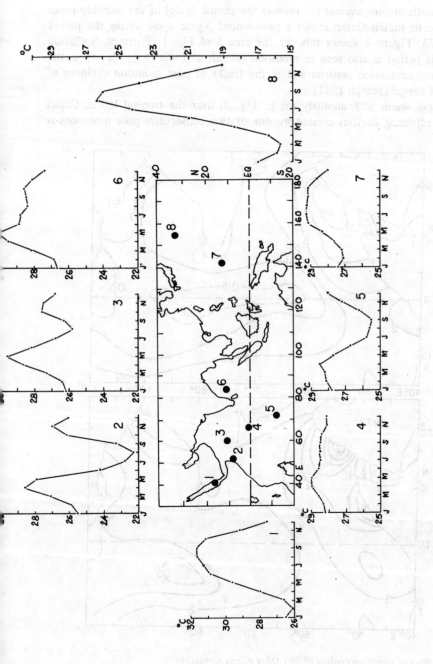

Fig. 1. Annual variation of monthly mean SST (in °C) at locations 1 – 8

Monsoon-Indian Ocean Interaction on a Seasonal Scale

Twelve-month moving average (to remove the annual cycle) of the monthly mean SST of north Indian Ocean shows a pronounced 3-year cycle during the period 1961–1973. Figure 4 shows this for the area 3 of Fig. 1. Triennial oscillation during this period is also seen in monsoon rainfall of India (e.g., Fig. 4), in the atmospheric circulation features and in the tracks of post monsoon cyclones of the Bay of Bengal (Joseph 1981).

The large warm SST anomaly (as in Fig. 3) over the tropical Indian Ocean (and the adjoining Pacific) created by one or two consecutive poor monsoons is

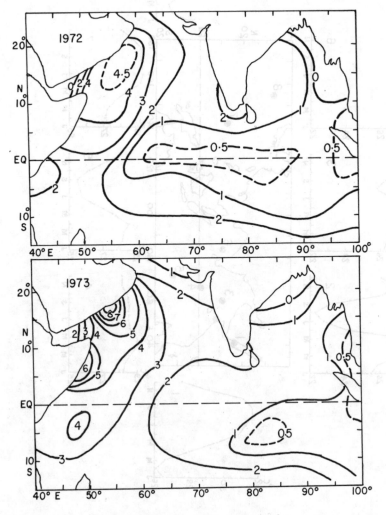

Fig. 2. Isolines of monsoon cooling of SST (May minus August) in °C

found to create anomalies in the atmospheric circulation over a large area over and around it, in much the same way as in the tropical Pacific Ocean, as seen from observational and GCM sensitivity studies (Horel and Wallace 1981; Keshavamurthy 1982). The atmospheric response at 200 hPa level to a large SST anomaly (warm) over the East Pacific tropics has been obtained by Sadler (1980) by subtracting the zonal wind of January 1972 from the zonal wind of January 1973. Over and to the west of the warm SST anomaly there is an easterly anomaly and the subtropical westerlies to the north and south of the anomaly have strengthened, as suggested by Bjerknes, and seen in the GCM sensitivity study of Keshavamurthy (1982). The warm SST anomaly created by the very poor monsoon of 1972 (persisting from October 1972 to May 1973) has changed the atmospheric circulation in the upper troposphere over South Asia in much the same way, as may be

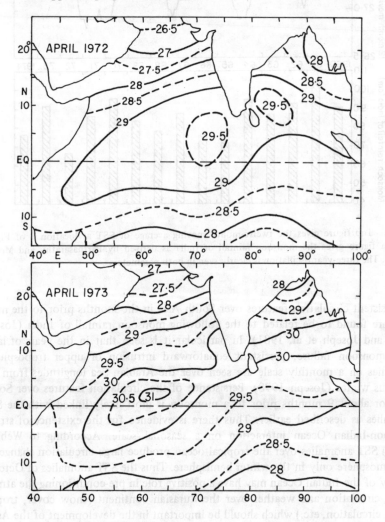

Fig. 3. Isotherms of sea surface temperature in °C

seen from Fig. 5, i.e., anomalous easterlies in low latitudes and increased strength
of the subtropical jet stream to its north in the upper troposphere (Joseph and
Pillai 1982).

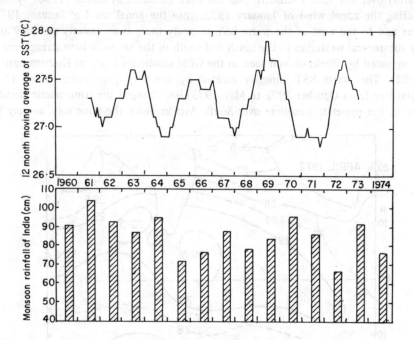

Fig. 4. Top figure gives the twelve-month moving average of SST at location 3 of Fig. 1.
Bottom figure gives the monsoon rainfall of India as derived by Parthasarathy and Mooley
(1978). The three-year oscillation is clearly seen in both parameters

Persistent circulation features over South Asia in the months prior to the monsoon are found to be related to the following monsoon rainfall of India (Joseph
1981, and Joseph et al. 1981). In particular it is seen that in the years of large
scale monsoon failure, persistent equatorward intrusion of upper tropospheric
westerlies on a monthly scale are seen over the Arabian Sea longitudes from the
previous winter (Joseph 1978). Persistence of the circulation features over South
Asia for about 9 months prior to a monsoon is most likely linked with the SST
anomalies as described earlier. Thus there is evidence for the existence of strong
monsoon-Indian Ocean interaction on a seasonal scale. According to Webster
(1982) SST anomalies over the tropical ocean produce large circulation changes in
the atmosphere only in the winter hemisphere. Thus the SST anomalies of October
to May of the Indian Ocean may have a positive role in pre-conditioning the atmospheric circulation and weather over the Eurasian continent (snow cover, tropospheric circulation, etc.) which should be important in the development of the Asian
summer monsoon.

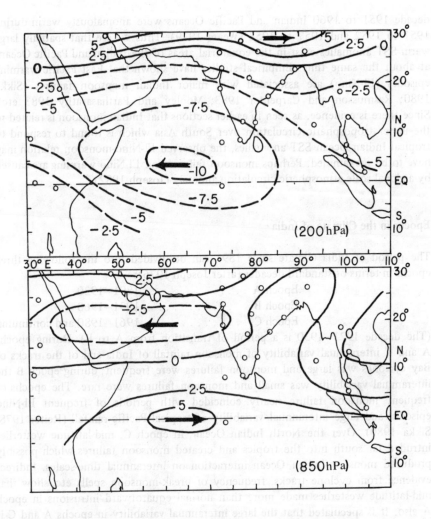

Fig. 5. The difference in monthly mean u at 850 hPa (bottom figure) and 200 hPa (top figure) for April 1973 minus April 1972 (isopleths at 2.5 m/s intervals, westerly in continuous lines and easterly in broken lines). Open circles show stations whose data have been used

Nearly Simultaneous Warming Episodes of Tropical Indian and East Pacific Ocean

Available observational studies have shown that the tropical Indian and East Pacific Oceans have warmed and cooled with only small phase lags (Joseph and Pillai 1984; Pan and Oort 1983). During the period 1961 to 1973 the warm anomalies occurred over the north Indian Ocean in the years of 1963, 1966, 1969 and 1972 respectively following a triennial cycle. During the same period tropical East Pacific Ocean also warmed and cooled with very little phase difference. During the earlier decade 1951 to 1960 Indian and Pacific Oceans were anomalously warm during 1951 to 1952 and 1957 to 1958 (Weare 1979). Thus we see that spatially large warm SST anomalies exist in the equatorial areas of the Indian and Pacific Oceans at about the same time. Empirical studies have shown that East Pacific warming episodes (El-Nino) are associated with major Indian monsoon failures (Sikka 1980; Rasmusson and Carpenter 1983; Mooley and Parthasarathy 1983 etc). Since there is evidence, as seen in earlier sections that Indian monsoon is related to the upper tropospheric circulation over South Asia which is found to respond to tropical Indian Ocean SST anomalies, the observed El-Nino-monsoon relation may have to be reexamined. Perhaps monsoon failures and El-Nino warming are caused by a global scale atmospheric circulation anomaly (Joseph 1982).

Epochs in the Climate of India

The period of India climate since 1891 can be divided into the following three epochs in terms of standard decades (after Joseph 1976):

Epoch A 1891–1920
Epoch B 1931–1960
Epoch C . . . . 1961–1980 and continuing

(The decade 1921–1930 is a period of transition from A to B). During epochs A and C interannual variability of monsoon rainfall of India and of the tracks of Bay cyclones was large and monsoon failures were frequent; during epoch B the interannual variability was small and monsoon failures were rare. The epochs of frequent monsoon failure nearly coincided with periods of frequent El-Nino episodes and large interannual variability of tropical Pacific rainfall (Reiter 1978; Sikka 1980). Over the North Indian Ocean, in epoch C, mid-latitude westerlies intruded far south into the tropics and created monsoon failures which possibly produced monsoon-Indian Ocean interaction on interannual time-scales. Indirect evidence from cyclone tracks, frequency of break-monsoon spells etc. show that mid-latitude westerlies made more than normal equatorward intrusions in epoch A also. It is speculated that the large interannual variability in epochs A and C is caused by ocean-atmosphere interaction.

Conclusions

Since large-scale ocean-atmosphere interactions are important in interannual changes of global atmospheric circulation in general and the Indian summer monsoon in particular, well-planned research programmes should be drawn up for empirical and theoretical GCM studies. The following areas require particular attention:

(a) The monsoon cooling of the north Indian Ocean, particularly the Arabian Sea, which causes SST anomalies, is not well understood.

(b) There is need for a GCM sensitivity study using a realistic SST anomaly over the Indian Ocean similar to the study by Keshavamurthy (1982) for the Pacific

Ocean. The available GCM simulations for the Indian Ocean area have used SST anomalies of very small spatial scale (Washington et al. 1978). A simulation with warm anomalies simultaneously in both Pacific and Indian Oceans as actually observed is also needed.

(c) The equatorial trough is likely to be influenced by SST anomalies. This and the interaction of the equatorial trough with the Asian summer monsoon need detailed study, particularly in view of the recently discovered 30–50 day mode of the equatorial trough during the monsoon season (Yasunari 1980; Sikka and Gadgil 1980).

Indian Ocean SST data of 1961–1973 has shown that one or two successive poor monsoons can create a spatially large and intense warm SST anomaly over the Indian Ocean, and this large warm anomaly existing during the pre-monsoon months of the following year changes the atmospheric circulation features over South Asia in such a way that India has copious rainfall during the monsoon of this year. Available rainfall climatology seems to support this hypothesis. Examining the long series of rainfall of India of more than a century as derived by Parthasarathy and Mooley (1978), it is found that there were only three occasions when the monsoon rainfall of India was less than normal by one standard deviation during two consecutive years; three such consecutive years have not occurred. The Indian Ocean perhaps has a stabilizing influence on the Indian monsoon. Unlike India, subtropical Africa is known to have experienced long spells of drought years, the recent disastrous drought in the Sahel during 1966 to 1974 being an example.

References

Bjerknes J (1969) *Mon Wea Rev* 97: 163–172

Horel J D, Wallace J M (1981) *Mon Wea Rev* 109: 813–829

Joseph P V (1978) *Ind J Meteorol Hydrol Geophy* 29: 412–418

Joseph P V (1981) Ocean-Atmosphere Interaction on a Seasonal Scale over North Indian Ocean and Indian Monsoon Rainfall and Cyclone Tracks–a Preliminary Study. MAUSAM 32: pp 237–246

Joseph P V (1982) in: *Proc. Int. Conf. Sci. Res. Monsoon Experiment,* Bali, Indonesia, 26–30 Oct. 1981, WMO–ISCU, p 1–71–1–74

Joseph P V, Pillai P V (1982) Air-Sea Interaction on a Seasonal Scale over North Indian Ocean II. Monthly Mean Oceanic and Atmospheric Parameters during 1972 and 1973 Pre-published Sci Rep 1982/2, India Meteorol Dep, Pune 5

Joseph P V, Pillai P V (1984) Air-Sea Interaction on a Seasonal Scale over North Indian Ocean I. Inter-Annual Variations of Sea Surface Temperature and Indian Summer Monsoon Rainfall. Paper presented at the WMO Symp Tropical Droughts New Delhi, 7–12 December 1981. MAUSAM 35 (1984), pp 323–330

Joseph P V, Mukhopadhyaya R K, Dixit W V, Vaidya D V (1981) Meridional Wind Index for Long Range Forecasting of Indian Summer Monsoon Rainfall, MAUSAM 32, pp 31–34

Keshavamurthy R N (1982) *J Atmos Sci* 39: 1241–1259

Mooley D A, Parthasarathy B (1983) *Indian Summer Monsoon and El-Nino.* PAGEOPH, 121: pp 339–352

Pan Y H, Oort A H (1983) *Mon Wea Rev* 111: 1244–1258

Parthasarathy B, Mooley D A (1978), *Mon Wea Rev* 106: 771–781

176

Rasmusson E M, Carpenter T H (1982) *Mon Wea Rev* 110: 354–384

Rasmusson E M, Carpenter T H (1983) *Mon Wea Rev* 111: 517–528

Reiter E R (1978) *Mon Wea Rev* 106: 324–330

Sadler J C (1980) *Mon Wea Rev* 108: 825–828

Sikka D R (1980) Some Aspects of the Large Scale Fluctuations of Summer Monsoon
Rainfall over India in Relation to Fluctuations in the Planetary and Regional Scale Circulation Parameters. Proc Ind Acad Sci (Earth Planet Sci) 89: 179–195

Sikka D R, Gadgil S (1980) *Mon Wea Rev* 108: 1840–1855

Washington W M, Chervin R M, Rao G V (1978) In: Krishnamurti T. N. (ed) *Monsoon Meteorology.* PAGEOPH 115, pp 1335–1357

Weare R C (1979) *J Atmos Sci* 36: 2279–2291

Webster P J (1982) *J Atmos Sci* 39: 41–52

Yasunari T (1980) *J Meteorol Soc Jpn* 58: 225–229

Characteristics of the Response of Sea Surface Temperature in the Central Pacific Associated with ENSO

Fu Congbin[1], H. F. Diaz[2] J. O. Fletcher[3]

Abstract–The zonal distribution of sea surface temperature (SST) in the equatorial Pacific ($4°N–4°S$, $120°E–80°W$) associated with ENSO has been studied by using the seasonal mean file of the Comprehensive Ocean-Atmosphere Data Set (COADS) for the period 1928–1983.

The warmest area in the western-central Pacific shows very little annual variation, but is very sensitive to the ENSO, displaying large variability during such episodes. The eastward migration of this warmest area ($28.5°C$ isotherm is used here as a criterion related to the strong tropical convection and heavy rainfall) is a common feature in the developing stage of almost all ENSO events since 1928, not only for the event of 1982. The extent of its eastward migration varies from event to event, and represents a large contribution to the interannual variability of zonal SST distribution in the equatorial Pacific, comparable to the behavior of the equatorial cold tongue in the east. This warmest water is not a passive factor, but an active one in the development of ENSO.

Eigenvector analysis of zonal profiles of SST in this area shows three major patterns. Pattern A is warm in the east and central areas with slightly below normal in the west (1972 type ENSO). Pattern B is warm in the east with normal in the central and slightly below normal in the west (1976 type ENSO) and Pattern C is nearly uniformly warm in the entire area (1963 ttpe ENSO). These three patterns account for 62% of the total seasonal equatorial Pacific SST variance about the mean ENSO profile. The patterns of SST profile mainly depend on the relative contribution of the warmest water in the west-central Pacific and the equatorial cold tongue in the east. The west-to-east SST gradient in these three patterns is appreciably different, which could prove useful in distinguishing the types of ENSO in the future.

Introduction

The evolution of sea surface temperature (SST) anomalies associated with ENSO (El Nino/Southern Oscillation) events has been a major topic in ENSO studies. The original meaning of El Nino dealt with the anomalous ocean current off the coast of Peru around Christmas time and the associated positive SST anomalies there (Wyrtki 1975). Therefore the so-called "El Nino" event has been defined mainly by the SST data from stations on the South America coast, such as Puerto Chicama ($7.7°S$, $79.3°W$), which has long been used as a representative eastern Pacific Station.

Since the 1970s the term "El Nino" has been popularized to refer to the SST anomalies in a wide area of the equatorial Pacific (Ramage 1975; Weare et al. 1976). After Bjerknes (1966) and Berlage (1966) demonstrated the linkage

1. The Institute of Atmospheric Physics, Academia Sinica, Beijing, China
2. NOAA/ERL, 325 Broadway, Boulder, CO 80309, U. S. A.
3. NOAA/OAR, Rockville, MD 20852, U. S. A.

between El Nino and Southern Oscillation, the SST anomalies in the equatorial Pacific became a principal characteristic of the ENSO phenomenon. Nevertheless, the main focus has remained in the eastern Pacific area.

Several investigators have analyzed the relationship of SST anomalies between the different longitudes from off the coast of South America to the central Pacific, suggesting a westward migration of the SST anomalies. For example, Hickey (1975) presented evidence that year-to-year fluctuations in SST at Christmas Island (20°N, 158°W) and Canton Island (2.5°S, 171°W) lag those at Galapagos (2°N, 92°W) by three and four months respectively. Barnett (1977) found evidence of about a half-year lag between SST at Tarawa (13°N, 172°E) and Christmas Island. Quinn (1979) found a 3-6 month time lag between the SST of the South American coast and SST at Canton Island. Rasmusson and Carpenter (1982) showed a 4-month time lag between the west coast of South America and the Hawaiian-Fiji track. Although its mechanism needs to be explored, the generally good agreement between different authors suggests the existence of a westward migration of SST anomalies during ENSO development.

The recent strong 1982 ENSO event, characterized as a "surprising one" (Rasmusson et al. 1983), has raised a number of questions about the "model" developed by Rasmusson and Carpenter (1982) based upon the composite features of 6 previous events since World War II. One of the major "surprises" of the 1982 event is "the lack of an early appearance and subsequent westward migration of positive SST anomalies in the extreme eastern equatorial Pacific." By using the longitude of 28°C isotherm as a reference, Rasmusson and Carpenter (1983) showed the *eastward* progression of the warmest water at the equator, while showing that the maximum SST anomaly moved relatively little during the event.

This leads us to ask: is the 1982 event an entirely different type of event or is it only another type of event which has occurred before but did not show up in the composite model? The SST record suggests two patterns of equatorial warming associated with ENSO events, one with the main center of anomalous warming near the date line, such as in 1963, and the other with the warming center mainly in the eastern Pacific, such as in 1976. The strong events, like 1972, and especially 1982, represent a mixing of these two patterns. (Fu and Fletcher 1985). This will be explored in more detail below. Rasmusson and Wallace (1983) have also suggested two types of ENSO events.

We suggest that the appearance of anomalously warm water in the west-central Pacific is crucial to the development of ENSO. This area is the counterpart of the so-called equatorial cold tongue. Together, they comprise the strong west-east thermal gradient which is believed to be associated directly with the east-west pressure oscillation and the Walker circulation (Bjerknes 1969). The assumption has long been made that the variability of zonal SST gradient is caused principally by the variation in the eastern Pacific; therefore, the equatorial cold tongue in the eastern Pacific has long been looked upon as a more active one compared with the area of warmest water in the west-central Pacific. However, as will be shown in Sections 2 and 3, this assumption is true on the mean annual scale, but is not true on an interannual scale. The interannual variability of seasonal SST anomalies in the central Pacific is of the same order as that in the eastern Pacific.

As is well known, the convective rainfall associated with this warmest area is one of the three major tropical heat sources (the other two are over Africa and South America) powering the general atmospheric circulation. The warm SST and large mountainous islands there are directly related to the vigorous tropical convection, equatorial rainfall and latent heat release in the atmosphere. It is our contention that this warmest area is perhaps not the passive factor it has long been thought to be compared with the equatorial cold tongue, but a rather active one in the development of ENSO, at least in some cases.

An attempt has been made to find a way of distinguishing the ENSO patterns based upon the zonal migration of this warmest area and the zonal SST distribution in the equatorial Pacific. Shukla and Wallace (1983) have suggested, based on numerical experiments, that " the simulated atmospheric response depends not only upon the position of the anomaly, but also upon the climatological mean SST distribution upon which the anomaly is superimposed." Observational studies (Garcia 1983) also show that the variation of tropical convection associated with the position of warmest area rather than with the position of maximum SST anomalies is very important in determining the atmospheric response, i.e., the location of the heaviest convection.

In this analysis we have used actual SST data which are described in the next section. Section 3 introduces the way we define the ENSO event and the comparison with other investigators, referring to the work about signal/noise in the Southern Oscillation index reported by Trenberth (1984). Section 4 gives the climatology of SST in the global tropics with emphasis on the characteristics of the warmest area in the west-central Pacific. Section 5 describes the interannual variability of SST in global tropics associated with ENSO and particularly the eastward migration of the warmest area which is given in Section 6. The result of eigenvector analysis of zonal SST profile are presented in Section 7 to show the main patterns of SST distribution associated with ENSO. The relationship between SST patterns and the equatorial westerlies and the tropical convection are studied in Section 8. The role of the warmest water in the west-central Pacific and the variation of east-west SST gradient in the equatorial Pacific related to it in the development of three types of equatorial warming are discussed in the summary, Section 9.

Data

Sea surface temperature and surface wind data used in this study come from the Comprehensive Ocean-Atmospheric Data Set (COADS) which was completed recently by a cooperative project among the Cooperative Institute for Research in Environmental Sciences (CIRES), University of Colorado; Environmental Research Laboratories (ERL), National Oceanic and Atmospheric Administration (NOAA); National Center for Atmospheric Research (NCAR); and National Climatic Data Center (NCDC) (Fletcher et al. 1983).

The data base contains seventy million individual marine reports over the world's oceans, observed primarily by the world's merchant fleets and also in the later years by ocean weather ships, buoys, and research vessels. These have

been edited and summarized statistically for each month. The period covered is
1854–1979, and the data is sorted by 2° latitude × 2° longitude boxes. Post-
1979 data derive from the telecommunication reports received from the Navy's
Monterey center, California.

With careful attention to a number of possible caveats (Slutz et al. 1984),
this data set is a unique one for the study of climate variation over the global
ocean, especially for the study of ENSO. The seasonal mean file COADS has
been used in this study because the characteristic time scale of ENSO is on the
order of year, with a mean recurrence interval of 3–4 years (Weare et al. 1976;
Julian and Chervin 1978; Fu and Li 1978; Cane 1983). Trenberth (1984) suggested in a discussion on signal *versus* noise in the Southern Oscillation that seasonal
values are the shortest averages effective in the monitoring of the SO.

After checking the variation in the number of observations within our area
of interest, we chose to study the period from 1928 to 1983.

Catalogue of ENSO Events Since 1928

ENSO events have been catalogued by several investigators (e.g., Quinn et
al. 1978; Weare et al. 1976; Van Loon and Madden 1981; Rasmusson and Carpenter 1983; Trenberth 1984). The strong events are recognized as such by
investigators, but there are some differences in their lists of ENSO events because
of the different criteria they use in defining the event. For instance, Quinn et al.
have documented events as far back as 1726, based on a number of sources, including fishery and marine bird reports, hydrological data in the Peruvian coast region,
SST data from the coast of Peru and Southern Ecuador, SST data over the equatorial Pacific, rainfall data in both of the above two regions, and the Southern Oscillation Index. Van Loon and Madden (1981) defined the event mainly according to
the rainfall in the equatorial region of the southern Pacific Ocean and the sea level
pressure in the tropical region from Darwin, Australia to Cocos Island. Horel and
Wallace (1981) used mainly the time series of the first eigenvector of SST in the
Pacific. Most lists agree on the time of the major events, however, there is some
disagreement regarding the timing of lesser ones.

The ENSO events catalogued in this paper are mainly based upon Fig. 1 which
includes SST data and Tahiti-Darwin SOI data. The SST index represents a departure of SST in the equatorial strip 0°–10°S from winter of 1927–28 through spring
of 1983. The reference period is 1950–79. The SOI data were obtained from the
Climate Analysis Center (CAC), NOAA/NMC. The index consists of the deviation
from the mean pressure difference of Tahiti minus Darwin.

Taking into account both SST anomalies and SOI (Fig. 1), 13 events are
catalogued as follows: 1930, 40, 41, 46, 51, 53, 57, 63, 65, 69, 72, 76, and 82.
Table 1 gives the comparison of our catalogue with those of several previous investigators. It shows that most of the events catalogued in this paper are the same as
those found in other compilations. Our list comes closest to that derived from the
figure given in the paper by Trenberth (1984), which is denoted by T in Table 1.

The years listed in Table 1 are defined as the onset years of ENSO events,

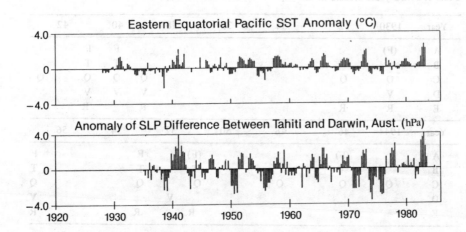

Fig. 1. Two indices of El Nino/Southern Oscillation. (a) Sea surface temperature anomaly 0° – 10°S (°C). (b) Southern Oscillation index defined by the sea level pressure difference between Tahiti and Darwin (from NOAA/CAC) (hPa)

using year (0), year (−1) and year (+1) to represent the onset year, the year before the onset and the year after the onset, respectively. The four seasons are grouped by DJF (December, January, February), MAM (March, April, May), JJA (June, July, August), and SON (September, October, November).

Climatology of Sea Surface Temperature in the Global Tropics

The annual cycle of sea surface temperature in the Pacific has been described by Wyrtki (1965). The annual cycle in tropical Atlantic and Indian Oceans, respectively have been developed by Merle (1978) and Weare (1979). Here, we describe the general climatological features of SST in the global tropics as a background to the discussion regarding interannual variability.

Figure 2 is a time-longitude section of seasonal mean SST in the 4°N–4°S latitude zone averaged for a 30–year period (1950–1979). It shows three areas with well-defined annual cycles of SST: (1) the Indian Ocean (especially in the western section) with an annual amplitude of about 2.8°C, warmest in northern spring and coldest in northern autumn; (2) eastern Pacific (east of ∼ 140°W) with an annual SST amplitude of about 3.7°C, warmest in northern spring and coldest in northern autumn; and (3) eastern Atlantic (30°W–10°E) with an annual amplitude of about 3.6°C, warmest in northern spring and coldest in northern summer. In contrast, the warmest area in the west-central Pacific has very little annual variation (amplitude ∼ 0.3°C) with the center of warmest water located about 140°–160°E through the year.

Table 2 shows the standard deviation of seasonal mean SST. The largest variance is located mainly in the eastern Pacific (east of the dateline) for every season

Table 1. ENSO years defined in this paper compared with some previous works

Year	1930	32	34	36	38	40	42
A	(F)					F	F
B						T	T
C	Q	Q			Q	Q	Q
D	V				V	V	V
E	R	R			R		R

Year	1944	46	48	50	52	54	56
A		(F)			(F)	F	F
B		T			T	T	T
C	Q	Q	Q		Q	Q	Q
D					V		V
E					R	R	R

Year	1958	60	62	64	66	68	70
A			F		F		F
B			T		T		T
C	Q		Q		Q		Q
D			V	V	V	V	
E					R		

Year	1972	74	76	78	80	82
A	F				F	
B	T		T		T	
C	Q	Q	Q	Q		
D	V		V	V		
E	R		R			

T=Trenberth (1984)
Q=Quinn et al. (1978)
V=Van Loon and Madden (1981)
R=Rasmusson and Carpenter (1983)
F=This paper
( ) data is sparse for this event.

Table 2. Comparison of interannual variance with the mean annual variation in 4 equatorial regions

Area	Annual Amplitude (A_m)	Standard Deviation (S_D)	S_D/A_m
100W	3.7°C	1.2°C	0.32
160E	0.3°C	0.9°C	3.00
40E	2.8°C	1.0°C	0.35
10W	3.7°C	0.5°C	0.14

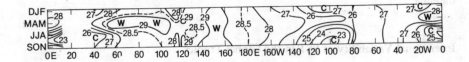

Fig. 2. Time-longitude mean SST sections along 4°N to 4°S in the global tropics. Mean reference period is 1950–1979

except northern spring (MAM), but it is only 1/3 of the annual amplitude in this area. In the warmest area in the west-central Pacific, the SST variance is lower than that in the eastern Pacific ($\sim 0.8°C$), but it is about 3 times that of the annual amplitude there. The Indian Ocean and western Atlantic have very small interannual variability (see Table 2). As will be shown in the next section, although the variance in the warmest area is less than that in the eastern Pacific, SST variability there, especially at the eastern edge of this area, is very sensitive to the ENSO event.

Figure 3 shows the zonal SST profile in the equatorial Pacific averaged for each

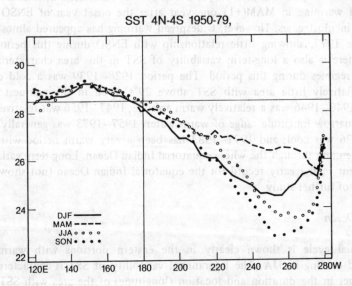

Fig. 3. Seasonal mean SST profiles in the tropical Pacific (4°N–4°S, 120°E–80°W). DJF (December–January–February); MAM (March–April–May); JJA (June–July–August); SON (September–October–November)

season. The maximum annual amplitude appears at about 100°W and decreases both eastward and westward. To the west of 150°W, the annual amplitude is reduced to less than 0.5°C. There is almost no seasonal signal in the area to the west of the dateline.

Interannual Variability of SST in the Global Tropics

Figure 4 is the Hovmoller chart (time-longitude cross-section) of seasonal mean SST for the period from 1928 DJF to 1983 MAM in the 4°N–4°S latitude zone of the global tropics. The isotherm interval is 1°C. The shaded area indicates the sea surface temperature above 29.0°C with the center above 30°C shaded darker. The black line area separates the areas of SST below 28°C with areas below 22°C being shaded darkest.

Since the actual SST is given here instead of SST anomaly, the well-defined annual signal in three regions, i.e., Indian Ocean, eastern Atlantic, and eastern Pacific illustrated in Fig. 1 are shown clearly in this figure, with the interannual variability superimposed on the annual cycle.

Indian Ocean

The annual cycle consists of warming in MAM and cooling in SON. Interannual variability is characterized by fluctuations of SST in both time duration and longitude range from year to year. One feature associated with ENSO is the additional warming in MAM(+1) one year after the onset year of ENSO which occurred in most cases. However, widespread warming has appeared almost every year since 1977, showing little relationship with ENSO during this period. Moreover, there is also a long-term variability of SST in this area characterized by different regimes during this period. The period 1928–1939 was a cold episode having relatively little area with SST above 29°C even in the warmest season (MAM), 1940–1946 was a relatively warm episode; 1947–1956 was relatively cold having a narrow longitude range of warm water; 1957–1973 was generally warm, 1974–1976 was cool; and 1977–1983 has been a very warm period with warm water extending through the whole equatorial Indian Ocean. Long-term variation is also present in the early record for the equatorial Indian Ocean (not shown) and is worthy of further study.

Atlantic Ocean

The annual cycle is shown clearly in the eastern portions with warming in MAM and cooling in JJA. The interannual variability of SST is characterized by the changes in the duration and location (longitude) of the area with SST above 28°C in the MAM(+1), one year after the onset ENSO year. In some cases, for instance, in MAM of 1973, the SST in the eastern Atlantic can be quite high, reaching about 29°C.

Eastern Pacific Ocean

The strong annual cycle in this area is illustrated by the westward extension

Time-Longitude Cross-Section
of SST 4°N—4°S

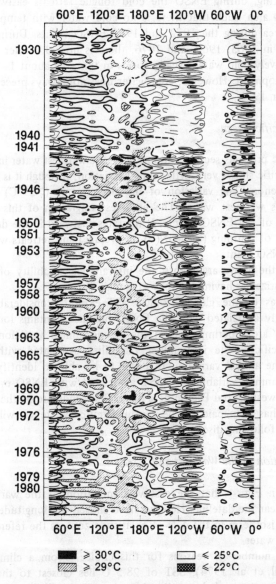

60°E 120°E 180°E 120°W 60°W 0°

■ ≥ 30°C   □ ≤ 25°C
▨ ≥ 29°C   ▩ ≤ 22°C

Fig. 4. Hovmoller chart of seasonal mean SST in the global tropics (4°N—4°S) for the period of DJF of 1928 to MAM of 1983. Interval of isotherm is 1°C. The shaded area indicates SST above 29°C with the center above 30°C shaded darkly. The lined area indicates SST below 25°C with the center below 22°C being hatched

of the so-called equatorial cold tongue which is expressed by the shaded region (<25°C) in Fig. 4. It develops strongly on average during SON and weakly in

MAM. It also shows strong interannual variability associated with the ENSO Generally speaking, during ENSO the cold tongue retreats eastward, increasing the SST in this area, such that there is no cold water with temperatures below 25°C in this area during the 1972 and 1982 ENSO events. During weak ENSO events, such as in 1963, 1969, and 1976, there is no cold water below 23°C in this area. Relatively cold water usually forms in this area about 1–2 years preceding ENSO development. However, cold water does not always precede some ENSO events, such as in 1982.

West-Central Pacific Ocean

As shown in the previous section and in Fig. 4, the warmest water in the equatorial west-central Pacific shows very little annual variation, although it is sensitive to the ENSO phenomenon. The variation of the shaded area (>29°C) represents the behavior of this warmest water. The east-west migration of this area is closely related to each of the ENSO events. It migrates eastward in the developing stage before the peak of warming (by SON(0) on average) and retreats westward before and after the ENSO event.

Based upon the above arguments, the interannual variability of global tropics could be summarized as having the following main features:

In most areas of the global tropics, SST exhibits considerable interannual variability mostly associated with the ENSO but also contains longer-term components, which are superimposed on the strong annual variation. Only in the west-central Pacific is the interannual variability associated with ENSO much stronger than the annual variation. In order to be able to identify early a signal related to interannual variability of SST associated with ENSO events, it is suggested that the west-central Pacific be a likely first candidate on the list. Relatively little attention has been paid to the central Pacific, where we will focus on our attention in the following discussion.

East-West Migration of the Warmest SST Area

It was shown in Fig. 4 that the east-west migration of the warmest SST area occurs mainly near the dateline. Here we have chosen the longitude of the eastern edge of 28.5°C isotherm in the 4°N–4°S latitude zone as the reference boundary for the warmest water.

There are a number of reasons for this choice. From a climatological perspective (Newell et al. 1974), SST of 28.5°C lies closest to the dividing line between the ascending and descending branches of the Walker circulation in the lower troposphere over the tropical Pacific (Plate 9.2, p 163; Newell et al. 1974); it is also closest to the boundary of cloud cover > 4 oktas (Plate 9.5, p 169); to the zero line of correlation of surface pressure anomalies with those at Djakarta from April to August (a representative measure of the Southern Oscillation) (Plate 9.11, p 177; Newell, et al. 1974; Berlage 1957); and to the dividing line between climatologically wet and dry areas in the tropical Pacific (Plate 9.3, p 165; Newell, et al. 1974). Therefore, the 28.5°C isotherm should be a useful reference boundary between areas of strong and weak tropical convection in the Pacific. It is expected that the east-west migration of 28.5°C could be a good indicator of the shift of the strong tropical convection area which is associated with the tropical heating patterns and therefore with the development of ENSO.

Figure 5 is the time sequence of the longitude of the eastern edge of 28.5°C isotherm averaged over the 4°N–4°S latitude zone by season since 1932. A noteworthy feature in Fig. 5 is that the eastward migration of the 28.5°C isotherm is a common characteristic of all of the 11 ENSO events in their developing stages, not only for the 1982 event. The extent of eastward migration varies from event to event. Generally speaking, in the strong events, such as 1940–41, 1957, 1965, 1972, and particularly 1982, the 28.5°C isotherm extends farther eastward (east of 155°W) and lasts longer (i.e., it remains to the east longer) than during the weak ENSOs.

It should be noted that during the 1982 event, as shown in Fig. 4, two areas of warmest water developed, one in the west-central Pacific and the other off the east coast of South America. This is the only documented occurrence of such warm water so far east (Fig. 4). It appears that the west-central Pacific area developed first; followed by another warm area to the east. However, there was only a narrow gap between these two warm areas (Fig. 4).

In Fig. 5, the solid line shows the edge of 28.5°C of the west-central warm area, whereas the dashed line depicts the warm water in the far eastern Pacific. As noted, such a far eastward extension of warm water (>29°C) has not been observed previously. The appearance of two warm water areas in the 1982 event might be related to the fact that the 1982 event was the longest and strongest in the modern record.

The westward retreat of this warmest area from two to four seasons prior to the onset of some ENSO events, such as 1940–41, 1957, 1963 1965, 1969, 1972, and 1976 (Figs. 4 and 5) seems to be another precursor of ENSO development. This appears to be associated with the development of strong anti-El Nino conditions in the eastern equatorial Pacific.

This evidence indicates that the east-west oscillation of the warmest area in the central Pacific is one of the major features present during the developing stage of ENSO and the oscillation of 28.5°C isotherm can be added to the growing list of useful monitoring indices (see Keen 1984).

Since the SST anomaly is generally larger in the eastern Pacific and the canonical model of El Nino shows an early appearance and subsequent westward migration of positive SST anomalies in the extreme eastern equatorial Pacific (Rasmusson and Carpenter 1982), the equatorial cold tongue has been looked upon as more significant in the development of ENSO than the warmest water in the west-central Pacific highlighted in this paper.

In recent years increased attention has been placed on the variation of this warmest area in the west-central Pacific. As pointed out by the Climate Research Committee of the US National Academy of Sciences (1983), this area of the ocean"contains the greatest amount of heat per unit mass of any part of the world's oceans. Thus, it has tremendous thermal inertia to help stabilize the ocean-atmosphere system. It has also the potential to release large quantities of both thermal

188

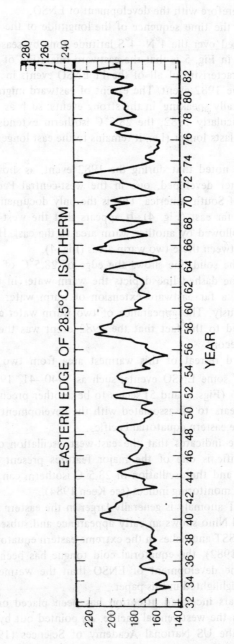

Fig. 5. Seasonal mean longitude of the eastern edge of 28.5°C isotherm in the equatorial Pacific Ocean

and potential energy that might be important to distant ocean areas and the overlying atmosphere." It also stated that, "While the surface temperature in this region appears to be less variable than in the eastern Pacific, the subsurface temperature (for example, the heat content of the mixed layer) displays large fluctuations that are well related to the SO."

We suggest that this warmest area also shows significant interannual variability associated with ENSO. This warmest area may be more significant, or at least no less significant, than the equatorial cold tongue in the eastern Pacific in the development of ENSO.

Variations of Zonal Profile of SST in the Equatorial Pacific Associated with ENSO

The large zonal SST gradient in the equatorial Pacific with relatively cool water in the east and warm water in the west is associated with the east-west seesaw of surface pressure in the Indian-Pacific Ocean, the so-called Southern Oscillation (Walker 1923, 1924) and the Walker circulation (Bjerknes 1969). According to Bjerknes (1969), a warming in the east with slight cooling or near normal conditions in the west is a major feature associated with ENSO development. This diminishes the west-east gradient and therefore weakens the Walker circulation, shifting the region of rising motion eastward. At the surface, the easterlies weaken or shift to westerly and strong easterlies develop at upper levels.

According to his hypothesis, the trigger of ENSO development is closely associated with changes in the eastern Pacific, although the variation of zonal SST gradient in the equatorial Pacific is not composed of only one type. Besides the variations in the equatorial cold tongue, the behavior of the warmest area also affects the SST gradient in this region.

In this section, the observed SST profiles in the equatorial Pacific during the past 9 ENSO events are described and then the principal patterns of variation of SST profile derived from eigenvector analysis of the SST data are shown.

Zonal SST Profile in the Developing Stage of ENSO Events

Figure 6 gives the zonal SST profiles in the equatorial Pacific ($4°N–4°S$, $120°E–80°W$) in JJA of onset year for 9 ENSO events, with the climatological mean during the period of 1950–1979 as a reference (dashed line). The $28.5°C$ line marked on each chart is used to indicate the extension of the warmest area described previously.

The season of JJA of the onset year (JJA(0)) is chosen to represent the developing stage of ENSO, because (1) positive SST anomalies widely cover the equatorial Pacific by July (Rasmusson and Wallace 1983); (2) the Southern Oscillation index generally reaches its minimum in November; and (3) the precipitation index at the Line Islands reaches a maximum in December (Rasmusson and Carpenter 1982). Therefore, in JJA(0), the SST pattern reaches a rather stable stage. As will be shown below, the SST pattern at this time can be regarded as representative of the maturing phase of the typical ENSO. However, the peak in terms of

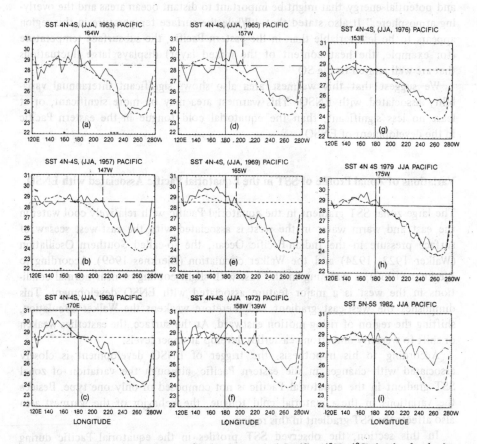

Fig. 6. Sea surface temperature profile in the tropical Pacific (4°N–4°S) in the developing stage of each ENSO event (season JJA). (a) 1953; (b) 1957; (c) 1963; (d) 1965; (e) 1969; (f) 1972); (g) 1976; (h) 1982. Dashed line is the climatological mean (1950–1979). Dotted-dashed line indicates 28.5°C level

area and strength comes in the subsequent fall and winter seasons. Our aim was to find some distinguishing features associated with different types of ENSO events based on the SST profile during this stage of development.

The differences shown in the profiles of Fig. 6 are related to the degree of eastward motion of the warm water indicated by the area above the 28.5°C line (dot-dashed line) and also to the re-distribution of SST in the equatorial Pacific. Table 3 lists the longitude of the eastern edge of 28.5°C and the temperature conditions in the West (120°E–160°E). Central (160°E–150°W) and East (150°W– 80°W) Pacific for each event.

The differences are summarized in Table 3; there appear to be three main patterns of SST profile:

Pattern A: Much warmer to the east of the dateline, near normal in the west and the warmest area extending to about 150°–160°W or farther east (i.e.

Table 3. Characteristics of zonal SST distribution in the equatorial Pacific ($4°W–4°S, 120°E–80°W$) during the past nine ENSO events for the season JJA (0)

Event	Eastern edge of 28.5°C	West	Central	East
1953	164°W	C	W	W–
1957	147°W	C	W–	W+
1965	157°W	C	W+	W
1972	158°W and (139°W*)	N	W+	W+
1982	148°W	N	W	W+
1963	170°W	W+	N	W-
1969	165°W	W	W	W-
1976	153°E	C-	N	W+1

West = $120°–160°E$; Central = $160°E–150°W$; East = $150°–180°W$

N, normal; N+, slightly above normal; N·, slightly below normal.

W, warm; W+, much warmer; W–, less warm.

C, cold; C+, much colder; C–, slightly cold.

* the same area above 28.5°C reached 139°W, but separated from the main center.

1957, 1965, 1972, and 1982).

Pattern B: Warm in the east, slightly below normal in the west, and near normal in the central. The warmest water is near its climatological position in the central Pacific (i. e., 1976).

Pattern C: Warm almost everywhere, especially in the west and central. compared with Patterns A and B. The warmest area extends to the east of the dateline, but not as far as in Pattern A (i. e., 1963, 1969).

Figure 7 depicts schematically the three patterns of SST profile and the climatological mean during the developing stage of ENSO events. Table 4 gives the maximum and minimum SST and the east-west SST difference for each type. In Pattern A, maximum SST increase slightly ($+0.7°C$), but the minimum SST increase greatly ($+1.7°C$), so the east-west gradient (ΔSST) decreases. This is particularly tyue of the 1982 event (Table 4). In Pattern B, the maximum SST decrease ($-0.4°C$) and minimum SST increase ($+1.5°C$) also resulting in a decrease of ΔSST, although smaller in magnitude. However, in Pattern C, the maximum SST increase greatly ($+1.2°C$) and the minimum SST increase slightly ($-0.3°C$), therefore the ΔSST increase substantially.

It is reasonable to expect that both the shift of this warmest area in the west-central Pacific and the change of east-west SST gradient for different patterns should result in significantly different responses of atmospheric circulation. This will be discussed in Section 8.

Cumulative ΔSST Profile During ENSO Period

Figure 8 shows the profiles of cumulative SST anomalies over the period from

192

Table 4. East-west sea surface temperature difference in the equatorial Pacific (4°W–4°S) for each type of SST profile (°C)

	Maximum (in west)	Minimum (in east)	ΔSST	Change
Climatological				
mean	29.2	23.8	5.4	–
1963	30.4	24.1	6.3	+0.9
1972	29.9	25.5	4.4	–1.0
1976	28.8	25.3	3.5	–1.9
1982	29.3	26.2	3.1	–2.3

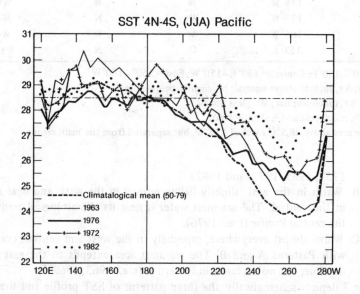

Fig. 7. Three patterns of SST profile summarized from Fig 6. *Thin solid line* (1963); *dark solid line* (1976); *dashed line with plus signs* (1972) and *dotted line* (1982). *Thick dashed line* is climatological mean (1950–79)

MAM(0), JJA(0), SON(0) to DJF(+1) for 7 ENSO Events. It illustrates the three patterns of equatorial warming described above: Pattern A, 1957, 1965, 1972; Pattern B, 1976; and Pattern C, 1963, 1969.

Of special interest is the peculiarity of the 1982 event. The corresponding cumulative curve for the period of MAM(0) to DJF(+1), shows a pattern similar to 1969, Pattern C. To illustrate the timing differences and the development of two warm areas, we show the accumulated SST anomalies for two period: SON (–1) to JJA(0) (1981 SON to 1982 JJA) and JJA(0) to MAM(+1) (1982 JJA to 1983 MAM). It clearly illustrates the two stages of equatorial warming (Fig. 8h). In the first period the warming is stronger in the west-central Pacific, whereas it is stronger in the east in the second period. These two stages comprise this extended (in time) event.

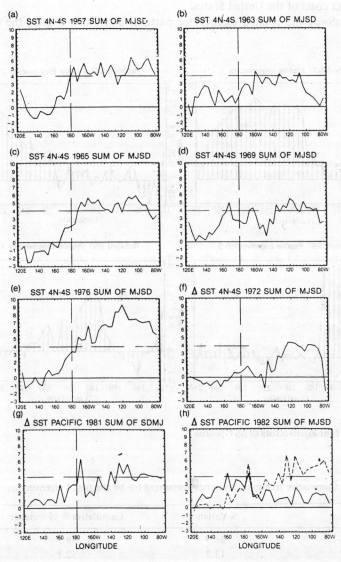

Fig. 8. The cumulatives SST anomalies during each ENSO period (MAM(0) to DJF(+1). (a) 1957; (b) 1963); (c) 1965; (d) 1969; (e) 1972; (f) 1976; (g) 1982; and (h) shows the cumulative SST anomalies for two periods: 1981 SON to 1982 JJA (*solid line*) and 1982 JJA to 1983 MAM (*dashed line*)

Results of Eigenvector Analysis of SST Profiles During the ENSO Events

We have applied eigenvector methods to the longitude profiles of SST, similar to the technique of McGuirk (1982), who applied it to precipitation variation along the west coast of the United States.

Figure 9 shows the rotated eigenvector patterns and Table 5 gives the cumu-

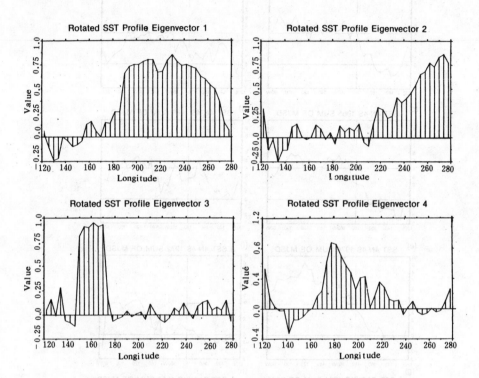

Fig. 9. First four eigenvectors of SST profile

Table 5. Cumulative fraction of variance accounted for by the first 4 eigenvectors

n	% Variance	Cumulative % of variance
1	38.9	38.9
2	13.2	52.1
3	9.7	61.8
4	7.4	69.2

lative fraction of variance accounted for by the significant functions (chosen as four on the basis of a Monte Carlo test). The first four factors account for more than 69% of the variance indicating that the SST profiles are characterized by large-scale variability. The most important factor, accounting for 38.9% of variance, is the anomaly to the east of 160°E, which shows strong warming in the central and the eastern Pacific, and slight cooling in the west. This factor is very close to Pattern A defined in Subsection C, such as the 1957, 1965 and 1972 events. The second factor (13%) is characterized by warming mainly in the east, with slight warming over the central Pacific and near-normal or slightly cool in the west. This factor is similar to Pattern B, represented by the 1976 event. The third (9.7%) factor shows the warming mainly in the west. The fourth (7.4%) shows warming mainly in the central Pacific. Pattern C seems to be a combination of factors 3 and 4.

In summary, the SST profiles for each ENSO event are a combination of these four major factors, each with different contributions. The three main patterns, A, B, and C, discussed earlier, describe the major features of the observed SST profile during these nine ENSO events.

Figure 10 shows a simplified sketch of these three patterns of equatorial warming

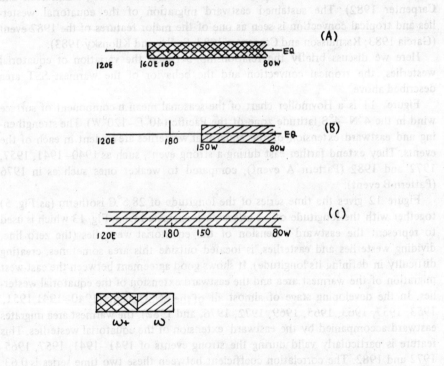

Fig. 10. A schematic map of three patterns of equatorial warming associated with ENSO events

associated with ENSO. Inspection of these figures brings two major questions to mind. First, what is the nature of tropical atmospheric circulation associated with these somewhat different patterns of equatorial warming? And second, what is the midlatitude response characteristic to each of SST patterns?

These will be the subjects of two other reports, but some initial findings will be shown in the next two sections.

Variation of the Equatorial Westerlies and the Shift of the Tropical Convection Area Associated with the East-West Migration of the Warmest SST Area

The anomalous equatorial westerlies in the western Pacific are thought to be a forcing factor in the development and eastward propagation of equatorial Kelvin waves which deepen the thermocline, depress the upwelling, and help to create and maintain the SST anomalies in the eastern Pacific (Knox and Halpern 1982; Cane 1983; Halpern 1983; Fletcher et al. 1983). It is also suggested that the tropical convective activity and rainfall are related to both sea surface temperature anomalies and the equatorial westerlies (Garcia 1983; Rasmusson and Carpenter 1982). The sustained eastward migration of the equatorial westerlies and tropical convection is seen as one of the major features of the 1982 event (Garcia 1983; Rasmusson and Carpenter 1983; Sadler and Kilonsky 1983).

Here we discuss briefly the relationship between the variation of equatorial westerlies, the tropical convection and the behavior of the warmest SST area described above.

Figure 11 is a Hovmoller chart of the seasonal mean u-component of surface wind in the 4°N–4°S latitude zone of the Pacific (40°E–120°W). The strengthening and eastward extension of the equatorial westerlies are evident in each of the events. They extend farther east during a strong event, such as 1940–1941, 1957, 1972 and 1982 (Pattern A event), compared to weaker ones such as in 1976 (Pattern B event).

Figure 12 gives the time series of the longitude of 28.5°C isotherm (as Fig. 5) together with the longitude of −2 m/s isotach extracted from Fig. 13 which is used to represent the eastward extension of the equatorial westerlies (the zero line, dividing westerlies and easterlies, is located outside this area sometimes, creating difficulty in defining its longitude). It shows good agreement between the east-west migration of the warmest area and the eastward extension of the equatorial westerlies. In the developing stage of almost all of the ENSO events (1940–1941,1951, 1953, 1957, 1963, 1965, 1969, 1972, 1976, and 1982), the warmest area migrates eastward accompanied by the eastward extension of the equatorial westerlies. This feature is particularly valid during the strong events of 1941–1941, 1957, 1965 1972 and 1982. The correlation coefficient between these two time series is 0.63 which

is significant at the 99% confidence level.

There is probably a positive feedback process operating between them. The anomalous equatorial westerlies help spread surface warm water eastward, reverse the normal upwelling of cooler water and increase near-surface convergence. This results in the eastward migration of the warmest SST and subsequent displacement of the active area of convection (the thermally-driven Walker circulation shifts eastward). The surface equatorial westerlies in the west tend to strengthen and extend farther eastward. The process could be self-amplifying during the developing stage of ENSO.

Figure 13 shows a time-longitude section of high reflective clouds (HRC) along

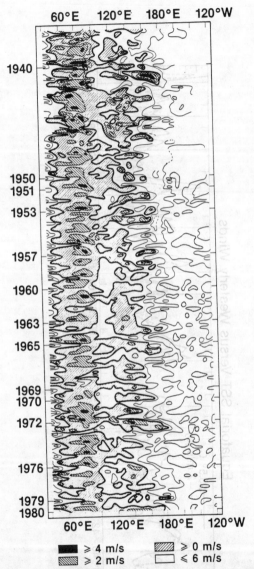

60°E   120°E   180°E   120°W

1940

1950
1951

1953

1957

1960

1963

1965

1969
1970

1972

1976

1979
1980

60°E   120°E   180°E   120°W

■■■ ≥ 4 m/s   ▨ ≥ 0 m/s
▨ ≥ 2 m/s   ☐ ≤ 6 m/s

Fig. 11. Hovmoller chart of seasonal mean u-component of surface wind in 4°N–4°S Latitude zone of Pacific. Shaded area indicates the region of equatorial westerlies

the equator from 120°E to 80°W for the three recent ENSO events (Garci 1983). The Hovmoller chart for the full data period (1971–1983) was not available at this writing. The heavy line with dots gives the seasonal mean longitude of 28.5°C isotherm in the equatorial Pacific (4°W–4°S) based on Fig. 4 (The data are plotted at the mid-month of the season, e.g., the mean of SON is plotted in October.)

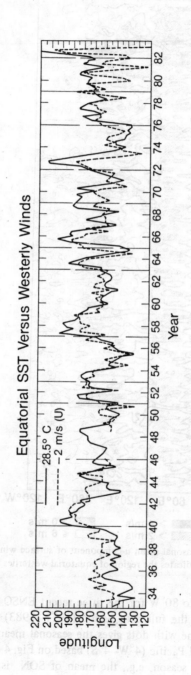

Fig. 12. Time series of longitude of the eastern edge of 28.5°C isotherm (copy of Fig. 5) and the longitude of −2 m/s isotach from Fig. 11

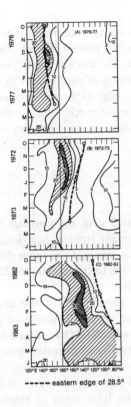

Fig. 13. Time-longitude sections of monthly percentage frequency of occurrence of highly reflective clouds from 0.5°N to 0.5°S and from 120°E to 80°W for three periods: **(A)** October 1976 – June 1977. **(B)** October 1972 – June 1973, and **(C)** October 1982 – June 1983. A 3% frequency of HRC corresponds roughly to a monthly rainfall of 100 mm, 10% to 175 mm, 20% to 300 mm, and 30% to 400 mm

This shows very good agreement between the eastward extension of the warmest area and the location of the strong tropical convection.

In the 1972 event (Fig. 13b), the warmest SST area migrated eastward with its edge at near 140°W. Correspondingly, the strong convection area shifted eastward, with the center near but east of the dateline within the area of warmest water. Figure 13(c), depicting the 1982 event, illustrates the far eastward migration of the warmest water and of the strong area of convection which was located near 140°W, about 60° longitude to the east of that in 1976.

The relationship can be partly explained by the fact that warm SST is favorable to the development of tropical convection, because the vertical gradient of pseudo-wet-adiabatic potential temperature of the tropical atmosphere depends highly on the sea surface temperature (Kraus 1973).

While it may be likely that a positive feedback process operates between sea surface temperature, surface wind, and the tropical convection during the developing stage of ENSO, sometimes the process fails or ceases. In the 1976 event (Fig. 13a), the warmest water was in the western Pacific, west of 160°E, and correspondingly the strong convection area was also located to the west of the dateline with the center near 160°E along the edge of the warmest water.

Understanding of both types of mechanisms — those that lead to full ENSO development, as well as those that are not necessary, if a greater measure of ENSO predictability is to be gained.

Summary

The convective rainfall associated with the warmest SST area in the west-central Pacific comprises one of the major tropical heat sources. It displays little annual variation, but large interannual variance in connection with ENSO events. This warmest area makes a contribution to the interannual SST variability comparable to that of the equatorial cold tongue.

The 28.5°C isotherm is a good index of the eastern edge of this warmest water. The east-west oscillation of the longitude of the eastern edge of 28.5°C isotherm is closely related to the ENSO. It has migrated eastward during the developing stage of all 12 events since 1928 and retreated westward in the decaying stage of most of the events.

The relative contribution of this warmest water and the equatorial cold tongue to the ENSO morphology varies from event to event. Three major patterns have been defined by analysis of zonal SST profiles: Pattern A (1957, 1965, 1972, and 1982) is warm in the central and eastern Pacific, both the warm area and the cold tongue have a significant impact to the pattern of SST anomaly; Pattern B (1976) shows warmth mainly in the eastern Pacific, and the cold tongue makes the dominant contribution; Pattern C (1963, 1969 and 1979) is warm mostly in the area from the central Pacific to the west of the dateline, with the effect of the warm area being larger than that of the cold tongue.

Acknowledgement. We are grateful to Professor C. Ramage for helpful discussions and his careful review of the manuscript, and R. J. Slutz, S. D. Woodruff and W. Otto for their assistance in preparing the data. Thanks are also due to O. Garcia for providing the HRC data which were very useful in interpretation of tropical convection. Special thanks to Ms. Julie McCaul for her patient typing of the manuscript.

References

Barnett T P (1977) *J Phys Oceanog* 7: 633–647
Berlage H P (1957) *K Ned Meteorol. Inst, Meded Verh* 69: 152 pp
Berlage H P (1966) *K Ned Meteorol Inst Meded Verh* 88:152
Bjerknes J (1966) *Tellus* 18: 820–829
Bjerknes J (1969) *Mon Wea Rev*, 97: 163–172
Cane M A (1983) *Science* 222: 1189 – 1195

Climate Research Committee, Board on Atmospheric Sciences and Climate, Commission of Physical Sciences, Mathematics, and Resources, National Research Council (1983) El Nino and the Southern Oscillation, A scientific plan. Natl Acad Press, Washington DC 20418, pp 72

Fletcher J O, Slutz R J, Woodruff S D (1983) Towards a comprehensive ocean-atmosphere data set. Trop Ocean Atmos Newslett 20: 13–14 [JISAO, Univ Washington, AK–40, Seattle, WA 98195]

Fu Congbin, Li K R (1978) *Ocean Select* 2(2): 16–21

Fu Congbin, Fletcher J (1985) *Sci Bull* 30: 1360–1364

Garcia O (1983) Equatorial Pacific convective activity during the last three ENSO events. Trop Ocean Atmos Newslett 21:31–34 [JISAO, Univ Washington, AK–40, Seattle, WA 98195]

Halpern D (1983) Variability of the Cromwell Current before and during the 1982 warm event. Trop Ocean Atmos Newslett 21:6–7 [JISAO, Univ Washington, AK–40 Seattle, WA 98195]

Hickey B (1975) *J Phys Oceanog* 5: 460–475

Horel J D, Wallace J M (1981) *Mon Wea Rev* 109: 813–829

Julian P R, Chervin R M (1978) *Mon Wea Rev* 106: 1433–1451

Keen R A (1984) Equatorial westerlies during the 1982 ENSO event. Trop Ocean Atmos Newslett 25: 10–11 [JISAO, Univ Washington, AK–40, Seattle, WA 98195]

Knox R A, Halpern D (1982) *J. Marine Res* Suppl 40: 329–339

Kraus E B (1973) *Nature* 245: 129–133

McGuirk J P (1982) *Clim Change* 4: 41–56

Merle J (1978) Annual and interannual variability of SST in the eastern tropical Atlantic Ocean. Proc 3rd Annu Climate Diagnotics Workshop, PB–298355

Newell R E, Kidson J W, Vincent D G, Boer G J (1974) The general circulation of the tropical atmosphere and interactions with extratropical latitudes, Vol 2. MIT, London, 371 pp

Quinn W H, Zopf D O, Short K S, Yang R J W K (1978) *Fish Bull* 76: 663–678

Quinn W H, (1979) Monitoring and predicting short-term climate changes in the South Pacific Ocean. Proc Int Conf Mar Sci Tech I. Catholic Univ Valparaiso, Valparaiso, Chile, pp 26–30

Ramage C S (1975) *Bull Am Meteorol So* 56: 234–242

Rasmusson E M, Carpenter T H (1982) *Mon Wea Rev* 110: 354–384

Rasmusson E M, Carpenter T H (1983) *Mon Wea Rev* 111: 517–528

Rasmusson E M, Arkin P A, Krueger A F, Quirroz R S, Reynolds R W (1983) The equatorial Pacific atmospheric climate during 1982. Trop Ocean Atmos. Newslett 21: 2–3 [JISAO, Univ Washington, AK–40, Seattle, WA 98195]

Rasmusson E M, Wallace J M (1983) *Science* 222: 1195–1202

Sadler J, Killonsky B J (1983) Meteorological events in the central Pacific during 1983 associated with 1982 El Nino. Trop Ocean Atmos Newslett 21: 3–5 [JISAO, Univ Washington, AK–40, Seattle, WA 98195]

Shukla J, Wallace J M (1983) *J Atmos Sci* 40: 1613–1630

Slutz R J, Woodruff S D, Jenne R L, Joseph D H, Steurer P M, Elms J D, Lubker S J, Hiscox J D (1984) Comprehensive ocean-atmosphere data set, Release 1, NOAA/ERL, 325 Broadway, Boulder, CO 80303, pp 137

Trenberth K E (1984) *Mon Wea Rev* 112: 326–332

Van Loon H, Madden R A (1981) *Mon Wea Rev* 109: 1150–1162

Walker G T (1923) *Mem Ind Meteorol* 24: 75–131

Walker G T (1924) *Mem Ind Meteorol Dept* 24: 275–332

Weare B C, Navato A R, Newell R E (1976) *J Phys Oceanog* 6: 671–678

Weare B C (1979) *J Atmos Sci.* 36: 2279–2291

Wyrtki K (1965) *Limnol Oceanog* 10: 307–313

Wyrtki K (1975) *J Phys Oceanog* 5: 572–584

The Connection Between the Surface Air Temperature in the Northern Middle Latitudes and the Sea Surface Temperature in the Tropical Pacific

Zhang Mingli[1], Zeng Zhaomei[1] and Pan Yihang[1]

Abstract – The correlation pattern between the SST in equatorial 130° W and surface air temperature in Northern Hemisphere has been analyzed. This pattern is associated with the anomalous climate in eastern Asia, especially related to the cold climate in summer over northeast China. In eastern North America, the climate distribution related to the SST fluctuation is different from that in east Asia. The difference between the teleconnection in Asia and North America can be traced back to the variation of polar vortex connected with the air temperature in China has been given.

Introduction

Recently, the abnormal climate appearing in most areas over the world during the El Nino period has motivated many meteorologists to study the effect of the variation of the sea surface temperature in the tropical Pacific on the global climate. One of the evident effects on the middle latitude areas is the cold weather prevailing over the eastern part of the Asian and North American continents during this period. A striking example is the extreme cold winter in the eastern United States during the 1976–1977 event. Horel and Wallace (1981) suggested that during the warm episode in the tropical Pacific, the PNA flow pattern prevails. Thus, the appearance of cold climate in the eastern United States can be related to the warm event in the tropical Pacific through the prevailing of the PNA flow pattern. In northeast China, the severe cold summer disaster happened to occur during the El Nino period (Research Joint Group on the Cold Climate in Northeast China 1979; Xu et al. 1982; Zhang and Zeng 1984). The time series of air temperature anomaly in northeast China show an opposite phase to the time series of sea surface temperature anomaly in the tropical Pacific. Usually, the abnormal cold climate prevails not only locally over northeast China, but spreads broadly over the Eurasian continent and connects to the abnormal climate in middle latitudes (Zhang et al. 1980). Thus, the cold summer disaster may be connected to the variation in the tropical Pacific through the atmospheric circulation in middle latitudes under the effect of air-sea interaction.

It is generally believed that the connection between the atmospheric motions in middle latitudes and the oceanic conditions in the tropical Pacific seems to be clearer in winter. Most of the studies on this topic are devoted to the situation in winter seasons, such as Horel and Wallace (1981), van Loon and Madden (1981). Having studied the significant teleconnection between the cold summer disaster in

1. Institute of Atmospheric Physics, Academia Sinica, Beijing, China

northeast China and the warm event in the tropical Pacific, here we pay more attention to this connection for the whole year instead of for the winter season alone.

In this paper, using the correlation maps between the SST in the tropical Pacific and the air temperature in the Northern Hemisphere, we will focus on the following problems: (1) the teleconnection of air temperature with the SST in the tropical Pacific, (2) comparison between the temperature patterns in Asia and in North America associated with the variation of SST in the tropical Pacific; (3) The variation of surface air temperature in China associated with the fluctuation of SST in the tropical Pacific.

The data series used in this study are: (1) The average SST anomalies centered at 2.5° S, 130° W from 1951−1980. It has been justified (Pan 1978; Pan and Oort 1983) that the equator at 130° W is a key region correlated with some dominant global atmospheric anomalies, and the time series of SST at equatorial 130° W may best represent those in the entire eastern tropical Pacific. (2) The surface air temperature anomalies for a 10°×10° grid in the Northern Hemisphere during 1969 − 1978 adopted from Zhang et al. (1982). (3) The 200 hPa geopotential height for a 5°×10° grid in the Northern Hemisphere during the 1969−1978 period adopted from Die Grosswetterlagen Europas, published by the Deutsche Wetterdienst. (4) The time series of surface air temperature at 45 stations, in which the series at 5 stations in North America and those at 4 stations in the Soviet Union are obtained from the Monthly Climate Data for the World (NOAA) during the period of 1966−1980. The remainders are obtained from the State Meteorological Administration of PRC during the period of 1951−1980.

The analyses in this study are primarily based on the monthly mean statistics. To depict the nature of teleconnection, we construct maps of correlation coefficient between the SST at the equatorial 130° W and the surface air temperature at each grid point of a 10°×10° grid in the Northern Hemisphere during 1968−1978. The length of sample we use to compute the correlation is 120 months through the ten years. The data series have been filtered by using a 7-month running mean method in order to remove the short-range disturbances.

The Teleconnection of the SST in the Tropical Pacific with the Surface Air Temperature in the Northern Hemisphere

Figure 1 denotes the contemporary correlation between the SST in equatorial 130° W and the surface air temperature in the Northern Hemisphere. The peak correlation centers of the SST in the equatorial 130° W with 500 hPa GPH in the Northern Hemisphere obtained from Zeng and Zhang (1985) are also plotted on Fig. 1 with dark dots indicating negative centers and dark triangles, positive centers. The centers in the two correlations show nearly a one to one correspondence. Horel and Wallace (1981) also obtained such a similarity between the correlation patterns derived both from 700 hPa height and surface air temperature with Southern Oscillation Index.

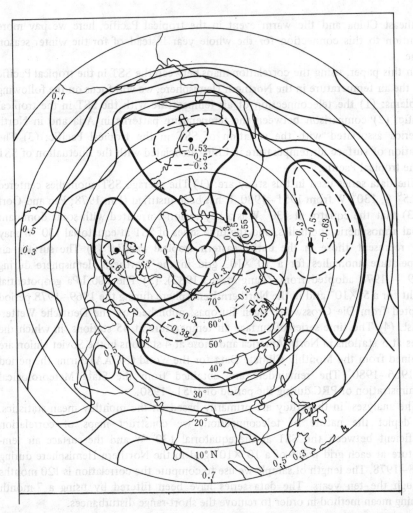

Fig. 1. Map of the correlation coefficient of the SST at the equatorial 130°W with the surface air temperature field in the Northern Hemisphere for 1969–1978 Numbers not attached to isolines indicate extreme values. Extensive areas with /r/=0.3 are statistically significant at the 95% level.

South of 20° N, the correlation coefficients are positive everywhere, and they are all above +0.5 except in north Africa and the western Pacific. This indicates that the temperature variation in the tropics is highly linked to the variation of the SST in the tropical Pacific. In the extratropics, the distribution of correlation coefficient displays a wave train with a chain of positive and negative centers alternatively arranged. This wave train can be compared with atmospheric planetary waves in the middle troposphere. The negative correlation centers correspond to the normal troughs. Among them the most significant center is the negative one located over eastern Asia with maximum coefficient of − 0.73, which corresponds to the major stationary trough over eastern Asia. The major positive center corresponds to the normal ridge along the western coast of North America with its coefficient reaching 0.55.

This situation implies that during the warm period in the tropical Pacific, the planetary-scale atmospheric disturbances in the middle latitudes are amplified. This will induce the air temperature to be below normal in the normal trough areas and above normal in the normal ridge areas. Thus, the temperatures along the zonal belt differ greatly and the climate tends to be abnormal, while during the cold period the temperature distribution inclines to be more even.

Comparison Between the Temperature Patterns in Asia and in North America as Associated with the Variation of Sea Surface Temperature in the Tropical Pacific

We have mentioned above that the distribution of correlation coefficient in middle latitudes displays a planetary wave pattern corresponding to the normal circulation pattern in the middle troposphere. However, if we examine the correlation pattern in detail, the disposal seems to be rather complex. The climate response along the major trough areas in both the Asian and North American continents are not the same. In Fig. 1, the negative center corresponding to the Asian trough locates at $50°$ N, $130°$ E. The center corresponding to the North American trough is south of $40°$ N.

Figure 2 indicates the correlation curves between the SST at equatorial $130°$ W and the surface air temperature at 12 stations along the eastern part of the Asian and North American continents during the 1966–1980 period. The curves on the left side are for the stations in Asia, and those on the right side are for the stations in North America. The curves on the same line are for stations almost located at the same latitude, and the curves in a row are put in an order ranging from north to south. These curves show a distinct contrast between Asia and North America. In Asia, the curves indicate large negative correlation coefficients north of $40°$ N. Their absolute values are larger than 0.4. As the latitude decreases, the correlation obviously drops and is not significant south of $40°$ N. The curves on the right show that the correlation pattern is quite different from that in Asia. North of $40°$ N in North America, the correlations are not significant at any lag in comparison with the considerable negative correlation in Asia. However, south of $40°$ N the correlation reaches -0.40.

In order to see the difference further, we compose the maps of average air temperature difference between seasons when the SST in tropical Pacific is relatively warmer and those when it is relatively colder.

Figure 3 is for the winter seasons (Dec.–Feb.), and Fig. 4 is for the summer seasons (Jun.–Aug.). Table 1 is the calendar of the warmer and colder seasons during 1969–1978.

The temperature difference map in winter seasons shows the evident contrast between Asia and North America. There are a pair of centers around the Pole. In Eurasia, the contour line of $-1°$ C encloses the whole northern part of the continent, while in the northern part of North America the temperature difference is

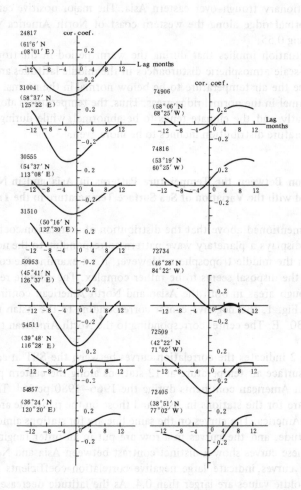

Fig. 2. The correlation coefficient curves of the SST at equatorial 130°W with the air temperature over 12 stations in Asia (left) and North America (right) for different lags during a 15-year period. A negative lag means that the SAT precedes the SST

positive. Figure 3 implies that the temperature pattern in Asia during the warm episode in the tropical Pacific is cold in the north and near normal in the south, and in North America, it is warm in the north and cold in the south.

The map in summer seasons shows features similar to the winter seasons. However, the values of temperature difference are much smaller. There is a negative center with temperature difference of − 1°C in northeast Asia, which implies that during the warm episode in the tropical Pacific, the cold climate tends to occur in that area. In North America, the difference is not significant in the summer seasons. Thus, the abnormal cold climate appearing in the eastern United States during the warm episode is only significant in winter seasons, but in northeast China, the cold

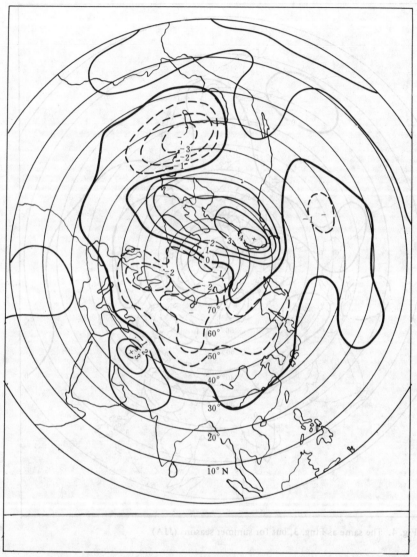

Fig. 3. The average surface air temperature difference between seasons during which the SST at equatorial 130°W is relatively warmer and the seasons when it is relatively colder, for 1969–1978 winters (DJF)

climate may occur both in winter and summer seasons.

From the composed mean 200 hPa charts during the warm and cold episodes (Figs. 5 and 6), respectively, we find that those features given above correspond to the different characters of polar vortex in the upper troposphere. During the cold period the mean polar vortex is relatively symmetric to the pole (Fig. 6). Its center is deflected to the western hemisphere. Its height reaches 10 920 GPM on 200 hPa level and the westerlies around the polar circle are stronger. During the warm

208

Fig. 4. The same as Fing. 3, but for summer seasons (JJA)

Table 1 Calendar of the SST warmer and colder seasons during 1969–1978

	Winter	Summer
Warmer season	1968–1969	1969
	1969–1970	1972
	1972–1973	1976
	1976–1977	1977
	1977–1978	
Colder season	1970–1971	1970
	1971–1972	1971
	1973–1974	1973
	1974–1975	1974
	1975–1976	1975
		1976

episode as the ridge along the western coast of North America develops and extends northward to the polar circle, the polar vortex obviously fills. Accompanied by this well-developed ridge in western North America, the contour line around the polar vortex shrinks along the 180° and 0° meridians and expands southward to the Asian and North American continents, so that the vortex is separated into two centers. The one located in north Asia deepens strongly and occupies the entire northern half of the continent. The contour line of 11 000 GPM extends southward to a distance of about 20 degrees latitude. However, in North America, since the ridge along the western coast is very strong, the other polar vortex center is relatively weaker. The 200 hPa geopotential height there increases 160 GPM in comparison with that during the cold period in tropical Pacific. The difference in character between the two vortices in Asia and North America may cause different climate variation in connected with the variation of SST in tropical Pacific. If so, what is the mechanism of interaction between them? As the differences are quite evident between the two periods, this question is interesting, and needs to be investigated further.

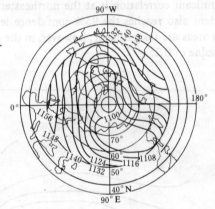

Fig. 5. The average 200 hPa GPH of Januaries during which the SST at equatorial 130°W is relatively warmer during the 1969–1978 period

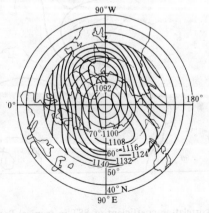

Fig. 6. The same as Fig. 6, but for the Januaries when SST is relatively colder

210

The Teleconnection with the Variation of Air Temperature in China

We have shown that the teleconnection of SST in the tropical Pacific with the climate in the Northern Hemisphere seems to be the most significant one in the midlatitudes. The authors have suggested that the PEA flow pattern in this area from Eurasia to the North Pacific may strengthen during the warm episode in tropical Pacific. The fluctuation of the flow pattern will influence the climate over eastern Asia. In this section, the connection of SST in the tropical Pacific with air temperature in China will be presented. This may provide some information for long-range forecasting. Figure 7 is the contemporary correlation map of air temperature in China which is obtained from the time series of air temperature in 34 stations over China during the period of 28 years through 1951–1978. Due to the larger samples than those in Fig. 1, the correlation coefficients are smaller. In China, the most significant correlation is at the northeastern area. In south China, the positive coefficient also reaches the 95% confidence level. The coefficients in eastern and western areas are insignificant; however, in the central area (105–115° E) there is a higher value of 0.2.

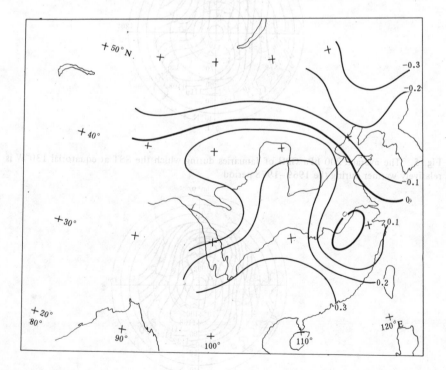

Fig. 7. Map of the correlation coefficient of SST in tropical Pacific with the surface air temperature in China for 1951–1978 period

Figure 8 indicates the lag months of the correlation coefficients. The variation of air temperature in south China lags 3—4 months to the SST in tropical Pacific, and the coefficient is 0.44. The coefficient is 0.28 along the Yangtze River and has a 7 month lag, while in northeast China it is − 0.37 and has a 0 to − 1 month lag.

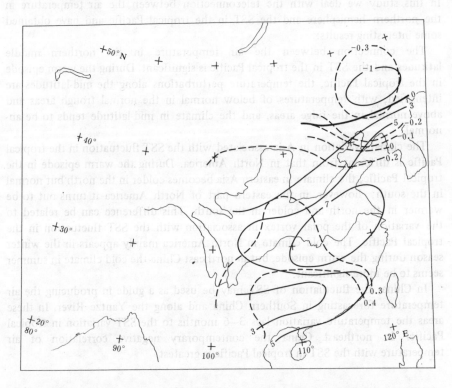

Fig. 8. Number of lag months of the correlation coefficient between SST in tropical Pacific and the surface air temperature in China (solid line) and the peak value of the correlation coefficient (dashed line) for 1951—1978 period

Thus the variation of SST can be used as a predictor in the forecasting of air temperature variation in several parts of China.

Through the analysis of the correlation map between the SST in tropical Pacific and the 500 hPa GPH, the authors (Zeng and Zhang 1985) have shown that the variation of 500 hPa height in mid-latitudes is simultaneous with that of SST in the tropical Pacific; however, it has 4—6 months lag in subtropics. The temperature variation in China has a similar teleconnection. It is an interesting phenomenon that the correlation lag in tropical and subtropical regions is behind that in mid-latitudes, although the spatial distance from the tropical Pacific is shorter in the tropics and subtropics. This may imply that for the response of the atmosphere to the tropical ocean, the teleconnection with the atmosphere in mid-latitudes has a quite different mechanism from that with the atmosphere in the tropics and subtropics.

212

Conclusions

In this study we deal with the teleconnection between the air temperature in the northern hemisphere and the SST in the tropical Pacific and have obtained some interesting results:

The connection between the air temperature in the northern middle latitudes and the SST in the tropical Pacific is significant. During the warm episode in the tropical Pacific, the temperature perturbations along the mid-latitudes are intensified, with temperatures of below normal in the normal trough areas and above normal in the ridge areas, and the climate in mid-latitude tends to be abnormal.

The climate variation in Asia associated with the SST fluctuation in the tropical Pacific is different from that in North America. During the warm episode in the tropical Pacific, the climate in eastern Asia becomes colder in the north but normal in the south; however, in the eastern part of North America it turns out to be warmer in the north and colder in the south. This difference can be related to the variation of the polar vortex in association with the SST fluctuation in the tropical Pacific. The cold climate in North America mainly appears in the winter season during the warm episode, but in northeast China the cold climate in summer seems to be more significant.

In China, the fluctuation of SST may be used as a guide in producing the air temperature forecasting in Southern China and along the Yantze River. In these areas the temperature variation lags 3–6 months to the SST variation in tropical Pacific. In northeast China the contemporary negative correlation of air temperature with the SST in tropical Pacific is greatest.

References

Horel J D, Wallace J M (1981) *Mon Wea Rev* 109: 813–829
Joint Research Group on the Cold Climate in Northeast China (1979) *Acta Meteorol Sin* 37: 44–58
Pan Y H (1978) *Sci Atmos Sin* 2: 246–252
Pan Y H, Oort A H (1983) *Mon Wea Rev* 111: 1244–1258
Van Loon H, Madden R A (1981) *Mon Wea Rev* 109: 1150–1162
Xue Z Y et al (1982) *Acta Oceanol Sin* 4: 169–173
Zeng Z M, Zhang M L (1985) Some statistical facts on the teleconnection between sea surface temperature in the eastern equatorial Pacific and 500 hPa geopotential height field in Northern Hemisphere for 1951–80 period. In this selection
Zhang M L et al. (1980) *Kexue Tongbao* 25: 1010–1013
Zhang M L et al. (1982) *Sci Atmos Sin* 6: 229–236
Zhang M L, Zeng Z M (1984) *Trop Ocean-Atmos Newslett* 23: 5–7

Regional Precipitation and Temperature Variability and Its Relationship to the Southern Oscillation

Henry F. Diaz[1], Fu Congbin[2]

Abstract – The western United States undergoes periods of prolonged dry and wet spells. Climatologically, the region is generally dry, with ample moisture relegated to the higher mountainous areas, where it accumulates in the form of snow during winter and early spring to be made available, as runoff, during late spring and early summer. The Yangtze River Basin of China experiences a dry/wet season regime, with the bulk of the precipitation occurring during the months of May to September (China Monsoon).

We have analyzed variations of seasonal precipitation in the western United States and the Yangtze River Basin (YRB) of China relative to the extreme phases of the Southern Oscillation sea level pressure and precipitation distribution from its High/Dry phase to its Low/Wet one. Using a set of 50 years, 30 Low/Wet events and 20 High/Dry ones, the winter and subsequent spring precipitation patterns in the United States, and summer (May-September) rainfall in YRB were categorized by various means. For China we have made use of the Atlas of Yearly Charts of Dryness/Wetness in China for the last 500 - year period. Analogous years were identified and composites of surface wind, sea level pressure, and sea surface temperature in the Pacific for these groups of years were produced.

A similar analysis was performed on winter and spring temperature in the United States based on the same set of Low/Wet, High/Dry ENSO years.

Introduction

Several diagnostic and numerical studies suggest that equatorial Pacific climate anomalies associated with the different phases of the Southern Oscillation may affect the weather in midlatitudes through the generation and interaction of planetary-scale waves (Horel and Wallace 1981; Hoskins and Karoly 1981; Webster 1981; Simmons 1982).

The present study examines some statistical evidence for the occurrence of anomalous seasonal (winter and spring) temperature and precipitation in the United States and summer precipitation in the Yangtze River Valley (YRV) of China associated with the extremes of the El Nino/Southern Oscillation (ENSO) phenomenon.

The frequency of occurrence by various class intervals of April to September rainfall in the YRV, as well as the frequency of occurrence of anomalous winter and spring temperature in the United States are compared during the high/dry and low/wet phases of ENSO (Van Loon and Madden 1981). The aim was to determine whether the resulting frequency distributions display significant departures from normality, which one would expect of a sample drawn randomly

1. NOAA/ERL, 325 Broadway, Boulder, CO 80303, USA
2. The Institute of Atmospheric Physics, Academia Sinica, Beijing, China

from the general population. We have also looked at the western Northern Hemisphere 700 hPa patterns associated with different regional temperature and precipitation patterns during recent ENSO years.

Data

Various data sets have been used in this analysis. A list of ENSO and anti-ENSO years was compiled from different sources (Quinn et al. 1978; van Loon and Madden 1981). The list was compared to the record of sea surface temperature data for the equatorial eastern Pacific taken from the comprehensive Ocean-Atmosphere Data Set (COADS; Slutz et al. 1984) and sea level pressure indices based on compilations by Berlage (1957), Trenberth (1984), and COADS data. Temperature and precipitation data for the U. S. were composed of both individual stations and divisional mean data published by the National Oceanic and Atmospheric Administration's National Climatic Data Center, Asheville, North Carolina. Rainfall data for China are derived from the Atlas of Drought and Flood (Central Meteorological Bureau 1982). Geopotential height data at 700 hPa were used to complete the circulation pattern of the North American Sector during various ENSO winter months to illustrate the type of circulation pattern present with certain temperature and precipitation regimes that developed during ENSO years.

Analysis

Figure 1 shows the winter temperature anomaly patterns for three ENSO winters in the United States, while Fig. 2 shows the precipitation anomaly in the eastern half of China during the same three ENSO episodes. The 1976 – 1977 winter is of a type associated with the so-called Pacific North American pattern, PNA (Wallace and Gutzler 1981, Horel and Wallace 1981). The other two winters are similar in that a general north-south anomaly gradient is present, with warmer than normal temperatures in the northern United States, and colder than normal along the southern U. S.

Note in Fig. 2 that southern and northern China are generally drier than normal in 1972 and 1963, while 1976 was wetter than normal in many of these areas. Although a number of similarities are present in these anomaly maps, there are also a number of discrepancies as well. For example, central China was quite wet in 1963, which was not the case in the other two years; the winter of 1963 – 1964 was cold in the southeast United States but was near-normal in 1972 – 1973.

Using the Flood/Drought Index developed at the Central Meteorological Bureau and the list of ENSO and anti-ENSO years given in Table 1, the seasonal precipitation for the years during and following each ENSO phase were composited in the region encompassing the Yangtze River Valley. Fig. 3 shows that there has been a tendency for greater than normal precipitation to occur in CRV following ENSO years, whereas a similar tendency toward wetter conditions prevailed during the anti-ENSO phase.

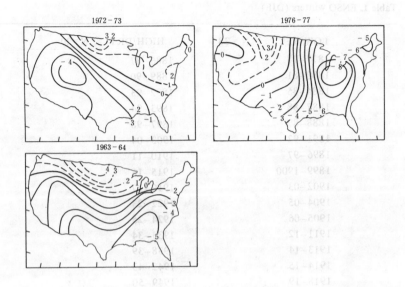

Fig. 1. The winter temperature anomaly patterns following three recent ENSO episodes

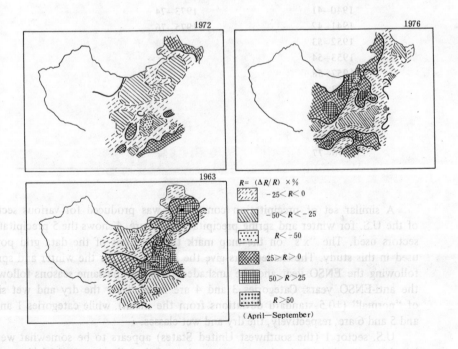

Fig. 2. April-September rainfall anomalies in China during three recent ENSO episodes

Table 1. ENSO winters (DJF)

LOW/WET N = 28	HIGH/DRY N =20
1877–78	1886–87
1880–81	1889–90
1884–85	1892–93
1887–88	1893–94
1888–89	1897–98
1891–92	1903–04
1896–97	1910–11
1899–1900	1915–16
1902–03	1916–17
1904–05	1922–23
1905–06	1924–25
1911–12	1933–34
1913–14	1938–39
1914–15	1942–43
1918–19	1949–50
1025–26	1955–56
1930–31	1961–62
1939–40	1970–71
1940–41	1973–74
1941–42	1975–76
1952–53	
1953–54	
1957–58	
1963–64	
1965–66	
1969–70	
1972–73	
1976–77	

A similar set of precipitation composites was produced for various sectors of the U.S. for winter and spring precipitation. Figure 4 shows the 5 precipitation sectors used. The "x's" on the map mark the locations of the data grid points used in this study. The shaded bars give the frequencies for the winter and spring following the ENSO year and the unshaded ones for the same seasons following the anti-ENSO years. Categories 3 and 4 are respectively the dry and wet sides of "normal" (±0.5 standard deviations from the mean), while categories 1 and 2 and 5 and 6 are, respectively, the dry and wet classes.

U.S. sector 1 (the southwest United States) appears to be somewhat wetter in winter, but it is distinctly wetter in spring, while following the high/dry years, the springs tend to be drier than normal. U.S. sector 2 (the northwest United States) displays a near-normal distribution , with possibly a greater tendency

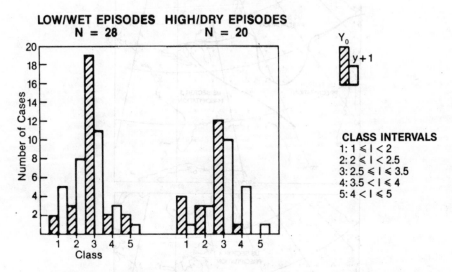

Fig. 3. Frequency distribution of precipitation classes in the Changjiang River valley of China during 28 ENSO episodes and 20 anti-ENSO events in the past 100 years

for the occurrence of drier winters following ENSO years, with an opposite tendency during anti-ENSO years. U.S. sector 2 springs are approximately normal with no occurrence of extreme seasons during these years.

U.S. sector 3 (the eastern United States) displays some shift toward drier winter seasons in both ENSO phases, while in spring there is a clear shift toward the drier end of the scale following both phases. U.S. sector 4 (the northern 2/3 of the contiguous States) shows no significant shifts one way or the other during the low/wet phase with, perhaps, some slight tendency toward wetter winters, but drier springs during the high/dry ENSO phase. The fifth sector (southern 1/3 of the U.S. tends to have wetter winters during the low/wet phase of ENSO, but drier winters during the high/dry phase.

In summary, for winter precipitation, the strongest signals occur in sectors 1 and 5 for greater frequency of occurrence of drier than normal winters following the high/dry (anti-ENSO) years. In the spring, the strongest signals occur in sectors 1 and 2 for wetter than normal low/wet years and in sector 5 for a tendency toward drier spring in connection with anti-ENSO episodes.

Figure 5 shows the regions used for testing whether a temperature signal following the extreme phases of ENSO is present in the winter and spring temperature in the United States. No clear signal is evident in sector 1 for winter or spring in either phase. However, following the low/wet phase, the data for sector 3 suggest a tendency for either colder than normal or warmer than normal winters to occur more frequently than normal winters. In the high/dry years, warmer than normal conditions tend to prevail.

218

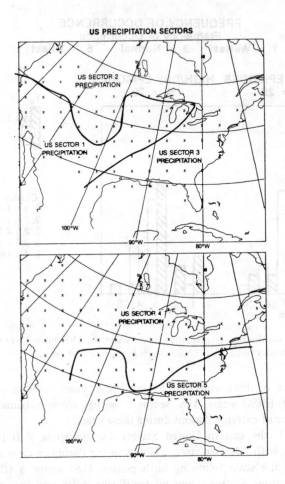

Fig. 4. Map showing location of data gridpoints and boundaries of U.S. percipitation sectors

A strong signal is also present for sector 4 (eastern 2/3 of the U.S.) with a strong tendency for colder than normal winters and springs to occur following the low/wet phase; an opposite tendency for warmer than normal winters following anti-ENSO conditions is also apparent. A normal distribution is evident in spring.

The patterns of winter temperature anomaly during ENSO years since 1950 have been classified into two basic patterns, although some seasons also possess secondary characteristics, given in parentheses in Table 2. It is obvious that two basic winter temperature patterns have been associated with the ENSO during its low/wet phase: The Pacific/North American (PNA) pattern and a zonal, north-to-south arrangement of the isopleths.

Similarly, we have classified the pattern of summer (April-September) rainfall in China into three types: Dry in most areas (denoted by "D" in Table 3), a single

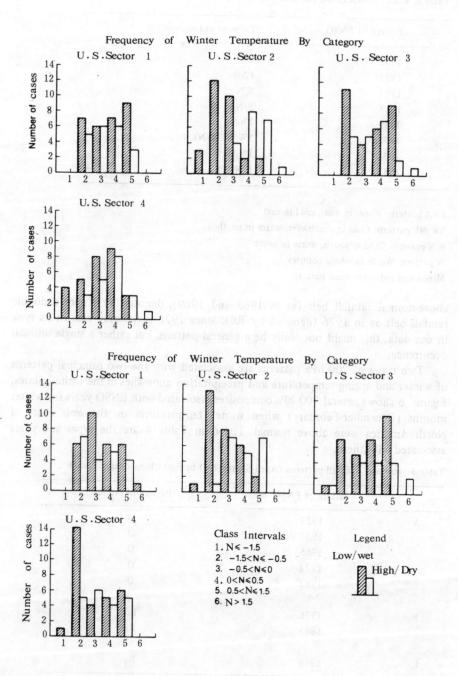

Fig. 5. Frequency distribution of winter and spring temperature classes in five different U.S. regions during 28 ENSO and 20 anti-ENSO episodes

220

Table 2. Winter temperature patterns [DJF(+1)] in·USA during ENSO

Pattern of ENSO	Patterns of temperature
1953	W
1957	PNA-
1963	S/N
1965	S/N
1969	PNA
1972	SW/NE(S/N)
1976	PNA
1982	S/N(W)

PNA pattern: Warm in west, cold in east
SW/NE pattern: Cold in southwest, warm in northeas
S/N pattern: Cold in south, warm in north
W pattern: Warm in whole country
Minus sign indicates weak pattern

above-normal rainfall belt (as in 1963 and 1969), denoted by Rs, or a double rainfall belt as in 1976 (denoted by Rd). Since 1976 is the only case of its type in our data, this might not really be a general pattern, but rather a single unusual occurrence.

Two types of 700 hPa patterns are associated with the two principal patterns of winter and spring temperature and precipitation anomalies in the United States. Figure 6 shows several 700 hPa composites associated with ENSO years for winter months (December-February) when winter temperatures on the west coast of North America were above normal. Listed in Table 4 are the types and years associated with them.

Table 3. Summer rainfall patterns (April-September) in East China during ENSO

	Pattern of ENSO	Pattern of rainfall
A	1953	D
	1957	D
	1965	D
	1972	D
	1982	D
B	1976	Rd
	1963	Rs
C	1969	Rs

D = dry in most areas
Rd = double rainfall belt
Rs = single rainfall belt

221

. They are all variations of the PNA pattern of Wallace and Gutzler (1981), with the position of the ridge over North America varying in longitude and in orientation. The Southeast United States is colder and wetter than normal and the Aleutian low is strengthened.

The precipitation composites, with the exception of the PAA pattern (Southeastern Alaskan precipitation above normal) do not display the PNA pattern. Rather, in these winters (see Table 4) the Aleutian low is displaced southeastward toward the North American continent and the downstream ridge and trough, normally associated with the PNA pattern, fail to develop.

Table 4. Type and years associated with 700 hPa composites in Figs. 6 and 7

Type	Description	Months of year		
TAA	Southeast Alaska winter, temperature above normal	Dec. 1952, Jan. 1958, Feb. 1964, Dec. 1976,	Feb. 1953, Dec. 1963, Dec. 1969, Jan. 1977,	Dec. 1953 Jan. 1964 Feb. 1970 Feb. 1977
TWA	Washington Coast winter temperature above normal	Jan. 1953, Feb. 1958, Dec. 1969. Feb. 1977	Dec. 1953, Dec. 1963, Feb. 1970,	Jan. 1958 Jan. 1964 Dec. 1976
TCCA	Central California Coast Winter temperature above normal	Jan. 1953, Jan. 1958, Jan. 1970, Feb. 1977	Dec. 1963, Feb. 1958, Feb. 1970,	Feb. 1954 Dec. 1969 Feb. 1973
TALLA	Winter temperature above normal in all regions	Dec. 1953, Feb. 1970,	Jan. 1958, Feb. 1977	Dec. 1969
PAA	Southeast Alaska winter precipitation above normal	Feb. 1953, Jan. 1958, Feb. 1977	Dec. 1953, Feb. 1964,	Feb. 1954 Dec. 1976
PWA	Washington Coast winter precipitation above normal	Dec. 1952, Feb. 1954, Jan. 1964,	Jan, 1953, Jan, 1958, Dec. 972	Jan. 1954 Feb. 1958
PCCA	Central California Coast winter precipitation above normal	Dec. 1952, Jan. 1973,	Feb. 1958, Feb. 1973	Jan. 1970

222

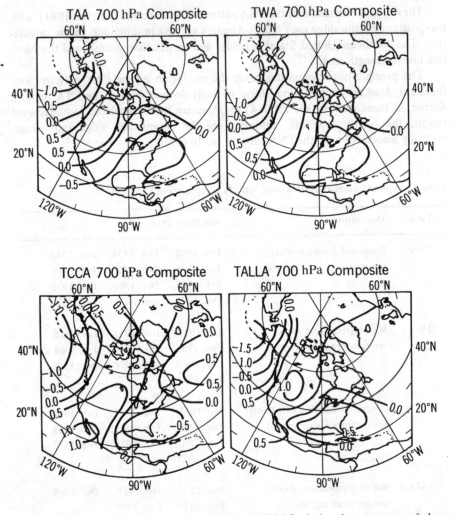

Fig. 6. Composite standardized height anomalies at 700 hPa during the occurrence of above normal winter-temperature anomalies in different areas of the west coast of North America

Conclusions

We have looked at the frequency distribution of winter and spring temperature and precipitation over the U.S. and summer rainfall in China during and following the occurrence of the extreme phases in the ENSO. The result can be summarized as follows.

Both extremes of the ENSO result in regionally anomalous conditions at monthly and seasonal time scales in both the U.S. and China. Sensitive areas in the U.S. following the low/wet ENSO phase include the southern one-third of the

United States, which tends to have colder and wetter winters and springs; and drier conditions at those times. During anti-ENSO years, the region near the Gulf Coast and the Southeast of the United States tends to experience dry winters and springs and warmer than normal temperatures.

In the CRV region of China wet weather tends to prevail during the year of peak anti-ENSO conditions, while dry conditions prevail the following year. The signal appears to be stronger in the high/dry than in the low/wet phase.

The atmospheric circulation pattern over the North Western Hemisphere displays the PNA pattern during many winter months following ENSO years. However, a second important pattern which actually has the opposite effect on the winter weather in the United States is also present. During these latter times, the Aleutian low develops southeastward of its normal position and the downstream ridge and trough develop poorly or not at all.

Obviously, from a predictive standpoint it is vitally important to ascertain which kind of development will occur in the North Pacific. We also feel that greater attention should be placed on conditions associated with the negative (high/dry) phase of ENSO, as in many instances the weather anomalies during these years are as strong as during the more notorious low/wet phase.

References

Berlage H P (1957) *K Ned Meteorol Inst Meded Verh* 69: 152 pp

Central Meteorological Bureau (1982) Yearly charts of dryness/wetness in China for the last 500−year period (In Chinese), pp 332

Horel J D, Wallace JM (1981) *Mon Wea Rev* 109: 813−829

Hoskins B J, Karoly D (1981) *J Atmos Sci* 38: 1179−1196

Quinn W H, Zopf D O, Short K S, Yang R J W K (1978) *Fish Bull* 76: 663−678

Simmons A J (1982) *Q J Met Soc* 108: 503−534

Slutz RJ, Woodruff S D, Jenne R L, Joseph D H, Steurer P M, Elms J D, Lubker S J, Hiscox J D (1984) Comprehensive ocean-atmosphere data set, release 1, NOAA/ERL, 325 Broadway, Boulder, CO 80303, pp 137

Trenberth K E (1984) *Mon Wea Rev* 112: 326−332

Van Loon, H, Madden R A (1981) *Mon Wea Rev* 109: 1150−1162

Wallace J M, Gutzler D S (1981) *Mon Wea Rev* 109: 784−812

Webster P J (1981) *J Atmos Sci* 38: 554−571

Some Statistical Facts of the Teleconnection Between Sea Surface Temperature in the Eastern Equatorial Pacific and 500 hPa Geopotential Height Field in Northern Hemisphere for 1951—1980 Period

Zeng Zhaomei [1] and Zhang Mingli [1]

Abstract — In this paper, through analysing the correlation coefficient fields between the monthly mean sea surface temperature anomalies in the key region (averaged over 10°S to 5°N along 130°W) and monthly mean 500 hPa geopotential height fields over the Northern Hemisphere for every decade of the past 30 years (1951—1980), the variations of the teleconnection between the tropical Pacific and atmospheric circulation have been examined. We have obtained some significant facts on statistics.

Introduction

Since Bjerknes (1966, 1969) provided strong evidence of links between the equatorial Pacific and the Northern Hemisphere westerlies, the large-scale tropical air-sea interaction and the effect of tropical ocean warming on the atmospheric circulation have become the most interesting topics which had received extensive discussing. In order to depict the connection of the global or hemispheric atmosphere with the tropical ocean, many meteorologists had focused on the correlation field between them. Yang and Wang (1979) had studied the correlation field between the average SST in tropical-subtropical eastern Pacific and the 500 hPa height field. They found that the variation of SST has some influence on the 500 hPa height field. Using Weare et al.'s (1976) SST index, Horel and Wallace (1981) had described a well-defined teleconnection pattern with 700 hPa GPH in the Northern Hemisphere in winter seasons. They demonstrate that the warm episodes in equatorial Pacific tend to be accompanied by the PNA flow pattern in the Pacific/North American region. With the aid of various correlation maps, Pan and Oort (1983) verified that the variation in the global atmosphere can be connected with the variation in the SST near the quator at 130°W, which can best represent the average temperature of the entire eastern equatorial Pacific. Thus, they suggested that the equatorial 130°W can be ·regarded as a key region for the global atmosphere.

In this paper, we will add more details on the connection between the equatorial Pacific and the atmosphere in the extratropics. In the following aspects this differs from the works of others: (1) As we had found some teleconnection of equatorial SST with the climate in summer seasons (Zhang and Zhen 1984a), we

---

1. Institute of Atmospheric Physics, Academia Sinica, Beijing, China

will not limit our study to the winter seasons, but extend them to the whole year. (2) Most other works describe a correlation pattern for a period of 10, 20, or 30 years. In order to study the short-range fluctuation of these correlation patterns,we separate the 30 years' range into three stages and calculate the correlation map for each stage respectively. (3) We use the SST series at equatorial 130°W to represent the SST in the eastern equatorial Pacific. (4) Since the 500 hPa height field has long been used as a basic tool in weather forecasting in China, we connect the SST in tropical Pacific with the 500 hPa geopotential height in our study.

The data sets used in this study are (1) the time series of monthly mean SST anomalies during 1950–1981, averaged over four grid points at 10°S, 5°S, 0° and 5°N, along 130°W, which are obtained from the Fishing Information (NOAA), and (2) the monthly mean 500 hPa GPH anomalies over the 10° N 10° gridpoint in Northern Hemisphere during the 1951–1980 period, which are adopted from the State Meteorological Administration of P.R.C.

The data series have been filtered by using the 7-month running mean method in order to remove the short period disturbance. The length of sample in statistics is 120 months. Using these data series, we have calculated correlation maps of SST in key regions with 500 hPa GPH field for 1951–1960, 1961–1970, and 1971–1980. To give a measure of the reliability of the individual correlation coefficient in this study, we pressure that values of $|r| \geqslant 0.3$ are significantly different from zero at the 95% confidence level (Pan and Oort, 1983).

The General Features of the Correlation Pattern in the Three Decades in 1951–1980

Figure 1 shows the contemporary correlation maps of the 500 hPa GPH field with the SST in key regions for each decade from 1951–1980. We have also calculated these correlation maps at various lags, but these maps are not shown here. Although the differences among the three decades are obvious, they still have some common features.

First almost all the correlation coefficients in the subtropics are positive and their absolute values are higher than those in the extratropics. Most of the correlation centers in the subtropical belt coincide with the semi-permanent centers of the subtropical high. The variation in the 500 hPa field usually lags to the variation in the tropical Pacific by 3–6 months. This result agrees well with that of others (Fu 1978; Chen 1982).

Second the correlation patterns in middle and high latitudes are characterized by a train of positive and negative centers which construct a planetary wave-like pattern just as the wave system of the atmospheric circulation in mid-troposphere in middle latitudes. If we compare these maps with the average 500 hPa height charts, we will find that the positive correlation centers locate near the normal ridge at the 500 hPa level, and the negative correlation centers locate near the normal trough at the 500 hPa level. Horel and Wallace suggested that during the warm episode in the tropical Pacific the PNA flow pattern prevails in the middle and upper troposphere between the Pacific and North American regions. If we examine the correlation pattern in the Pacific/Eurasian region, we consider another dominant flow pattern exists during the warm episode, which we may call the PEA flow pattern. Under this condition, the major trough over eastern Asia to the western Pacific develops deeply, the subtropical high in the Pacific Ocean may be strengthened and the broken high frequently tends to establish in the central part of the Eurasian continent. However, to inspect the whole correlation pattern, we have the impression that either the PNA or the PEA flow pattern is only the truncated part of the whole wave-like correlation pattern. This correlation pattern implies that during the warm episode in the tropical Pacific the atmospheric planetary wave in the middle latitude tends to be amplified and both the troughs and ridges strengthen in their normal location. Thus the climate in a broad area will be abnormal during the warm episode. The variation of 500 hPa height in middle latitude seems to be simultaneous with the variation of SST in tropical Pacific.

Lastly in the arctic circle, the correlation coefficients are mainly positive, However, the correlation is less considerable than in both the subtropical and middle latitude areas. The variation of atmosphere in the arctic area leads to the variation in tropical ocean by 2–3 months, and their correlation coefficient reaches 0.6.

As a whole, the correlations display a seesaw distribution along the meridian, that is, the correlations are positive in the subtropics and polar regions, and mainly negative in mid-latitude regions. The time lag of maximum correlation is not the same among the different latitudes. In the midlatitudes, it nearly equals zero, in the polar region to 2–3 months and in the subtropics to 4–6 months.

In addition, the most significance and steady peak correlation center in midlatitude are at eastern Asia, the Aleutian area and the north-western coast of North America.

The Differences Among These Three Decades

Returning to Fig. 1, the three correlation maps corresponding to the three decades have some evident differences. The correlation coefficients in 1961–1970 seems to compose three belts around the zonal belts. The negative correlation belt is located in midlatitude. However, in 1971–1980, the correlation coefficient in midlatitude is cut off by a train of correlation peak centers alternatively with negative and positive. The peak values of these correlation centers are higher than thoes in 1961–1970. Thus in 1971–1980 the correlation pattern is more significant in wave-like type. In 1951–1960, the correlation field is even more different. It is neither a zonal type nor a wave-like type. It seems to be rather in disorder with the correlation center. The most evident difference with the other two decades is that there is a significant positive center is north Asia, and the negative one which locates in north Asia in the other two decades shrink northward.

According to the correlation maps, the atmospheric circulation connected to the variation of SST in the tropical Pacific seems to have evident differences among these three decades. In the 1970s, the atmospheric wave system in midlatitudes tends to amplify the meridional exchange intensity as the tropical Pacific Ocean grows warmer. Thus both the PNA and PEA flow pattern seem to be more signif-

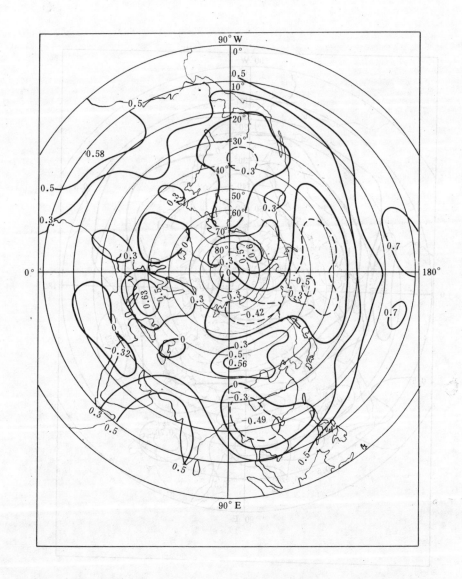

Fig. 1. Maps of the contemporary correlation coefficient of the sea surface temperature anomalies in the key region with the 500 hPa GPH in the Northern Hemisphere for each decade in 1951–1980 (a) 1951–1960

228

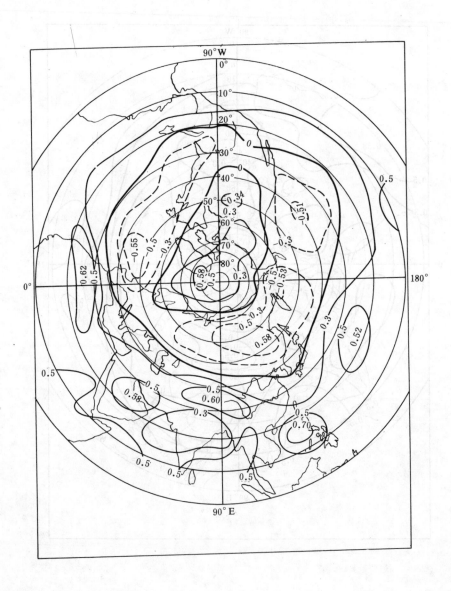

Fig. 1. (b) 1961–1970

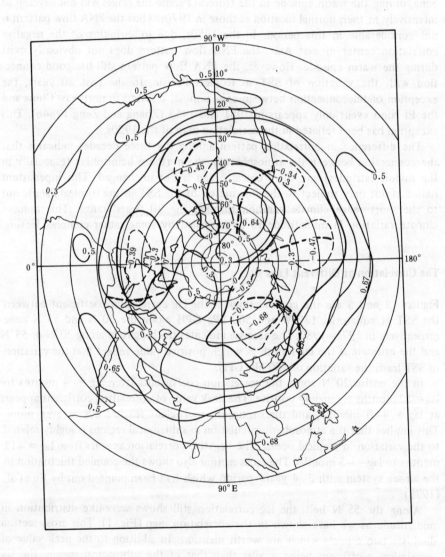

Fig 1. (c) 1971–1980

icant during this period, and the abnormal climate will frequently occur in the eastern part of the two continents. We have identified (Zhang and Zeng 1984b) that the temperature variation in the Northern Hemisphere with the spectral analysis for each decade in this century and found that the short-range variation is most significant in the 1970s. It means that the abnormal climate tends to occur in the 1970s. In the 1960s, as the negative coefficient link is a belt around the midlatitude zone, during the warm episode in the tropical Pacific the ridges will not develop so intensively in their normal location as those in 1970s. Thus the PNA flow pattern is not considerable in this period. In the 1950s, due to shrinking of the negative correlation center in east Asia, the PNA flow pattern does not obviously exist during the warm episode. However, the PNA flow pattern still has good connection with the variation of SST in tropical Pacific. In the past 30 years, the exception of teleconnection between cold summer disasters in northeast China and the El Nino event only appears in 1951 and 1953 (Zhang and Zeng 1984b). This exception has been refleted in the correlation map of the 1950s.

The difference in correlation pattern among the three decades indicates that the connection between the atmosphere in the northern hemisphere (especially in the middle latitudes) and the tropical ocean is not unchanged. This inpersistent nature is not only subject to the fluctuation of the SST in the tropical Pacific but to the short range climate varation through the 30 years' range. This induces climate variation in midlatitude to be governed by some rather complex factors.

The Correlation at Different Lags

Figures 2 and 3 are the cross-sections of the lag correlation coefficient between the SST at equatorial 130°W and 500 hPa GPH along the 20°N and 55°N zone respectively in 1971—1980. The abscissa indicates the longitude along 20°N or 55°N and the ordinate is the lag month in which positive value reflect that the variation of SST leads the variation of 500 hPa GPH.

In the entire 20°N zone, the correlation coefficients from lag =- 4 months to lag = 12 months are mainly positive. The peak value of correlation cofficient appears at lag = 4—6 months; and the maximum coefficient reaches 0.8 or even more. This implies that the atmospheric circulation in subtropical regions is highly related to the variation in tropical ocean. The negative correlation appears from lag = −12 months to lag = −5 months. The cross-section also shows the coupled fluctuation in the air-sea system with 3—4 years' period which had been pointed out by Fu et al. (1978).

Along the 55°N belt, the lag correlation still shows wave-like distribution in midlatitude as we have shown in the correlation map (Fig. 1). This cross-section discloses two aspects which are worth mention. In addition to the peak value of correlation coefficient being smaller than that of the subtropical region, the lag months of the variation of atmosphere in the middle latitude decrease in comparison with that in the subtropics. The peak value along the western coast of North America appears at lag = 1 month, in west Europe is at lag = 0 month and in eastern Asia is at lag = −1 to -2 months. The distribution shows that the variation

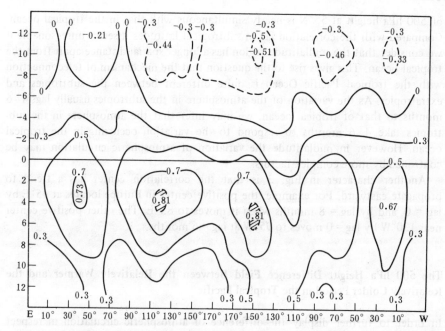

Fig. 2. The cross-section of the lag correlation coefficient between the SST at equatorial 130°W and 500 hPa GPH along 20°N in 1971–80. The abscissa indicate the longitude along 20°N and the ordinate is the lag month in which positive values reflect that the variation of SST leads to the variation of 500 hPa GPH

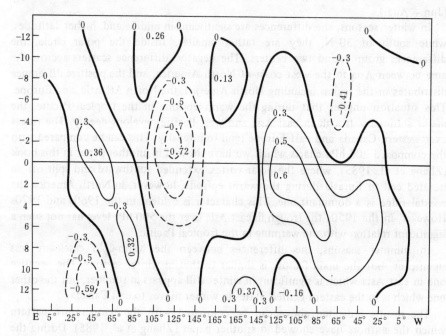

Fig. 3. As in Fig 2 but for 55°N

of 500 hPa height at 55°N is nearly simultaneous with that in the tropical ocean. Compared with the correlation lag in subtropical latitudes it seems unreasonable if we consider that the midlatitude region has a longer spatial distance apart from the tropical ocean. This gives rise to the question that the mechanism of teleconnection with the tropical Pacific Ocean is quite different between the subtropics and extratropics. As the variation of the atmosphere in the subtropics usually lags 4–6 months to that of tropical ocean, we may infer that the atmosphere in the sub-tropics take 4–6 months to respond to the variation occurring in the tropical ocean. However in midlatitude the variation of atmospheric circulation may be controlled by more complex factors.

Another character in Fig. 3 is that the correlation center has a trend to propagate eastward. For example, the positive center in Europe locates at 45°E by lag = 0, and by lag = 8 months when it moves to 85°E. The other positive center near 130°W by lag = 0 moves to 95°W at lag = 6 months.

The 500 hPa Height Difference Field Between the Relatively Warmer and the Relatively Colder Period in the Tropical Pacific

In order to further display the difference of atmospheric circulation in respect to different oceanic warming, we composed maps of 500 hPa height difference between the periods of the SST in equatorial 130°W are warmer and colder for winter season and summer season respectively. Figure 4 is the map for winter seasons (Dec.–Feb.) during the 1951–1980 period and Fig. 5 for summer seasons (Jun.– Aug.).

In winter seasons, the differences are significant in middle and higher latitudes, while south of 30°N, they are rather smaller. Inside the polar circle, the differences group round two centers. The negative difference scatters accross the area between Asia to the west coast of North America, and the positive difference distributes in the region including North America, the North Atlantic, and Europe. This situation implies that during the warm episode in the tropical Pacific, the major stationary trough in east Asia and west Pacific develops deeply. The ridges over western Canada and east Europe tend to intensify. This can be compared with the composed 200 hPa charts which we have showed in another paper in this book (Zhang et al. 1985), where the polar vortex extended southward and split off an isolated cell in Eurasia during the warm episode, however, in North America the coastal ridge is a dominant one, This character is evident in the 1960s and 1970s However, in the 1950s the trough in east Asia over the 500 hPa level has not such a significant relation with the warming in the tropical Pacific.

In summer seasons, the differences between the warm-cold cases are less significant, but the main feature is similar to that in winter seasons. The negative one in east Asia which is significant in winter still appears in summer, but the other one which is in the eastern United States in winter moves to east Canada.

The results indicated above can be compared with the temperature pattern which the authors have showed in another paper (Zhang et al. 1985). During the warm period in the tropical Pacific, the climate in eastern Asia is cold in the north

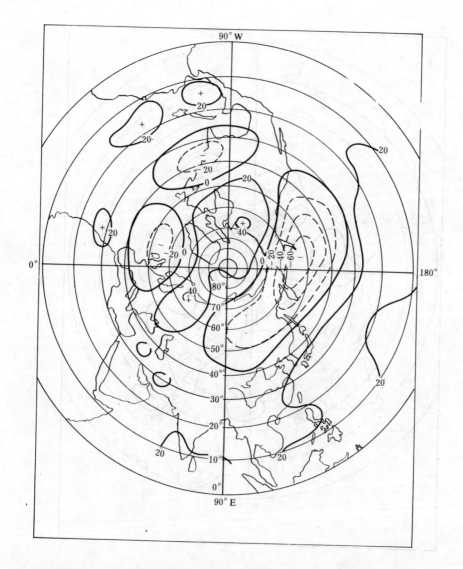

Fig. 4. The averate 500 hPa GPH difference between the DJF seasons that SST (130°W) was relatively warm and the seasons that it was relatively cold for the period 1951–1980

and near normal in the south corresponding to the negative correlation with 500 hPa height in northeastern Asia, and in eastern North America, it is warm in the north and cold in the south corresponding to the positive correlation with 500 hPa height in the most northern area of North America. The connection of climate in northeast Asia is significant both in winter and in summer, but in the eastern United States it is only significant in winter.

234

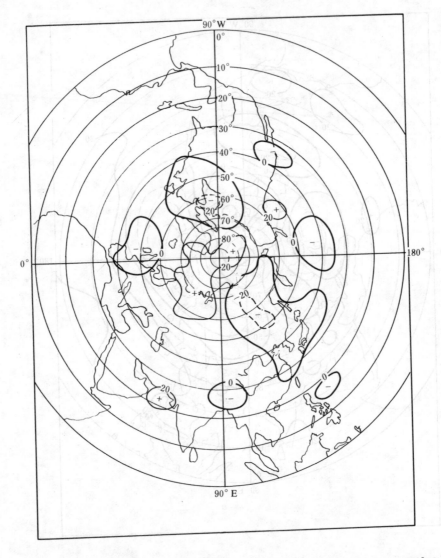

Fig. 5. The average 500 hPa GPH difference between the JJA seasons that SST (130°W) was relatively warm and the season that it was relatively cold for the period 1951–1980

Summary and Discussion

According to the analysis above, we can obtain the following conclusions:
(1) The correlation between SST in the tropical Pacific and 500 hPa height in the Northern Hemisphere shows a positive coefficient in the subtropical and polar

region and a planetary wave-like distribution of coefficient in the middle and higher latitude area. This implies that during the warm episode in the tropical Pacific, the atmospheric planetary wave seems to be amplified, and the major stationary trough and ridge are deepened in their normal position. This will lead to the abnormal climate prevailing in most area of midlatitudes.

(2) The correlation pattern is not persistent throughout the whole recent 30 years. In the 1960s, the correlation coefficient links to a negative belt in midlatitudes, but in the 1970s, the negative coefficients in midlatitudes are cut off by some positive centers, and the wave-like distribution is more significant than in the other two decades. In the 1950s, the correlation centers scatter and are less considerable than those in the 1960s and 1970s. The persistance of the correlation character showing the short-range climate fluctuation may be controlled by more complex factors.

(3) The lag correlation shows 4–6 month lag in the subtropics, and nearly 0 month lag in midlatitudes. We may infer that the variation of atmsophere circulation in the subtropics responds to the variation of SST in the tropical Pacific, however, in midlatitudes the connection of atmospheric circulation with the tropical Pacific has different mechanism.

(4) The 500 hPa height pattern associated with the warm episode in the tropical Pacific shows a deeply developed major trough in east Asia and a strengthened ridge in west Canada. This may lead to the different climate pattern connected with SST in the tropical Pacific between Asia and North America.

Acknowledgement. The authors wish to thank Profs. J. P. Chao and Y. H. Pan for helpful discussions and their valuable suggestions.

References

Bjerknes J (1966) *Tellus* 28: 820–828
Bjerknes J (1969) *Mon Wea Rev* 97: 163–172
Chen L T (1982) *Sci Atmos Sin* 6: 148–156
Fu C B et al. (1978) *Ocean Select* 2: 16–21
Horel J D, Wallace J M (1981) *Mon Wea Rev* 109: 813–829
Pan Y H, Oort A H (1983) *Mon Wea Rev* 111: 1244–1258
Weare B C, Navato A R, Newell R E (1976) *J. Phys Oceanogr* 6: 671–678
Yan J C, Wang L Y (1979) *Acta Meteorol Sin* 37: 40–48
Zhang M L, Zeng Z M (1984a) *Trop Ocean-Atmos Newslett* 23: 5–7
Zhang M L, Zeng Z M, (1984b) *Sci Sin* (Ser B) 27: 722–733
Zhang M L, Zeng Z M, Pan Y H (1985) In this selection

Abnormal Heavy Rainfall in Changjiang Valley and the Correlation with Extraordinary Oceanic Features in 1983

Zhang Yan[1], Li Yuehong[1] and Bi Muying[1]

Abstract

An abnormal heavy precipitation, one of the four serious flood events in the last sixty years, occurred at Changjiang Valley, Central China, May—July, 1983. This abnormally wide area and long duration of heavy rainfall was directly associated with the abnormal changes of the Subtropical High (S. H.) of the NW Pacific, while the abnormal changes of the. S.H. were correlated with the abnormal warming of SST of the tropical east Pacific and the appearance of El Nino in 1982—1983 at the highest intensity in this century, and were reasonably correlated with the unusual variations of Southern Oscillation, SST anomalies of the Pacific Ocean, and the anomalies of the cloudiness over these areas. The intensities of these anomalies were quite consistant. This has once again confirmed that a teleconnection exists between several parts of the Pacific Ocean and the atmospheric circulation. A chain reaction between the abnormal weather in China and the East and the West Pacific (the tropical and the subtropical Pacific) has been revealed. That may give some information for long-range weather forecasting.

1. Weather and Climate Research Institute, Academy of Meteorological Science, Beijing, China

East Asian Cyclogenesis and the Southern Oscillation

Long Baosen[1] and Howard P. Hanson[2]

Abstract — Wintertime cyclogenesis over the East China Sea and the surrounding area exerts a profound influence on the climate in this region. In a climatological study of the frequency and position of this cyclogenesis, we have found a significant link to variations of the Southern Oscillation Index (SOI) and El Nino events. This appears to be a manifestation of the West Pacific teleconnection pattern discussed by Horel and Wallace (*Mon. Wea. Rev,* 109: 813*ff*) on the synoptic scale in which low (high) values of the SOI, associated with high (low) equatorial sea-surface temperature, are also associated with greater than (less than) the average frequency of cyclone formation over the region, We infer the physical basis for this is the preferential formation of an anomalous 500 hPa low (i.e., strengthening and broadening of the East Asian Trough) in the vicinity of $155°E$ during the Southern Oscillation/El Nino (ENSO) events via the West Pacific teleconnection.

Introduction

The influence of El Nino/Southern Oscillation (ENSO) events on the global atmospheric circulation and the associated weather have been one of the recent focuses of air-sea interaction research, and a number of interesting studies have shown that the ENSO phenomena have a significant effect upon the middle latitude climate. In this paper the number of cyclones that formed within the area $20°-40°N$ and $110°-150°E$ (Fig. 1) as well as within the primary center of activity (shaded area) in each year from 1899 to 1962 were counted (based on U.S. Weather Bureau historical weather maps) with emphasis on wintertime. The time series of these events are compared with corresponding time series of the equatorial Pacific sea surface temperature (SST) and other ENSO indices. Significant relationships between the frequency of cyclones and ENSO indices have been found.

There is the evidence (e.g., Bjerknes 1972; Horel and Wallace 1981) that a teleconnection exists between the Eastern Asian atmospheric circulation and the ENSO phenomena. The extropical cyclogenesis that occurs over the East China Sea and surrounding area and the propagation of the resulting cyclones bring severe weather to this region. We suggest that the anomalies of the general circulation are directly responsible for the anomalies of cyclone formation that occur in the area studied here.

Statistical Relationships Between Cyclogenesis and ENSO

Figure 1 presents a geographical frequency of cyclones formed within the area

---

1. First Institute of Oceanography, State Oceanic Administration, Qingdao, Shandong, China
2. Cooperative Institute for Research in Environmental Sciences, University of Colorado/-NOAA, Boulder, CO 80309, USA

20°–40°N, 120°–150°E for the months October–April, 1899–1982. It can be seen that there are two maximum centers of frequency with the primary one between the islands of Taiwan and Japan. This area is just on the southern side of an entrance region to the upper-level subtropical jet stream and the region where disturbances on the surface subtropical front are rather active. Figure 2 shows the seasonal variability of frequency of cyclone formation. The frequency reaches a maximum in April both for all 64 years and for the 48 non-ENSO years. In ENSO years the maximum appears earlier during the winter months. During the rest of the year the cyclone formation is almost the same as that in the non-ENSO years. It seems that there is a tendency for earlier cyclone formation during the ENSO years. As indicated by some papers (e.g., Horel and Wallace, 1981), the ENSO phenomena have a significant effect on the middle latitude climate only in the wintertime; this can also be inferred from Fig. 2.

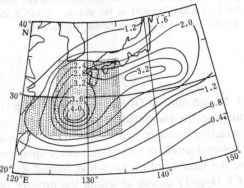

Fig. 1. Average number of winter maritime cyclogenesis events, per 5° square, for the mouths October April, 1899–1962. Hatched area is used in the text as a primary center area of cyclone formation

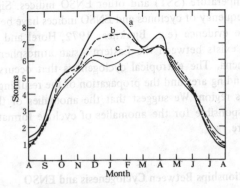

Fig. 2. Averaged seasonal cycle during strong ENSO years (curve a for 8 strongest ENSO years); all (16) ENSO years (curve b); all (64) years (curve c); and non-ENSO (48) years (curve d)

Both yearly time series of frequency of cyclone formation for the entire study area and the area of primary maximum center are given by Fig. 3 (A, B).

These two time series look quite similar to one another. The curves C, D, and E are, respectively, the yearly equatorial Pacific SST time series, the time series of yearly Darwin sea surface pressure (DAR) and the Southern Oscillation Index (SOI). Comparing (A,B) with C, D and E, one can find positive correlation relationships among (A, B), C and D, and negative ones between (A,B) and E. Correlation coefficients between B, C, and B, E, are respectively 0.33 and −0.31 (these are significant at the 95% level). During the ENSO years, the cyclones occurred more than usual except for 1899–1900 and 1911–1912 [vertical lines refer to the ENSO years defined by Rasmusson and Wallace (1983)], as can be seen from Fig. 3 and Table 1. The correlation coefficients among these time series of variables are presented in Table 2.

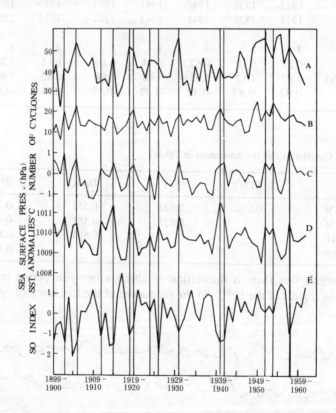

Fig. 3. Time series of: wintertime total cyclones in study area (A) and primary center area (B); yearly averaged equatorial Pacific SST anomalies (C) Darwin sea surface pressure anomalies (D) and Wright's (1975) SOI (E)

The frequency of cyclone formation and the Darwin sea surface pressure have also been analyzed on the basis of monthly means subjected to a twelve-month running average to subtract the seasonal variation. These smoothed time series are shown in Fig. 4. The peaks of cyclone formation appear behind those of Darwin sea surface pressure. The phase lag is computed to be about two months.

Table 1. Anomalies of: Storm formation in the area of maximum formation (STM'), equatorial pacific SST, Darwin surface pressure (DAR) and Wright's SOI during ENSO episodes

ENSO year S	1899-1900	1902-1903	1905-1906	1911-1912	1914-1915	1918-1919	1919-1920	1923-1924
STM'	-5	6	8	-3	1	3	6	1
SST(°C)	0.94	1.14	0.98	0.44	0.91	0.46	0.62	0.30
DAR(hPa)	1.13	1.46	0.90	1.10	2.30	1.00	0.58	0.36
SOI	-1.41	-1.10	-0.69	-0.87	-1.08	-1.04	-1.20	-0.09

ENSO Year	1925-1926	1930-1931	1940-1941	1941-1942	1951-1952	1953-1954	1957-1958
STM'	3	3	7	1	5	9	3
SST(°C)	0.33	0.80	0.62	0.74	0.80	0.65	1.24
DAR(hPa)	0.80	1.12	1.95	1.48	0.68	0.62	0.97
SOI	-1.42	-0.83	-1.12	-1.72	-0.17	0.14	-0.81

Table 2. Correlations of the Anomalies in Table 1

	STM'	SST	DAR	SOI
STM	0.672	0.531	0.351	-0.319
STM'		0.638	0.581	-0.413
SST			0.901	-0.832
DAR				-0.914

It is obvious that there is fluctuation of about a two-year period in the 1930's and in the middle of the 1950's in the time series of the frequency of cyclone formation.

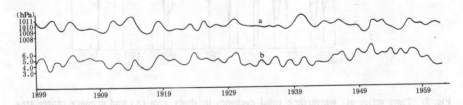

Fig. 4. Twelve-month running averaged time series of monthly averaged frequency of cyclone formation (b) and Darwin sea surface pressure (a)

These fluctuations at a two-year period can also be found in the power spectral analysis of cyclone formation (Fig. 5) as a band from $(3 \text{ yr})^{-1}$ to $(1.7 \text{ yr})^{-1}$. There is also a lower frequency band that is broader from $(3 \text{ yr})^{-1}$ to $(8 \text{ yr})^{-1}$ centered at $\sim (0.53 \text{ yr})^{-1}$ and containing more power. This analysis was made by Fast Fourier Transforming (FFT) a "pseudomonth" time series of cyclone formation, where a "pseudomonth" (p-month) is six weeks, i.e., there are 8 p-months per year. This is discussed in more detail by Hanson and Long (1984). Figure 6 shows the cross spectrum of the high-pass-filtered (with a cutoff at $(8 \text{ yr})^{-1}$) time series of cyclone formation and Darwin sea surface pressure; the cross spectrum is dominated by the peak at $\sim (5.3 \text{ yr})^{-1}$. We therefore conclude from this spectral analysis that the variations in the frequency of cyclone formation and the Darwin sea surface pressure are consistent with each other. Darwin sea surface pressure leads by 0 to 2 p-months with the correlation coefficient maximum (0.346) at 1 p-month (1.5 months).

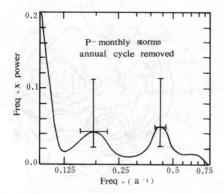

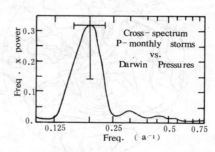

Fig 5. FFT power spectrum of interannual cyclone variability

Fig. 6. The cross-spectrum of high-pass filtered cyclone series and Darwin pressure anomalies series

Discussion of the Physical Basis for the Relationships

A number of valuable studies have shown that there are significant correlations between ENSO and the general atmospheric circulation and teleconnections in the Northern Hemisphere geopotential field during wintertime. Figure 7 (from Wallace and Gutzler 1981) shows a sea level pressure teleconnection pattern between two centers located at (65°N, 170°E) and (25°N, 165°E). As can be seen from these one-point correlation maps, a negative correlation exists between the centers. This pattern is also present in the 500 hPa geopotential height field and is called the "Western Pacific Pattern" (Fig. 8, ibid). It is characterized by a deeper Aleutian low (particularly at the latitude west of date line) and a stronger jet stream over Japan accompanied by a stronger 500 hPa geopotential height field over Western Pacific. The West Pacific pattern is analogous to and in phase with a similar pattern in the West Atlantic (Wallace and Gutzler 1981). Figures 9 and 10 (ibid) show the two extremes of the East Asian Trough with a positive value of WA (the index of the Western Atlantic pattern) and with a negative value of WA. As can be seen, a deeper Aleutian low (Fig. 9a) is accompanied by a deeper and wider East Asian Trough (Fig. 10a) and a weaker Aleutian low (Fig. 9b) is accompanied by a weaker and narrower East Asian Trough (Fig. 10b). Looking carefully at these two figures, one can find that during the time with positive value of WA (Fig. 10a), the East Asian Trough is located near 120°E (somewhat to the west of the normal climatological position), there is an evident trough over the mainland of China, and the Siberian anticyclone is stronger than normal and is associated with a deeper Aleutian low (Fig. 9a). During the time with negative values of WA (Fig. 10b), the East Asian Trough is located near 140°E (somewhat to the east of the normal position), there is no evident trough over the mainland of China, and the Siberian anticyclone is weaker and is associated with a weaker Aleutian low (Fig. 9b).

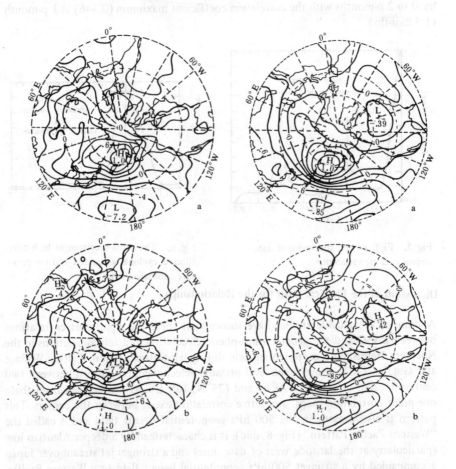

Fig. 7. One-point correlation maps for sea level pressure for base grid points (a) 65°N, 170°E and (b) 25°N, 165°E, based on 15 winter data set 1962–1963 to 1976–1977 (from Wallace and Gutzler 1981). Contour interval is 0.2.

Fig 8. One-point correlation maps for 500 hPa geopotential height field for base grid points (a) 60°N, 155?E; (b) 30°N, 155°N (from Wallace and Gutzler 1981)

From the above discussion a conclusion emerges that, as the pressure at the sea surface and 500 hPa geopotential height field in the Western Pacific rise, the Aleutian low deepens and is accompanied by a deeper East Asian Trough, and vice

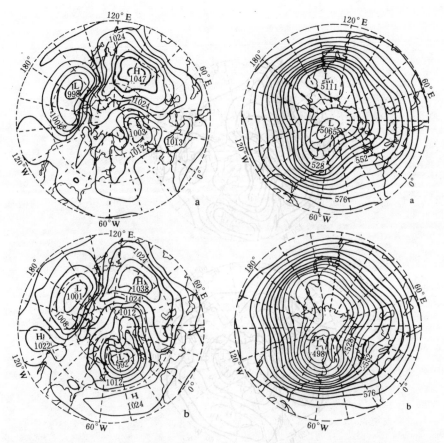

Fig. 9. The strong Western Atlantic (WA) pattern, surface pressure field. (a) Positive values of WA correspond to a strong Aleutian Low; (b) negative values of WA correspond to a weak Aleutian Low (from Wallace and Gutzler 1981)

Fig. 10. The corresponding pattern at 500 hPa. (a) positive values of WA correspond to a strong East Asian Trough, and (b) vice versa (from Wallace and Gutzler 1981)

versa. There is further evidence that the deepening of the East Asian Trough is related to ENSO events. Figure 11 (Horel and Wallace, 1981) shows that warmer SST in the equatorial Pacific is accompanied by negative 700 hPa height anomalies in a broad belt across Northern Canada and the western Pacific, and negative anomalies over the southeastern United States. This study also indicates that the shapes of patterns derived by correlating ENSO indices with the Northern Hemisphere sea surface pressure, 300 hPa and 1000–700 hPa thickness fields are, for the most part, quite similar to the correlation pattern in Fig. 11.

We have shown in Section II that when ENSO indices, such as the equatorial Pacific SST and Darwin sea surface pressure, are above normal, especially during ENSO years, the frequency of cyclone formation over the East China Sea is higher than usual. The discussion above indicates that as the two kinds of indices rise,

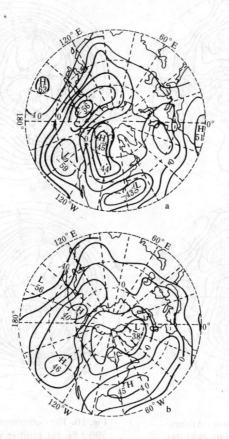

Fig. 11. Correlation coefficients between 700 hPa geopotential height at grid points poleward of 20°N and (a) Southern Oscillation Index, based on 28 year data set 1951–1978 (from Horel and Wallace 1981)

the East Asian trough deepens in association with the deeper Aleutian Low. Therefore, we suggest that the deepening and the broadening of the East Asian Trough associated with ENSO events is directly responsible for the increase in cyclones formed over the East China Sea and the surrounding area during ENSO episodes.

The vorticity equation can be used to explain which factors have an influence on cyclogenesis. Starting with the vorticity equation and the thermal wind equation, one can derive

$$\frac{\partial \zeta_0}{\partial t} = -[u\frac{\partial \zeta}{\partial x} + v\frac{\partial \zeta}{\partial y}] - \frac{g}{f}\nabla^2 \left\{ -[u\frac{\partial (z-z_0)}{\partial x} + v\frac{\partial (z-z_0)}{\partial y}] \right.$$
$$\left. +\frac{R}{g}\ln[\frac{P_0}{P}][\gamma_d-\gamma]\omega+Q/C_P \right\}.$$

The local variation of the surface vorticity ($\partial \xi_0 / \partial t$) depends on the upper level vorticity advection (first term on right-hand side), the thickness advection from the surface to the upper level (second term — representing the baroclinicity of the atmosphere), the stability of the air (i.e., the non-adiabatic vertical advection, third term), and the heating from diabatic processes (e.g. latent heat release). Q/C_p. As the East Asian Trough becomes stronger during ENSO episodes, upper level systems over the area studied here will present more positive vorticity advection to the weather systems at the surface and increase the potential for cyclone formation, while the other terms remain relatively constant (we will discuss this equation in some detail elsewhere). The results of the Air-Mass Transformation Experiment of 1974 and 1975 have shown (e.g., Chen et al. 1983) that latent heat release, moisture and heat fluxes from the ocean surface and larg-scale forcing due to upper level systems are three physical factors important to oceanic cyclogenesis. This is basically consistent with the equation discussed above. The mechanism for the anomalies of the general atmosphere is beyond this paper, although in general the dynamical mechanism has been thought to be the dispersion of the Rossby waves triggered by ENSO phenomena.

Concluding Remarks

First, positive correlation relationships exist between the frequency of cyclone formation off the east coast of China, the tropical Pacific SST and the Darwin sea surface pressure. The phase of Darwin sea surface pressure leads by about two months with a maximum correlation coefficient (0.346) at 6 weeks.

Second the power spectral analysis of frequency of cyclone formation has shown two spectral bands, one centered at $\sim (1.7 \text{ yr})^{-1}$, the other centered at $\sim (5.3 \text{ yr})^{-1}$ with more power and broader bandwidth. A spectral peak at $\sim (5.3 \text{ yr})^{-1}$ has also been found in the cross-spectrum of the frequency of cyclone formation and Darwin sea surface pressure. The spectral analyses show that both time series of frequency of cyclone formation and Darwin sea surface pressure have about a 5-year period.

Finally, the deepening and broadening of the East Asian Trough related to ENSO events are directly responsible for the more cyclone formation. This can be explained using the vorticity equation.

Acknowledgements. We would like to express our appreciation for the Scientific Exchange Program between NOAA and the National Bureau of Oceanography of the PRC for the opportunity to work together, and to A. S. Frisch of NOAA/ERL/WPL for facilitating this particular exchange. We thank R. Slutz and E. M. Rassmusson for the HSSTP and ENSO index data. H. P. Hanson is supported by the Equatorial Pacific Ocean Climate Studies program of NOAA and the National Science Foundation through grant No. ATM 82-09115.

References

Bjerknes J (1972) *J Phys Oceanogr* 2: 212–217
Chen, T-C, Chang C. -B. Perkey D. J. (1983) *Mon Wea Rev* 111: 181–1829

246

Hanson, H. P. Long B. (1984) Climatology of cyclogenesis over the East China Sea.
   Submitted to *Mon Wea Rev.*
Horel, J. D. Wallace J. M. (1981) *Mon Wea Rev* 109: 813–829
Rassmusson, E. M. Wallace J. M. (1983) *Science* 222: 1195–1201
Wallace, J. M. Gutzler D. S. (1981) *Mon Wea Rev* 109: 784–812
Wright, P. B. (1975) An index of the Southern Oscillation, Report CRU RP4, Climate Res
   Unit, Univ. East Anglia, 22 pp

Equatorial Oceanic Thermal and Rossby Waves Associated with El Nino Events

Chao Jiping [1] and Ji Zhengang [2]

Comprehensive Abstract

Observations have shown that there is a cold water pool in the equatorial Pacific Ocean, especially in the eastern part of the ocean, where the depth of the mixed layer is less than that in the west area.

By using a linear oceanic mixed layer model, the influences of the horizontal gradients of sea surface temperature (SST) and the variation of the mixed layer depth upon the tropical oceanic waves are investigated. The gravity wave, intertial oscillation and Kelvin wave are all filled out in this model, with only the Rossby wave remaining.

Two nondimensional parameters which represent the influences of large-scale inhomogenities of three-dimensional sea temperature are important in this model. They are

$$S_x = \frac{1}{\Delta \overline{T}_Z} \cdot \frac{\partial \overline{T}}{\partial x} - \frac{1}{\overline{D}} \cdot \frac{\partial \overline{D}}{\partial x}$$

$$S_y = \frac{1}{\Delta \overline{T}_Z} \cdot \frac{\partial \overline{T}}{\partial y} - \frac{1}{\overline{D}} \cdot \frac{\partial \overline{D}}{\partial y},$$

where $\Delta \overline{T}_2$ is the vertical temperature difference between the sea surface and the surface of the thermocline, $\overline{T}$ is the large-scale climatic sea temperature field, and $\overline{D}$ is the characteristic depth of the mixed layer.

The theoretical analysis shows that the equatorial Rossby wave will be modified and a kind of slow thermal wave is revealed when S_x and S_y are taken into account. When the Coriolis parameter $f=0$, this kind of new wave disappears. While only by letting $\beta = \partial f / \partial y = 0$, the thermal wave does exist, even though Rossby wave is filled out in this case. An interesting result is that the thermal wave propagates eastward, which is opposite to that of the classical equatorial Rossby wave.

Two numerical experiments are made in this study:

1. $\tau_x = \tau_y = 0$, $S_x =$ const., $S_y =$ const.

It is assumed that the anomalous windstress is neglected and the large-scale thermal structure of the ocean has no change under the perturbation fields. In this case, the governing equation is the same as the one which has been investigated

---

1. National Research Centre for Marine Environmental Forecasts, SOA, Beijing, China
2. The Graduate School, University of Science and Technology of China, Beijing, China

by the theoretical analysis.

By considering the contribution of different S_y to the variation of SST anomaly, we have the following results under the time development.

(1) $S_y < S_{yc}$

Here S_{yc} is a certain critical value. In this case, the influences of the large-scale inhomogeneities of the ocean thermal structure are not imporant. The SST anomaly (SSTA) propagates westward about 7×10^3 km in 240 days. It shows that the Rossby wave is still dominant.

(2) $S_y = S_{yc}$

There are no wave phenomena and the perturbation field is not be able to propagate or to spread out in this case.

(3) $S_y > S_{yc}$

In this case the perturbation field propagates eastward about 7×10^3 km after 240 days. As mentioned above, the Kelvin wave has been filled out in this model. Therefore the eastward propagating of the SSTA could be attributed to the thermal wave which propagates eastward.

2. $\tau_x \neq 0$, $\tau_y = 0$, $S_y = S_y(t)$.

When the windstress anomaly τ_x in latitude and the time variation of the large-scale oceanic thermal structure are taken into account, there is not much change in the results compared with the experiments mentioned above, except that the perturbation fields increase about $1°C$ after 240 days.

The results of theoretical analysis are supported by the numerical experiments.

As a first approximation, it seems to be true that both these two waves obtained in this study may be used to explain the observation facts that the SSTA can usually propagate in both direction during the El Nino events. This physical process provides a possible explanation for the El Nino event in 1982—1983.

The main argument of this paper is that the thermal wave can have an effect which is somewhat similar to the Kelvin wave. If we consider not only the effect of the Kelvin wave, but also the effect of the eastward propagating thermal wave, we may obtain a better understanding of El Nino phenomenon. The thermal wave can only appear under a special environment. The most favorable condition is the tropical ocean with the presence of a cold water pool. Thus the effect of thermal wave on the El Nino events is emphasized as a special phenomenon in the tropical area.

The East Asia Monsoon and the Southern Oscillation, 1871–1980

Guo Qiyun [1]

Abstract – Winter and summer monsoon intensity index are defined for the period from 1871 to 1980 on the base of data set of sea level pressure ($50°S-70°N$) in January and July. A new defination of the Southern Oscillation Index (SOI) is given. The variations of monsoon intensity index and SOI are investigated and compared each other.

Introduction

It is well known that the monsoon plays an important role in forming the climate in China. The variations of its intensity and duration to a great extent control the climatic anomalies over China, especially in its eastern part. Recently, the summer and winter monsoon variations are investigated in relation to the climatic anomalies in China (Guo Qi-Yun 1983, in press a), using a 30-year data set from 1951 to 1980. The correlation coefficients between summer monsoon intensity and rainfall have shown that the rainfall observed in the North China Plain and South China is larger than that in the Changjiang River valley when the summer monsoon is strong. At the same time the Changjiang River valley suffers from drought. On the contrary, when the summer monsoon is weak the opposite picture appears, plentiful rainfall occurs in the Changjiang River valley and droughts take place north and south of it. The correlation coefficients between winter monsoon intensity and temperature show that the winter temperature has a close connection with the winter monsoon intensity. When the winter monsoon is strong the temperature is low, and vice versa. This is true especially in East China, thus suggesting that the essential step in research of the climatic variations and their causes is to examine the monsoon variations. Unfortunately, the scarcity of a long series of monsoons and the uncertainty of its definition have greatly prevented the development of reserch, so most of it is restricted to case studies only. Therefore, an effort is made in using the longer series to form a longer series of winter and summer monsoon intensity index.

However, it was indicated in a previous paper that the Asian monsoon is linked closely with general atmospheric circulation, the atmospheric circulation in the Southern Hemishphere usually exerts some influence on the Asian monsoon. For example, it has been proved that the drought/flood in the Changjiang River valley has a high correlation with Southern Oscillation (SO) (Tu Chang-Wang 1954) and the development of summer monsoon in China is connected with the cold air activity over Australia (Guo Qi-Yun 1965). Recently, a possible mechanism has been suggested, which links the summer monsoon with planetary atmospheric circulation (Guo Qi-Yun, in press b). These investigations indicate the

---

1. Institute of Geography, Academia Sinica, Beijing, China

importance of expanding the scope of research to a global, and finally to hemispheric extent, if we are interested in the cause-effect relation and the long-term variation.

Therefore, a series of global sea level pressure charts in January and July from 1871 to 1980 are used to construct a long set of winter and summer monsoon intensity indices and Southern Oscillation Index (SOI). On the base of these data set the long-term variations of monsoon, the relationship between monsoon intensity indices and SOI are examined.

Data and Method Analysis

The monsoon is considered to develop on the base of the sea-land thermo-contrast and the index describing the monsoon must be a representative one for a great part of East Asia. Thus, the monsoon intensity index is defined as following

$$I_{w(s)} = \sum_{i=1}^{5} (P_{110°E} - P_{160°E})_i, \quad (i, \; 1=10°N,......5=50°N,)$$

where P is the monthly mean sea level pressure; the indexes I_w or I_s represent the winter or summer monsoon intensity.

Concerning the SOI, the definitions are various from one author to another. Generally, the pressure difference between a pair of stations, one located in the Indian Ocean Australia such as Darwin, and the other located in the Southern Pacific Ocean such as Tahiti/Easter Island is adopted as SOI (Quinn and Burt 1972; Trenberth 1976). In order to look for a more representative index of SO, the sea level pressure data set is used, which is available at $10° \times 10°$ grid points within the area $10°-360°$ longitude and $50°S-70°N$ latitude. Considering the SO mainly reflects the opposite trend of pressure variation between Southern Pacific Ocean and Indian Ocean, six points mean pressure in the area $100°-120°W, 20°-30°S$ is taken to represent the Southern Pacific Ocean pressure. Then the correlation coefficients between these six points mean pressure and the pressure elsewhere over the Southern Hemisphere are calculated. In Fig. 1 the correlation coefficients are given with the isopleth in 95%, 99%, and 99.9% significance level. It is shown that the positive correlation predominates in the Southern Pacific Ocean and the

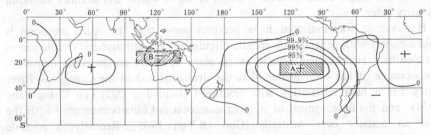

Fig. 1. Correlation coefficients between sea level pressure averaged for the area indicated in the text and the pressure elsewhere over the Southern Hemisphere

negative in the Indian Ocean. Zero line is almost along the date-line. This is in full agreement with Walker's finding. However, this calculation gives the geographical distribution of correlation in detail, and one can use it to find a better definition of SO. Two areas are identified as the maximum of positive and negative correlation. The Pacific one is called the A region limited to $90°-130°W$ and $20°-30°S$. The other in the Indian Ocean is named the B region located in $100°-140°E$ and $10°-20°S$. Finally, SOI is defined as the difference of mean pressures between the A region and the B region. The foot nortes 1 and 7 denote the SOI for January and July. Thus, four long series are obtained, namely I_w, I_s, SOI_1 and SOI_7, which are the basic data used in the present paper.

Variations of Monsoon Intensity and Southern Oscillation

Figure 2a gives the variation of I_w from 1971−1980. The break lines are average of I_w for different periods. Three epochs during 1761−1870 can be identified,

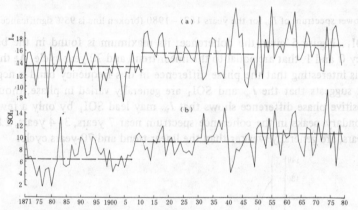

Fig. 2. The variations of I_w(a) and SOI_1(b) from 1871 to 1980

befor the 1910s the I_w remained lower, from the 1910's to the 1950's, it varied around an intermediate value and after the 1950's came to a high value period but with strong fluctuation. Generally, the I_w increased for the whole 110-year period. This characteristic is well illustrated by the power spectrum of I_w (Fig. 3), which is calculated from the 110-year series with a maximum 35 year lag. In the spectrum, the linear trend predominates, and the quasi-biennial oscillation (QBO) fits the 95% significance level. This is in good agreement with the QBO found in climatic variations in China (Reed 1965) and over the world (Wang Shao-Wu and Zhao Zong-Ci 1978).

The long-term variation of SOI_1 in this Fig. 2b shows a trend as strong as I_w. The 110-year period can also be divided into three phases. In the first phase SOI_1 was weakened, and in the third phase SOI_1 was intensified significantly. The intermediate phase from 1909 to 1950 was characterized by an increasing trend. Besides, the linear trend QBO is also prominent in the spectrum of SOI_1 (Fig. 4). The only difference, between the spectra of I_w and SOI_1 is the 3.9 year peak, which did not appear at all in I_w.

The cross-spectrum analysis is carried out to compare the variations of I_w

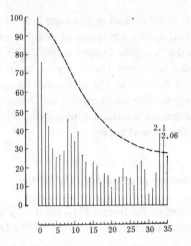

Fig. 3. Power spectrum of I_w for the years 1871–1980 (broken line is 95% significance level)

with SOI_1. Figure 5 gives the coherence, its maximum is found in the band of frequency 0 and 1, that are equal to the linear trend and 70-year cycle. At the same time, it is interesting that the phase difference in this frequency band is negligibly small. It suggests that the I_w and SOI_1 are generally varied in phase, though the small positive phase difference shows that I_w may lead SOI_1 by only a few years. The secondary peaks in the coherence spectrum near 7 years, 3–4 years, 2.8 years and 2 years, are generally weaker than the linear trend and 70-years cycle.

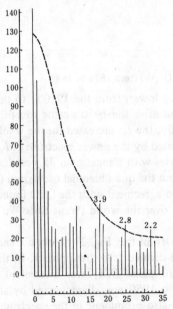

Fig. 4. Power spectrum of SOI_1 (for the years 1871–1980)

The variation of I_s (Fig. 6) shows that the 110-year period started with high I_s, decreased near the turn of the 20th century, kept in high I_s from the 1920s to

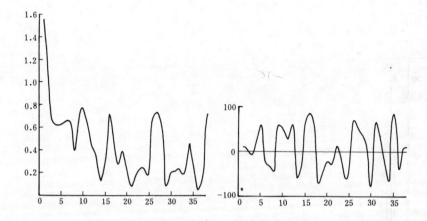

Fig. 5. Coherence (a) and phase spectrum (b) between I_w and SOI_1

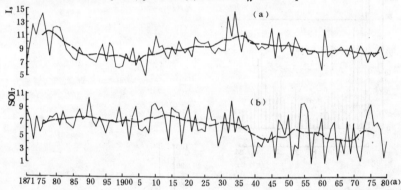

Fig. 6. The variations of I_s (a) and SOI_7 (b) from 1871 to 1980 (broken line is 10 year running mean)

the 1940s, and decreased again after about 1950. That is to say within this period the I_s accomplished nearly two cycles. But the series is too short to give a definite indication of the length of cycle. However, the spectrum of I_s (Fig. 7) throws some light on the matter by showing that the length of cycle may be close to 70 years. Takahahi has found a 70-years cycle in storm damage in Japan(Takahashi 1984). Of course, considering the lower resolution in low frequency band of spectrum, it cannot give a definite answer to the exact length of the cycle. But we believe that the predominated cycle of I_s is rather shorter than that shown in Fig. 2a.

The curve of SOI_7 in Fig. 6b illustrates that the intensity of SOI_7 decreased gradually for the whole 110-year period. Especially, after 1938 it weakend in a significant way, the 15 relative minima in the SOI_7 curve fall fully into this time interval. The spectrum of SOI_7 (Fig. 8) indicates the predominance of the linear trend and QBO. The 3—4 year cycle is not as strong as in January, but there is a relative by weak peak around this frequency band. It suggests that the 3—4 year cycle found recently in ENSO does exist at least for 100 years or mor.

The cross-spectrum analysis between I_s and SOI_7 is given in Fig. 9. It shows that the maximum of coherence is found in the high frequency band. The phase

254

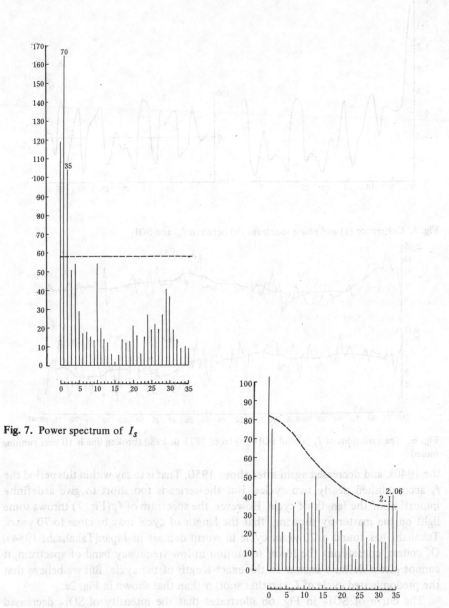

Fig. 7. Power spectrum of I_s

Fig. 8. Power spectrum of SOI_7

spectrum indicates that there is nearly no phase difference. Figure 9 shows that the in-phase variation can be also observed in I_w and SOI_1. It means that the QBO is in-phase not only in July but also in January. In the low frequency band, the I_s has no close correlation with SOI_7. The second peak in coherence appears in the frequency band of a 3—4-year cycle. Considering that the 3—4-year cycle is not strong enough, we can consider that this peak may occur mainly from the 3rd—4th year of SOI$_7$.

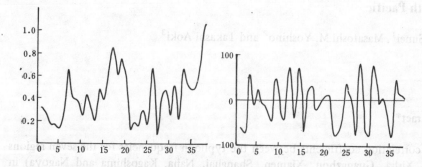

Fig. 9. Coherence (a) and phase spectrum (b) between I_s and SOI$_7$

Conclusions

From the above investigation, one can conclude that
1. The variation in both the monsoon and the Southern Oscillation has an obvious long-term trend; although one cannot exactly give the length of the cycle, it is probably longer than 100 years.
2. There is a prominent QBO in the variations of all circulation indices, except for I_s. This is in good agreement with obvious QBO in climatic variations in China and around the world.
3. For the whole 110-year period examined, the winter monsoon circulation intensified gradually. At the same the SOI$_1$ was enhanced as well.
4. The winter monsoon index changes nearly in-phase with SOI$_1$. The maximum of coherence is found in the low frequency band. But in July the maximum of coherence between I_s and SOI$_7$ fallsn in the high frequency band into the QBO band.

References

Guo Qi-Yun (1983) *Acta Geogr Sin* 38 3: 207–217
Guo Qi-Yun (In press a) The variation of Winter monsoon intensity during 1951–1980 in East Asia and its influence on the climate in China
Tu Chang-Wang (1954) Collected scientific papers METEOROLOGY 1919–1949. Science Press, Beijing, pp 349–393
Guo Qi-Yun (1965) *Collected Papers of Geography* No 9. Science Press, Beijing pp 43–55
Guo Qi-Yun (in press b) Research on the possible machanism connected the variations of summer monsoon in East Asia to the planetary atmospheric circulation
Quinn, W. H., Burt M.V. (1972) *J. Appl Meteorol*, II 616–628
Trenberth, K. (1976) *Q. J. R Meteorol Soc* 102: 639–653
Reed, R.J. (1965) *Bull Am Meteorol. Soc.* 46: 374–387
Wang Shao-Wu, Zhao Zong-Ci`(1978) *Symp.Climatic Change* China. Science Press, Beijing, pp 117–130
Takahashi, K. (1984) Beijing Int. Symp.Climate, collected abstracts. pp I–2

Long-Range Changes in the Correlation Between Typhoon Frequency for the Seven Regions in East Asia and Sea Surface Temperature in the North Pacific

Xie Simei[1], Masatoshi M. Yoshino[2] and Takashi Aoki[3]

Abstract*

The correlation coefficients between the typhoon frequencies for the seven regions (i.e., Xisha, Guangzhou, Xiamen, Shanghai, Naha, Kagoshima and Nagoya) in East Asia and the monthly mean sea surface temperature (SST) from 2 years before to 1 year after are calculated in this paper. The results show that significant correlation appears in the period from the summer of 2 years before to the summer of 1 year before. During this period negative correlations are located in the northwestern part of the North Pacific and positive correlations are located in the southeastern part of the North Pacific. It is found that the positive correlations change into the negative ones in the Equatorial East Pacific in the same year (simultaneous correlation). The high level correlation region 1 year after is found in the area of the Alaska Current. It is also suggested that there are interannual variations of about 1, 1.5, and 2 years for the changes in correlation intensity.

1. National Research Center for Marine Environmental Forecasts, SOA, Beijing, China
2. Institute of Geoscience, University of Tsukuba, Japan
3. Meteorological Reasearch Institute, Tsukuba, Japan

* For details see Acta Oceanological Sinica, Vol. 4 No. 3 pp. 382–394

Section III

Land Surface Processes Related to Climate Variation

Regional Circulation Anomaly Characteristics over Eastern Asia in Relation to the Boundary Forcings

Zhu Baozhen[1] and Song Zhengshan[1]

Abstract − The regional characteristics of climate over Eastern Asia is that the anomalies of 500 hPa circulation are generally small compared with those in other areas in the Northern Hemisphere. The possible physical explanations are as follows: (1) The geographical position of Eastern Asia with respect to the dynamic downstream effect of SSTA in the Atlantic and Pacific Oceans; (2) The "see-saw" pattern induced by the entire polar cap ice cover and the dynamic downstream effect of sea ice anomalies in the Davis Strait and the Bering Sea; (3) The dominant zonal type circulations over Eastern Asia produced by the blocking effects of the Tibetan Plateau.

Introduction

As the general goal of this volume is to make a comparison of climate between China and other regions in the world, We emphasize that one of the regional characteristics of climate over Eastern Asia is that the anomalies of midtropospheric circulation are generally small compared with those in other regions in the Northern Hemisphere.

We consider the monthly time scale and define an anomaly as a deviation from the monthly mean climatic normal. The regional characteristics of climate described above can be found quantitatively from the distribution of wintertime standard deviation of monthly mean geopotential height (Fig. 1). Two main high centers are located over the Pacific and Atlantic Oceans, and the smallest values in East Asia. The purpose of this paper is to find the possible physical causes of these regional characteristics.

Responses of Monthly Mean Anomalies to SSTA

The geographical position of East Asia is close to the Western Pacific Ocean and on the opposite side of the Atlantic Ocean. But it is located on the upstream of the western Pacific and downstream of the Atlantic. It is reasonable to consider that the response to SSTA depends on the geographical dynamical factors. In the first part of this paper, the primary aim was to see whether worthwhile correlations are well defined with Atlantic or Western Pacific.

We consider SSTA anomalies to be significant on monthly mean charts, if they cover a large extent ($\geq 20°$ long. in mid-latitudes) and the magnitude of the anomaly centers is large enough ($\sim 2°$ C). The ten winter months with such a signi-

---

1. Institute of Atmospheric Physics, Academia Sinica, Beijing, China

ficant positive SSTÁ in the Atlantic and the six winter months in West Pacific were first determined from monthly SSTA charts during 1949–1979 (Table 1). We found that the months of warm SSTA in the Atlantic area did not overlap with the months of warm SSTA in the West Pacific.

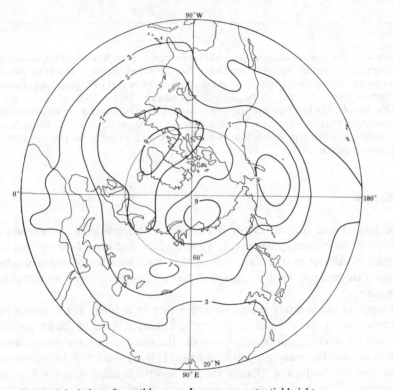

Fig. 1. Standard deviation of monthly mean January geopotential height

Table 1. Months of wintertime warm SSTA

Atlantic		West Pacific	
Jan.	1951	Jan.	1954
Jan.	1952	Jan.	1956
Jan.	1953	Jan.	1973
Feb.	1951	Feb.	1956
Feb.	1952	Feb.	1979
Feb.	1953	Dec.	1952
Dec.	1951		
Dec.	1952		
Dec.	1953		
Dec.	1969		

Then the corresponding composite mean 500 hPa geopotential height anomalies were constructed. The simultaneous relations between circulation anomaly and SSTA during winter seasons are then studied by comparing them. The 500 hPa composite map for warm SSTA in the Atlantic Ocean (Fig. 2) shows strong positive anomalies over the Atlantic Ocean with negative anomalies to the north over Greenland. The second positive center is located over the Ural region together with a weak negative center in the south and relatively strong negative anomalies over the West Pacific. It is very interesting that the patterns of the atmospheric circulation during this warm episode in the Atlantic are similar to the PNA pattern induced by a warm episode in Equatorial SSTA in the Pacific Ocean reported by Horel and Wallace (1981).

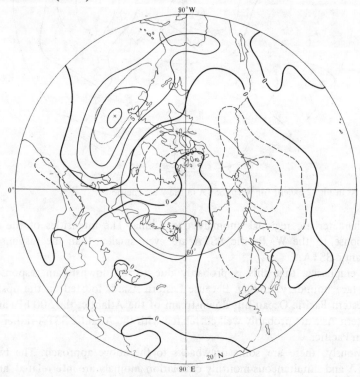

Fig. 2. Composite mean 500 hPa geopotential height anomalies for warm SSTA in the Atlantic. Contour interval is 2.5 dam. Dotted line indicates a sketch of warm SSTA

The composite map for the warm SSTA in the Western Pacific Ocean (Fig. 3) shows huge positive anomalies over the Pacific Ocean with a negative center over western coast of North America, a second pair of centers over Eastern America and the Atlantic Ocean, and a third couple of centers over Europe.

It is very interesting:
(1) that the anomaly over the Western Coast of N. America in the case with the warm SSTA in the W. Pacific is stronger than in the Atlantic case; and
(2) that both of the two composite maps show small deviation values and therefore

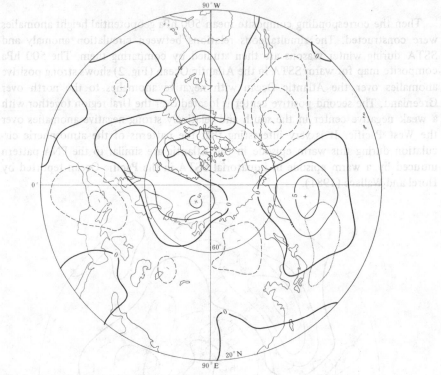

Fig. 3. As in Fig. 2 except for warm SSTA in the West Pacific

a characteristic regional anomaly over China. The intensities of the anomaly response to the W. Pacific SSTA are even smaller than the response to the Atlantic SSTA.

We emphasize that this is probably due to the downstream response to the sea surface boundary forcing. Because Eastern Asia is located on the upstream of the Western Pacific Ocean but downstream of the Atlantic, the 500 hPa anomalies in Eastern Asia are probably well correlated with the Atlantic SSTA rather than the Western Pacific.

Obviously, there are some drawbacks to the above approach. The two fields of SSTA and simultaneous monthly circulation anomaly are interrelated, and therefore with this approach it has been difficult to interpret the cause and effect between them. However, the rate of change of SSTA is much slower than the accompanying atmospheric circulation, so we may consider SSTA as a boundary forcing the circulation anomalies, and that the SSTA results in the anomaly of the atmosphere.

Hoskins and Karoly (1981) used a linearized primitive equation model on a sphere to study the steady solutions for the thermal forcings. For an isolated circular thermal forcing at $45°$ N, the balance is by zonal advection with the warmest air to the east of the source. There are wave trains of alternating positive and negative geopotential height anomalies emanating from the heat source. The distribution of geopotential anomalies shown above has a strong qualitative resemblance to their results.

The Possible Effects of Sea Ice Cover

Sea ice in the arctic area plays an important role in the climate anomalies. It acts to exchange sensible and latent heat between the atmosphere and the ocean, and also has a significant effect on the radiation budget of the atmosphere through the ice albedo feedback. Because the sea ice anomalies are themselves forced to a large extent by atmospheric anomalies, it has been difficult to argue that a climate feature that correlated with an ice anomaly is the cause rather than an effect. However, the ice anomalies do have high monthly persistence, it is reasonable to consider it as a surface forcing to the climate anomalies. Therefore we also examined whether the regional characteristic anomaly over East Asia described above is linked to the sea ice anomaly.

First, we compared the responses of climate to a significant heavy ice extent in Davis Strait, the Bering Sea, and the entire polar cap.

Table 2 shows the months of wintertime sea ice extremes (1953–1977) taken from Johnson (1980) and Overland and Pease (1982). To examine sea ice extremes, the five Januaries of max. sea ice area over the Davis Strait and five Januaries and five Februaries of max. ice area over the Bering Sea were first determined. Then the 500 hPa anomaly patterns were averaged for the heavy ice forcing in different seas.

Table 2. Months of wintertime sea ice extremes

Davis Strait	Bering Sea	Entire polar cap
Jan. 1957	Jan. 1958	Jan. 1958
Jan. 1972	Jan. 1970	Jan. 1959
Jan. 1973	Jan. 1971	Jan. 1971
Jan. 1974	Jan. 1976	Jan. 1972
Jan. 1975	Jan. 1977	Jan. 1973
	Feb. 1958	
	Feb. 1961	
	Feb. 1964	
	Feb. 1976	
	Feb. 1977	

Figures 4 and 5 are the 500 hPa geopotential height anomalies for heavy sea ice in the Davis Strait and Bering Sea respectively. The chart for heavy sea ice anomaly (SIA) in the Davis Strait shows a strong negative anomaly over the Greenland area with a positive anomaly extending over Western Europe. The second negative center is located in central Asia with two relatively weak positive center over East Asia. The chart for heavy SIA in the Bering Sea shows a strong negative center over the Aleutian Islands with a positive center over Western America and a second negative area extending from the Norwegian Sea to Florida.

It is worth pointing out that both of the two 500 hPa anomaly charts show

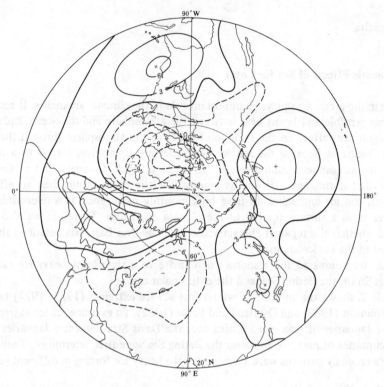

Fig. 4. Composite mean 500 hPa geop. height anomalies for SIA in Davis Strait. Contour interval is 3 dam

small deviation values over China, and the intensity of the anomalies corresponding to Bering Sea heavy ice is even smaller than that corresponding to the Davis Strait case, like the response to SSTA of the Western Pacific and Atlantic. This suggests that the downstream response of the atmosphere to SIA could be another possible dynamical interpretation.

Figure 6 is the 500 hPa anomalies for heavy sea ice years over the entire polar cap. The pattern is quite different form those of Figs. 4 and 5. A distinguished zonal type anomaly constrasts to the wave types of Figs. 4 and 5. A huge circumpolar negative belt is at high latitudes only with a very narrow positive region extending over North Europe. The two broad positive areas south of the negative belt form a "see-saw" pattern. All of the zonal type anomalies are probably favorable to the formation of minor anomaly characteristics over East Asia.

This is roughly consistent with the numerical experimental results of Herman and Johnson (1978). The variations of the fixed sea ice boundary in GCM cause a statistically significant difference in the monthly mean model climate. Figure 7 is reproduced from Herman and Johnson. It is the difference between the 300 hPa gopotential heights corresponding to max. and min. ice conditions respectively. The results indicate

(1) that the anomaly centers are located in the Atlantic and Pacific Ocean, and the anomalies are usually small over East Asia; and

Fig. 5. As in Fig. 4 except for SIA in the Bering Sea

(2) that the anomaly pattern is rather zonal, with a belt of negative anomalies at high latitudes when there is a maximum ice cover.

All of these are qualitatively consistent with our results. However, these authors report a difference between geopotential heights corresponding to the max. and min. sea ice, the intensity of the anomalies higher than ours.

The Dynamic Effects of the Tibetan Plateau

From a statistical study of the Northern Hemispheric daily circulation types of the mid-troposphere, it is found that zonal-type circulations prevail in Eastern Asia (Luo et al. 1984).

The large-scale circulations over the northern hemisphere were divided into 3 types (W, C and E) by Гирс (1971). Features of W and C type circulation are that the zonal type circulation predominates in East Asia, but the meridional one in North America. Features of E-type circulation are different, meridional type circulation develops both in Asia and North America.

Using climatological data in 1891—1966 we find that the total number of days per year for the W-and C-type circulations is generally 180—270 days and the maximum may be more than 300 days in a specific year, while the total number of

266

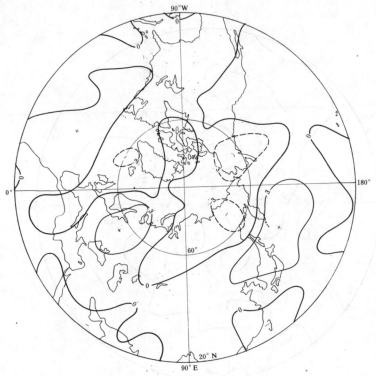

Fig. 6. As in Fig. 4 except over the entire polar cap

E-type circulation is only 90—180 days per year (see Fig. 8). This demonstrates that in East Asia zonal type circulation is much more frequent than meridional type. This is mainly due to the fact that large amplitude troughs generally show filling as they move from the west to China as a result of the blocking effect of the Tibetan Plateau. Staff members (1958) gave the first account of the observational fact. We will discuss the dynamic effects of orography on atmospheric circulation by using numerical and theoretical models.

Firstly, the numerical model used in this paper is a three-level primitive equation model in δ-coordinates. Orography, surface friction, and the horizontal diffusion effects are included.

We made a study on 48 hours numerical prediction with and without orography for a real case starting from 12 GMT, April 19, 1974. Figure 9 shows 48-hour 500 hPa forecast. It is seen from Fig. 9 that the intensity of the forecasted trough over the Tibetan Plateau without mountain effects is too strong. It extends into middle Xingjiang (Fig. 9a). However, the forecasted trough with orographic effects is weakened markedly and the speed of the trough slows down, leading to the formation of rather zonal type circulation with short waves (Fig. 9b), which is in agreement with observations.

Secondly, a theoretical approach was also made by Zhang and Zhu (1984) to study the formation of the short wave along the northern slope of the Tibetan plateau.

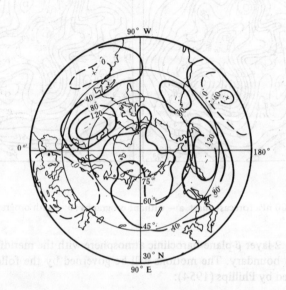

Fig. 7. The 300 hPa geop. height difference (gpm) corresponding to max. minus min. ice conditions by Herman and Johnson (1978)

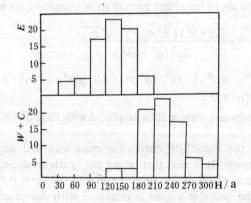

Fig. 8. Frequency distribution for E and W+C circulation types

The Tibetan plateau is the topography with a major axis oriented in west-east direction and ·a sharp north-south slope, so we consider small-amplitude, linear

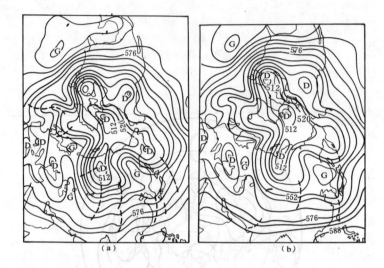

Fig. 9. 48 h 500 hPa forecast chart. a — without orography; b — with orography

motions in a 2-layer β-plane baroclinic atmosphere with the meridional slope effect at the lower boundary. The motions will be governed by the following system of equations used by Phillips (1954):

$$\left(\frac{\partial}{\partial t} + \bar{u}_1 \frac{\partial}{\partial x}\right) [\Delta\phi'_1 - \lambda^2 (\phi'_1 - \phi'_3)] + [\beta + \lambda^2 (\bar{u}_1 - \bar{u}_3)] \frac{\partial\phi'_1}{\partial x} = 0$$

$$\left(\frac{\partial}{\partial t} + \bar{u}_3 \frac{\partial}{\partial x}\right) [\Delta\phi'_3 + \lambda^2 (\phi'_1 - \phi'_3)] + [\beta - \lambda^2 (\bar{u}_1 - \bar{u}_3) + H'] \frac{\partial\phi'_3}{\partial x} = 0$$

where $H' = 2f_0 a h'/H$ $h' = \partial h/\partial y$, the constant north-south slope of large-scale mountains. From the above linearized perturbation equations, we have the complex wave velocity

$$C - \bar{u} = \frac{(2\beta + H')(a^2 + \lambda^2) \pm \sqrt{A}}{2a^2 (a^2 + 2\lambda^2)}$$

where

$$A = 4\beta^2 \lambda^4 - 4\alpha^4 u_T^2 (4\lambda^4 - \alpha^4)b + 4H' [\beta\lambda^4 - \alpha^2 (\alpha^2 + 2\lambda^2)(\lambda^2 - \alpha^2) u_T]$$
$$+ H'^2 (\alpha^2 + \lambda^2)^2$$

Evidently, the perturbation wave will be amplified with time if $A < 0$, and neutral if $A < 0$.

Figure 10 gives the instability curves for cases with and without orographic slope effect respectively. It is seen that, along the northern slope, the presence of orography destabilizes the flow for short waves and stabilizes it for longer waves, and further that the unstable region is increased with decreased slope gradient. It is also noticed that baroclinic instability and orographic instability have different preferred length scales, with the maximum growth rate for baroclinic instability occuring at wave length 6000 km, and that for orographic instability near 3000 km. But in the case of the southern slope, an entirely different situation occurs. As the dashed line in Fig. 9 shows, the unstable regions are strongly reduced, thus the orographic instability hardly occurs on the southern slope. Thus we emphasize that the formation and development of the short wave disturbances along the northern periphery of the plateau may be partly described as a kind of orographic slope instability.

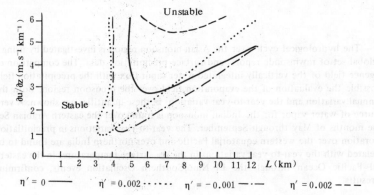

Fig. 10. Instability curves with and without mountain slope

From the above numerical experiment and theoretical approach, it can be concluded that the predominate zonal type circulation and short wave development over China are closely connected to the dynamical blocking effect of the Tibetan Plateau. This is most favorable for the formation of small monthly anomalies over China.

Summary

In summary, the possible explanations of the rather small anomaly characteristics of geopotential height over Eastern Asia may be as follows:
(1) the blocking effects of the Tibetan Plateau;
(2) the downstream dynamic effect of geographical characteristic position of China with respect to the SSTA of the Atlantic and Pacific Oceans; and
(3) the feedback effects of the geographical polar ice cover.

References

Herman G. F., Johnson, W. T. (1978), *Mon Wea Rev*, 106 12: 1649–1664
Horel, J. D., Wallace J. M. (1981), *Mon Wea Rev* 109 4: 813–829
Hoskins, G. J., Karoly D.J. (1981) *J Atmos Sci* 38, 6: 1179–1196
Johnson C. M. (1980), *Mon Wea Rev* 108 11: 1774–1781
Luo Meixia, Zhu Baozhen, Zhang Xuehong (1983): *Sci Atmos Sin* 7, 2: 145–152 (In Chinese)
Overland, J. E., Pease C. H. (1982), *Mon Wea Rev* 110 1: 5–13
Phillips. N. A.(1954), *Tellus* 6, 3: 273–286
Staff members (1958), Academis Sinica. *Tellus* 10, 1: 58–75
Zhang Shuhua, Zhu Baozhen (1984) The dynamic effect of Tibetan Plateau slope and cumulus convection on the large-scale perturbation. Collected papers of QXPMEX, vol 2. Science Press, Beijing, pp 245–252 (In Chinese)

Variability of the Hydrological Cycle in the Asian Monsoon Region

A. H. Oort[1]

Abstract — The hydrological cycle over the Asian monsoon region is investigated by using a 15-year global set of rawinsonde reports and surface precipitation data. The comparison of the divergence field of the vertically integrated water vapor flux with the precipitation field makes possible the evaluation of the evaporation field over the monsoon region. Both the normal annual variation and the year-to-year variations in these quantities are shown. A very strong source of water vapor for the Indian monsoon is found over the eastern Arabian Sea during the months of May through September. The year-to-year variations in precipitiation and evaporation over the western equatorial Pacific and over northern India are found to be well correlated with the year-to-year variations in the sea surface temperatures in the eastern equatorial Pacific Ocean through the El Nino/Southern Oscillation events, confirming previous results.

Introduction

Water substance is one of the major elements in the climatic system on earth. It passes through the various subsystems of the climatic system, such as the atmosphere, oceans, cryosphere, biosphere, and land surfaces, in a continuous cycle of evaporation, condensation, and precipitation. In a sense, the hydrological cycle links the climatic system together. Water is also, of course, of crucial importance to the existence of life on earth.

In this paper, we will explore what can be learned about the hydrological cycle using principally the aerological observations from the global rawinsonde network. In other words, we will explore the atmospheric branch of the hydrological cycle and try to infer from this some new facts about the less-known terrestrial branch. Our main emphasis will be on the hydrological cycle over the monsoon region of Southeast Asia and its seasonal and interannual variations. We will show that the aerological method can be used successfully to define the sources and sinks of water for the monsoon. For a more extensive account of the global aspects of the hydrological cycle the reader is referred to Peixoto (1973) and Peixoto and Oort (1983).

The geographical distribution of the rawinsonde stations used in our studies is given in Fig. la. The figure shows good coverage over the continents and the oceans in the Northern Hemisphere, except for the southeastern North Pacific and the tropical Atlantic Oceans. Large data-void areas are found over the southern oceans, especially south of 30°S. The monsoon region of Southeast Asia and its subdivision into various subregions to be used later in this paper are also indicated on the map.

---

1. Geophysical Fluid Dynamics Laboratory/NOAA, Princeton University, Princeton, New Jersey 08542 USA

For a further description of the data and their reduction, see Oort (1983).

The meteorological network of surface stations where precipitation is measured is given in Fig. 1b. It shows a coverage over land that is even more dense than the upper air network.

Formulation of the Problem

The water cycle in the atmosphere can be studied using the principle of conservation of water substance for a vertical column of unit area extending throughout the depth of the atmosphere:

$$\frac{\partial W}{\partial t} = - \operatorname{div} \mathbf{Q} + E - P,$$ (1)

where $W = \int_0^{p_0} q\,dp/g$ = precipitable water in the column,

$$\mathbf{Q} = (Q_\lambda, Q_\phi) = (\int_0^{p_0} qu\,dp/g, \int_0^{p_0} qv\,dp/g)$$

= integrated flux vector of water vapor,
E = evaporation rate at the earth's surface,
P = precipitation rate at the earth's surface.
The other symbols used are:
g = acceleration due to gravity,
p = pressure,
p_0 = surface pressure,
q = specific humidity,
u = zonal wind component (positive if eastward),
v = meridional wind component (positive if northward).
In Eq. (1), some small terms involving the rate of change of liquid water with time, and its advection are neglected (Peixoto 1973). Figure 2 illustrates the balance between the various terms in Eq. (1) in a schematic way.

If we average Eq. (1) in time over a period (T) of a month or longer the time rate of change becomes small compared to the other terms (less than a factor of 10), and Eq. (1) can be written in the form:

$$\operatorname{div} \overline{\mathbf{Q}} = \overline{E} - \overline{P} ,$$ (2)

where

$$(\overline{}) = \int_Q^T ()\,dt/T = \text{time average.}$$

Eqation (2) shows that the computation of the divergence field of the integrated water vapor flux in the atmosphere also provides an estimate of evaporation-minus-precipitation rates ($E-P$). This last parameter ($E-P$) forms the link between the atmospheric and terrestrial branches of the hydrological cycle. The terrestrial branch includes river and subterranean runoff over land, net fresh water fluxes in the oceans, and water storage in both land and oceans.

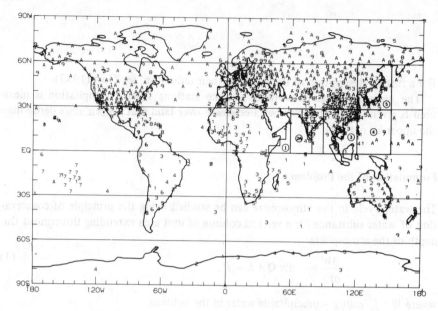

Fig. 1a. Distribution of upper air stations used in the mean January analyses at 500 hPa during the 1964–1973 period, and the number of years of observations available, ranging from 1 to 10 (=A) years. The total number of stations over the globe for January was 1093. Several geographical subregions in the Southeast Asian monsoon area are also shown for later reference

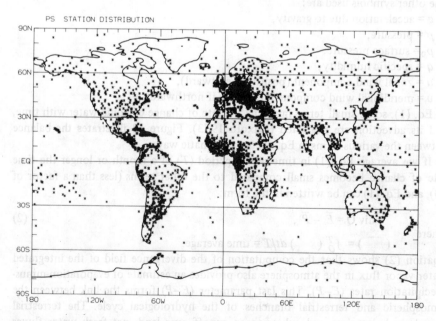

Fig. 1b. Typical distribution of surface stations available for precipitation analyses, such as those by Jaeger (1976)

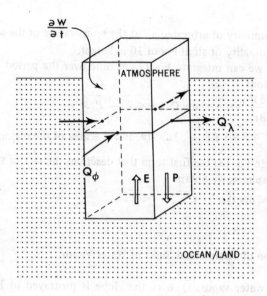

Fig. 2. Schematic diagram of the terms in Eq. (1) describing the atmospheric branch of the hydrological cycle. W is the precipitable water in the column, $Q\lambda$ and $Q\phi$ are the eastward and northward components of the vertically integrated flow of water vapor in the atmosphere E is the surface evaporation rate, and P the surface precipitation rate.

If for a certain region of the earth's surface $E > P$, the region acts as a source of atmospheric water vapor, whereas if $E < P$ the same region would act as a sink of atmospheric water vapor.

Knowing the water vapor flux divergence field from atmospheric data and the precipitation field from the (dense) meteorological surface station network, one can, in principle, compute the field of evaporation. This approach to obtain estimates of evaporation is attractive because the large scale evaporation rate cannot be measured directly over land by any known method. The only direct and reliable determinations of evaporation have been made so far in a few locations by measuring the vertical velocity (w) and humidity fluctuations simultaneously

$$E = \rho \, \overline{w'q'} \, ,$$

where ρ = density of the air. These measurements are usually made from meteorological towers over land or in shallow water. Over the oceans, evaporation can be estimated from ship reports using a bulk aerodynamic method. However, this last approach also involves many uncertainties.

$$E = \rho \, c_E \, \overline{V(q_s - q_a)} \, ,$$

where c_E = exchange coefficient for water vapor,

V = horizontal wind speed at about 10 m height,

q_s = specific humidity of saturated air at the temperature of the sea surface,

q_a = specific humidity of air at about 10 m height.

As a final step we can integrate Eq. (1) in time over the period T and in space over a specific region:

$$< \frac{\partial W}{\partial t} > = - < \text{div } \overline{Q} > + < \overline{E} - \overline{P} > , \tag{3}$$

where $\quad <(\quad)> = \iint\limits_{A} (\quad) \, dx \, dy$ and A = area of the region.

Again, neglecting in Eq. (3) the first term that describes the rate of storage of water vapor in the atmosphere, we obtain

$$< \text{div } \overline{Q} > = < \overline{E} > - < \overline{P} > \tag{4}$$

Global Distribution of Sources and Sinks of Water Vapor

The net flow of water vapor, $\overline{Q}$, over the globe is protrayed in Figs. 3a and 3b for the months of January and July, respectively. The flow is indicated by both streamlines and arrows. The number of barbs on the shaft of an arrow gives the strength of the local water flux.

The overall impression one gains from studying Fig. 3 is one of zonal flow, westward in the tropics and eastward in middle latitudes. As expected, this latitudinal distribution resembles the distribution of the wind in the lower troposphere because the air below 500 hPa contains more than 90% of the water vapor in the atmosphere, There are seasonal shifts in these zones, northward during Northern Hemisphere summer and southward during Northern Hemisphere winter, with summer-winter differences on the order of 10° to 15° latitude. The main departures from zonal symmetry are found over the Southeast Asian monsoon region, as we shall discuss later.

. Looking now in more detail at the regions where the streamlines diverge or converge, we notice that the source regions (div Q > 0 or $E > P$) occur mainly over the subtrpoical oceans, while the sink regions (div Q < 0 or $E < P$) are found in the Intertropical Convergence Zone (ITCZ) near the equator and (less clearly) over the cyclone tracks in middle and high latitudes.

The zonally averaged values of the actual divergence of the Q fields for the year and the extreme seasons are plotted in Fig. 4. This figure shows clearly where the sources and sinks of water vapor are found and how they shift latitudinally with the seasons. The strongest source region with a net evaporation-minus-precipitation rate of more than 100 cm/yr is found in the subtropics of the winter hemisphere, and the strongest sink (on the order of 60–80 cm/yr) in the ITCZ of the summer hemisphere, We should add that the profiles south of about 40°S are unreliable because of the sparseness of the rawinsonde stations at these latitudes, as is clear from an inspection of Fig. 1a.

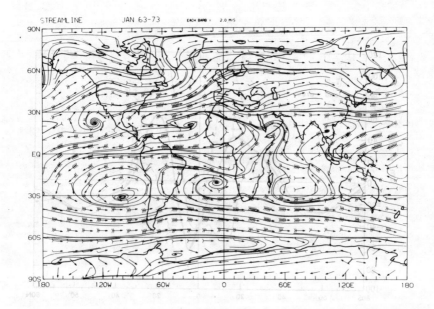

Fig. 3a. Global distribution of the total flow vector of water vapor, Q, during January and some corresponding streamlines; each barb represents an average transport, $\overline{Q}/p_0$, of $2\,\mathrm{ms}^{-1}\mathrm{gkg}^{-1}$

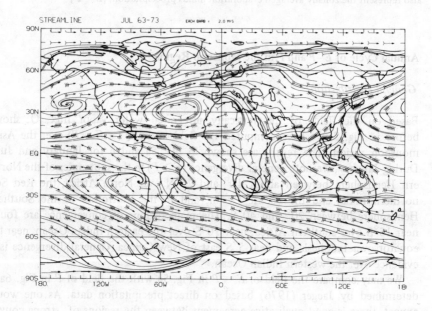

Fig. 3b. As in Fig. 3a, except for July

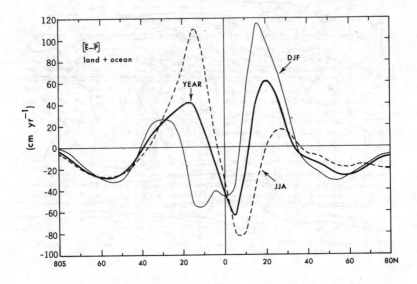

Fig. 4. Meridional profiles of the zonally averaged divergence of the water vapor transport, [div $\overline{Q}$], in cm yr^{-1} for the annual mean (thick solid), northern winter (thin solid) and northern summer (dashed) from Peixoto and Oort (1983). According to Eq. (2), the profiles also represent the zonally averaged evaporation minus proecipitation, [$\overline{E} - \overline{P}$]

Annual Cycle of E-P and P in the Monsoon Region

General Distribution

Based on the fields of the vertically integrated water vapor flux, Q, shown before in Fig. 3, we have computed the field of div Q ($\cong E - P$) over the Asian monsoon region. The results are shown in Fig. 5a and b for January and July. During January, we note general divergence ($E > P$) in the subtropics of the Northern Hemisphere between about 10° and 25°N over North Africa, the Red Sea, northern India, and the Pacific Ocean, and in the subtropics of the Southern Hemisphere between 20° and 40°S. The main regions of convergence are found near 15°S over southern Africa, near 10°S over the eastern Indian Ocean, near the equator over Indonesia and near 5°S east of New Guinea. General divergence is in evidence over the western Indian Ocean.

We may compare the field of $E - P$ in Fig. 5a with the field of P in Fig. 6a as determined by Jaeger (1976) based on direct precipitation data. As one would expect, there is good qualitative agreement between the regions of strong convergence ($E < P$) and the precipitation ($P > 0$) maxima. Adding the two fields together we obtain in Fig. 7a a map of the inferred evaporation rate, E, for January. Since E

is a positive definite quantity, no negative areas should appear in Fig. 7a, provided the data were perfect. In other terms, the negative areas give some indication of how large an error was made in the evaluation of the evaporation rate. There does not appear to be an easy, acceptable way to avoid the regions with negative evaporation rates. Many further investigations have to be made on the nature of the uncertainties in estimating the basic quantities, div Q and P. Obviously, there is a problem over the eastern Indian Ocean near the longitude of 80°E where our water vapor convergence is too intense ($E \ll P$) compared with Jaeger's precipitation values. We should point out that there is a lack of both upper air and surface data south of the equator between about 60° and 90°E. Nevertheless, overall the January evaporation map seems reasonable with strong evaporation over central Africa, the western Indian Ocean, northern India, Sumatra, Malaysia, the Philippines, New Guinea and over the Pacific Ocean near the latitudes of 15°N and 10°S. Our Fig. 7a shows much more intense centers and higher values than the evaporation maps derived using bulk aerodynamic methods by, e.g., Budyko (1956).

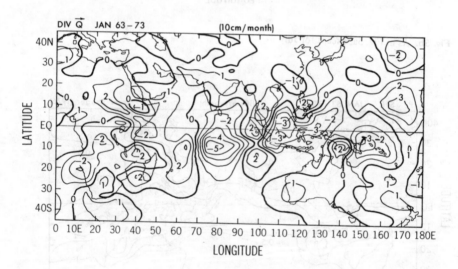

Fig. 5a. Map of the div $\overline{Q}$ or $\overline{E} - \overline{P}$ field over the Asian monsoon region during mean January conditions in units of 10 cm/month

Of more interest than the January maps are probably the summer monsoon maps of $E - P$, P and E displayed in Figs. 5b, 6b and 7b. Again, there is overall agreement between the centers of strong convergence in Fig. 5b and Jaeger's (1976) precipitation maxima in Fig. 6b, especially over Southeast Asia and the New Guinea area. However, there is disagreement over the Somali area where we find sizeable convergence and low precipitation amounts, and also near the west coast of India where we find strong divergence ($E > P$) and high precipitation rates. These discrepancies must be mainly due to the sparse distribution of the rawinsonde stations (see Fig. 1a) and possibly of the rainfall stations (Fig. 1b) as well as due to some

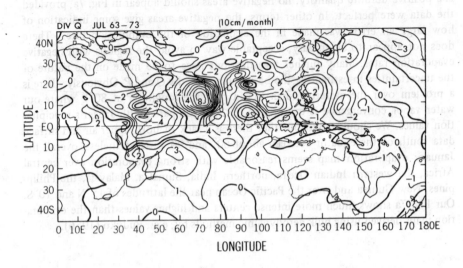

Fig. 5b. As in Fig. 5a, except for July

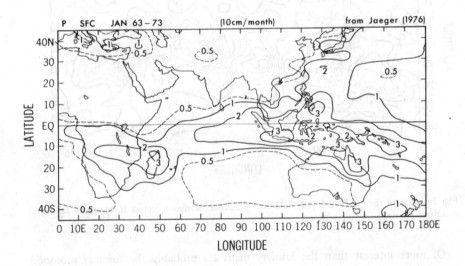

Fig. 6a. Map of the $\bar{P}$ field over the Asian monsoon region during mean January conditions in units of 10 cm/month after Jaeger (1976)

spatial smoothing in our analysis procedure. Nevertheless, some very interesting features show up:

(1)   A very strong center of divergence of water vapor and evaporation over

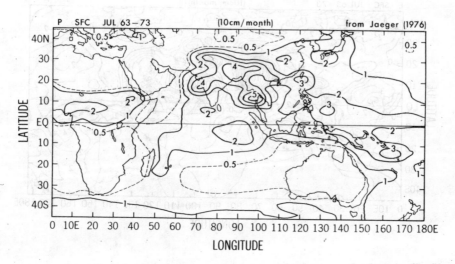

Fig. 6b As in Fig. 6a, except for July

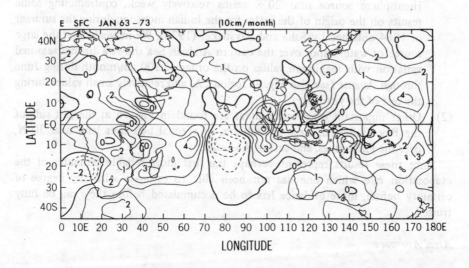

Fig. 7a. Map of the $\overline{E}$ field over the Asian monsoon region during mean January conditions in units of 10 cm/month, computed as the sum of the $\overline{E} - \overline{P}$ and $\overline{P}$ fields in Figs. 5a and 6a. The areas where the computed values of E are negative must be in error, because the evaporation rate, E, is a positive, definite quantity

the eastern Arabian Sea (see Figs. 5b and 7b). Apparently, this region is the dominant source of wate vapor for the Indian monsoon, confirming an earlier conclusion of Pisharoty (1965). In comparison, the Southern

280

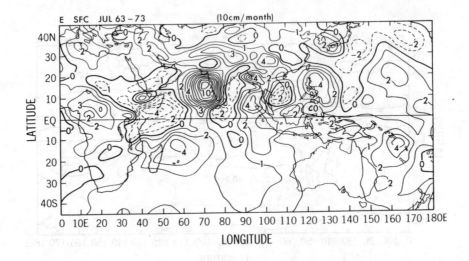

Fig. 7b. As in fig. 7a, except for July

Hemisphere source near 20°S seems relatively weak, contradicting some results on the origin of the water for the Indian monsoon during the summer of 1964 obtained by Saha and Bavadeker (1973). We may add that the large source of water vapor over the eastern Arabian Sea shows a distinct seasonal variation with maximum values on the order of 100 cm/month during June through August (see also Fig. 9a region 2A) and relatively small values during the rest of the year.

(2) Other important evaporation areas are found in Fig. 7b at the east side of the Bay of Bengal, west of Sumatra, in a zonal belt over land near 30°N, over the Philippines, and east of Australia.

Since these results represent perhaps the first more reliable estimates of the evaporation rate which one has not been able to estimate with any degree of certainty before, more evidence has to be accumulated before they can be fully trusted.

Areal Averages

To better study the various hydrological regimes and their seasonal variation we have broken the monsoon region down into seven subregions as shown in Fig. 8. Averaging the −div Q and P fields over these regions for the 12 calendar months yields the profiles shown in Fig. 9. One consistency requirement is again that $P - E < P$. Obviously, this condition is violated over the west Arabian Sea during the May–November (half-year) and over the western Bay of Bengal during the November–January season. For the other regions and months the variations in $P - E$ and P seem reasonable and consistent. The highest rainfall rates are found over northern

India (region 2C) during summer. For the year as a whole, the western Bay of Bengal, the South China Sea and the Philippine Sea areas are the wettest areas and show the strongest convergence rates.

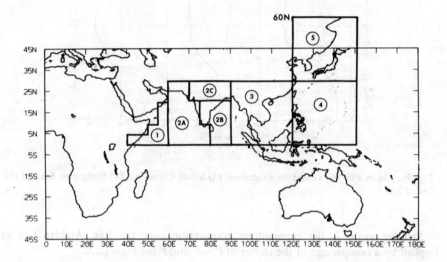

Fig. 8. Breakdown of the monsoon region into seven subregions. The areas covered by regions 1, 2A, 2B, 2C, 3, 4, and 5 are 4.41, 7.86, 4.82, 3.49, 13.39, 13.39, and 9.80×10^{12} m^2, respectively

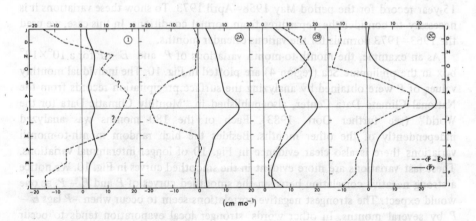

Fig. 9a. Annual cycle of $\bar{P}$ (solid) after Jaeger (1976) and $\bar{P} - \bar{E}$ (dashed) from our 1963 – 1973 data for the monsoon subregions (1) west Arabian Sea, (2A) east Arabian Sea, (2B) west Bay of Bengal, and (2C) northern India, in units of cm/month. If for a certain subregion or month $\bar{P} - \bar{E} > \bar{P}$ the values are incompatible; either one or both values must be in error since E should be positive

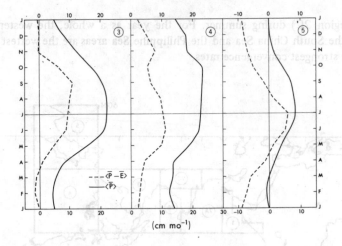

(cm mo^{-1})

Fig. 9b. As in Fig. 9a, except for subregions (3) South China Sea, (4) Phillippine Sea and (5) Sea of Japan

Striking again is the tremendous evaporation over the East Arabian Sea as implied by a comparison of the curves of $P - E$ and P for region 2A.

Interannual Variations in $E–P$ and P in the Monsoon Region and Connections with ENSO Events

The year-to-year variations in monsoon activity can also be studied from our 15-year record for the period May 1958–April 1973. To show these variations it is necessary to consider the departures from normal conditions. In this case, we used the 1963–1973 normals for the various calendar months.

As an example, the month-to-month variations of P and $E - P$ for a $10° \times 10°$ box in the philippine Sea (region 4) are plotted in Fig. 10. The individual monthly values of P were obtained by analyzing the surface precipitation records from the National Climate Data Center, also published in "Monthly Climatic Data for the World" (see further Oort, 1983). Each of the 180 months was analyzed independently of the other months. Besides the high random month-to-month variations there is also clear evidence in Fig. 10 of longer interannual variations. These last variations are more evident in the smoothed curves in Fig. 10. We notice a strong negative correlation between the smoothed curves of P and $E - P$, as one would expect. The strongest negative correlations seem to occur when $-P$ lags $E - P$ by several months. In other words, stronger local evaporation tends to occur when the local precipitation decreases and weaker evaporation when the precipitation increases. This effect can probably only be explained when taking into account the large-scale interactions with other regions, such as would occur during variations in the east-west Walker Circulation. In fact, the interannual signals in region 4 must be directly associated with El Nino/Southern Oscillation (ENSO) events. In this

connection, the sea surface temperature (SST) in the eastern equatorial Pacific
Ocean near 130°W can be thought of as a key parameter characterizing the ENSO
phenomenon (see e. g., Pan and Oort, 1983). The month-to-month SST variations
for the key region are reproduced in Fig. 11a. The smoothed SST curve is almost
perfectly correlated with the smoothed curve of $-P$ over the Philippine Sea as shown
in Fig. 11b. Apparently, the warmer SST in the key region goes together with
decreased rainfall in the western tropical Pacific and vice versa. This effect is
connected with a weakened east-west Walker Circulation over the equatorial Pacific
Ocean during ENSO events as was demonstrated by, e.g., Rasmusson and Carpenter
(1982).

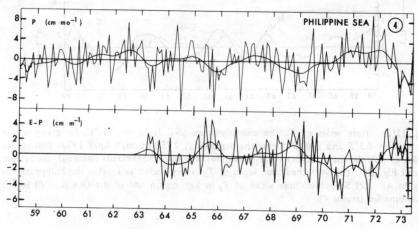

Fig. 10. Time series of the monthly mean departures from climatology of P (top) and $E - P$
(bottom) for region 4 in units of cm/month for the period May 1958 through April 1973.
There is a clear negative correlation between the two curves after smoothing. The normal
annual cycle was removed in each time series. The markings along the abscissa indicate the
January of a particular year. The smoothed lines were obtained by applying a 15-point
Gaussian type filter with weights 0.012, 0.025, 0.040, 0.061, 0.083, 0.101, 0.117 and 0.122
at the central point.

For our present purpose, it is encouraging to note that there is a good correspondence (although with some lag) shown in Figs. 11b and 11c between the $-P$
curve computed from surface precipitation records and the $(E - P)$ curve computed
from the upper air divergence, even on an interannual time scale.

The same analysis was also performed for the other six monsoon regions. In
general, a similar correlation was found between $E - P$ and $-P$ except over the
Arabian Sea, where evaporation and not precipitation seems to dominate the
hydrological cycle. Regarding the connection with ENSO events we found, besides
the one described above for region 4, only a clear relation for the region 2c over
northern India, in the sense of a decreased rainfall during ENSO years.

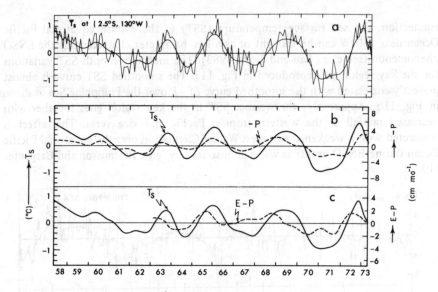

Fig. 11(a) Time series of T_S, the monthly-mean SST departure, in °C, for a key region centered at 2.5°S and 130°W during the period May 1958 through April 1973. This record indicates well the major ENSO events (maxima) and anti-ENSO events (minima). See further legend Fig. 10. **(b)** Smoothed time series of T_S in key region and $-P$ in the Philippine Sea (region 4). **(c)** Smoothed time series of T_S in key region and of div $Q(= E - P)$ in the Philippine Sea (region 4).

Summary and Conclusions

Extensive tests of the applicability of the aerological approach to estimate the $E - P$ field from the divergence of the vertically integrated moisture transport, Q, were described.

Indications are that the results are reasonable and compatible with the direct analyses of the P field by Jaeger (1976) based on surface precipitation records. Combining the $E - P$ and P fields we obtained estimates of the evaporation field, E, over the monsoon region. We suggest that this is the most promising approach to estimate the evaporation rate over both land and oceans. The approach commonly used over the oceans is to estimate $\overline{E}$ using a bulk aerodynamic formulation. However, this method cannot be used over land because of the extremely complex physiography of the land surface, and it is also unreliable over the oceans because of temporal and spatial sampling problems in the ship reports. Of course, sampling problems also play a serious role in the aerological approach. For example, Rasmusson (1968) has shown that over the data-rich area of North America only area integrals over a minimum area of 2×10^6 km^2 show good correspondence with direct surface estimates of runoff and water storage. Therefore to get the full benefit of the aerological approach for regional studies significant improvements in the rawinsonde network have to be made in the region considered.

The eastern half of the Arabian Sea was found to be the most important source of water vapor for the Indian summer monsoon, confirming Pisharoty's (1965) conclusions.

A Strong inverse relationship between the year-to-year SST variations in the eastern equatorial Pacific Ocean and the rainfall (and $P - E$) variations over the philippine Sea and northern India was also confirmed in the present results.

Acknowledgement. I would like to thank the organizers of the Beijing International Symposium on Climate for their great hospitality, and Tony Broccoli, Frank Bryan and Syukuro Manabe for their review of this paper.

References

Budyko, M. I. (1956) *The heat balance of the earth's surface* (in Russian) Gidrometeorologi-cheskoe Izdatel' stvo, Leningrad, 255 pp (Translated from Russian by N.A. Stepanova, Office Tech Serv U S Dept Commerce, Washington, D C 1958)

Jaeger, L., (1976) *Monthly precipitation maps for the entire earth* (in German). Ber Dtsch Wetterdienstes, Vol 18, No 139, 38 pp

Oort, A. H. (1983) *Global atmospheric circulation statistics, 1958–1973.* NOAA Professional paper No 14 U S Government Printing Office, Washington D C, 180 pp aud 47 microfiches

Pan Y. H., Oort, A H (1983), *Mon Wea Rev* 111: 1244–1258

Peixoto, J. P. (1973) *Atmospheric vapour flux computations for hydrological purposes.* Rep No 20, WMO, Geneva, Switzerland, 81 pp

Peixoto, J. P., Oort, A. H. (1983), The atmospheric branch of the hydrological cycle and climate in *Variations of the global water budget.* Reidel, London, England, pp 5–65

Pisharoty, P. R., (1965), *Evaporation from the Arabian Sea and the Indian southwest monsoon.* Proc Symp Met Results, I I O E, Bombay, pp 43–54

Rasmusson, E. M. (1968) *Mon Wea Rev* 96: 720–734

Rasmusson, E. M., Carpenter, T. H. (1982) *Mon Wea Rev* 110: 354–384

Saha, K. R., Bavadekar, S. N. (1973) *Q J R Met Soc* 99: 273–278

Climatic Effect of Snow Melting in High Latitudes in a One-Dimensional Air-Land Model

Ji Jinjun[1]

Abstract— It is shown that there is a clear relationship between the large-scale anomalous snow extent in middle and high latitudes (and Tibetan plateau) and the planetary circulation in Northern hemisphere, and the time of season turning. The aim of this chapter is to investigate the snowmelting process and the local short-term climatic effect of anomalous snow depth by using a one-dimensional air-land coupling model, which involves both heat and moisture balances between the atmosphere and the sublayer of soil. The tmospheric component of the model is constructed by radiation-turbulent and radiation transfer equations and the equation of air humidity. For the sublayer of soil, the equations of soil temperature and moisture are used. The coupling of both components is realized through the budgets of heat and water vapor at the surface. First, simulations of annual cycles of temperature and moisture in the atmosphere and soil, as well as the components of heat and water balances at the surface were performed. Secondly, comparing the experiments with anomalous snowdepths to those in normal case, the time-lag climatic effect was examined. The results indicated that for the experiments with a larger snow depth than normal the soil moisture content is more abundant, the temperature in soil and air are lower. The moisture exchange process is more active, the rates of evapotranspiration and precipitation enhanced, while for those of less snow depth, the soil is drier and the temperature higher, the intensities of evapotranpiration and precipitation weaken. The above climatic effect responds more sensitively to the change in moisture budget process and would last at least to summer.

Introduction

Many authors have investigated the effect of snowcover on climate, especially the feedback between high albedo of snowcover and surface temperature. However, most of them studied the equilibruim state climate using different complicated models, such as sensitivity of climate system to changes in solar radiation. Only a few works (Namias 1963, Hahn and Shukla 1976 and the others) have concerned the influences of snowcover on short-term climate.

Seasonal snowcover responds rapidly to atmospheric dynamics on the time scale of days and longer, however, the time-lag effect of anomalous snowcover would last for a long time. It is shown· clearly that there is a relationship between the large-scale anomalous snow extent in middle and high latitudes (and Tibet plateau) and the planetary circulation in Northern hemisphere, and thus the time of season turning. For example, the large snowcover in winter and early spring postpones the coming of summer circulation. Seasonal heat storage in snow is small; indeed, the climate effect of snow melting is realized through the changes in

---

1. Institute of Atmospheric Physic Academia Sinica, Beijing, China

albedo and soil moisture.

Yeh et al. (1983) have performed a set of experiments on the climatic effect of sudden removal of snowcover in spring with a simplified GCM and the results indicated the anomalous temperature and moisture distributions and corresponding circulations would remain till August. Ji (1984) studied theoretically the time-lag feedback of large-scale precipitation by employing an air-land coupling model and found that a kind of diabatic planetary wave exerted by heat and water vapor budgets between air and land surface is responsible to this time-lag effect.

The attempt of this chapter is to investigate the snow melting process and its local short-term climatic effect of different snow depth. Rather than utilize a complex model we will seek an understanding of the thermal aspect of climatic effect of anomalous snowdepth, a one-dimensional air-land model which involves both heat and moisture exchanges in the system has been developed.

Description of the Model

The radiation-turbulance equation is adopted to determine the atmospheric temperature. The parameterization scheme of radiation transfer in the model used by Kibel (1943) and Charney (1975) is described as follows. The atmosphere is transparent for insolation, and water vapor in air is the only absorptive medium for infrared radiation. The absorptive coefficient of longwave radiation is independent of spectrum. The governing equations can be given

$$C_p\rho\,\frac{\partial T}{\partial t} = C_p\rho\frac{\partial}{\partial z}\,(K\frac{\partial T}{\partial z}) + \alpha_1\rho(A+B-2E) \tag{1}$$

$$\frac{\partial A}{\partial z} = \alpha_1\rho_w\,(A-E) \tag{2}$$

$$\frac{\partial B}{\partial z} = \alpha_1\rho_w(\dot{E}-B) \tag{3}$$

where T, ρ are air temperature and density respectively. C_p is the specific heat of air at constant pressure. A, B are the downward and upward longwave radiation fluxes respectively. α_1 is absorptive coefficient of longwave radiation, E blackbody emission and equal to σT^4, σ Boltzman constant. We assumed that distribution of water vapor density ρ_w reduces exponentially with height

$$\rho_w = \rho_{w_o}\,e^{-z/h}$$

where $\rho_{\omega o}$ is water vapor density at the ground, and h the equivalent height of water vapor distribution. The coefficient of vertical turbulent exchange of heat K varies with height.

The variation of water vapor in air is determined by evapotranspiration and precipitation which occurs when the mixing ratio of the air q exceeds to percent of saturated value q_s, the remainer is precipitated automatically

$$\int_0^\infty \rho\frac{\partial q}{\partial t}\,dz = M_E - P \tag{4}$$

where M_E is the rate of evapotranspiration from the surface and P the rate of precipitation.

At the top of the atmosphere, the downward longwave radiation flux vanishes. The upward longwave radiation from the surface is blackbody emission and the temperature at 20 km height is maintained with time. Thus, the boundary conditions of Eqs. (1)–(3) are

$$A = 0 \qquad z = \infty \qquad (5)$$

$$B = E \qquad z = 0 \qquad (6)$$

$$T = T_\infty \qquad z = \infty \qquad (7)$$

The lower boundary condition of (1) is a heat balance equation which will be described later.

The time-dependent soil temperature and moisture are governed by the equations

$$\frac{\partial T_s}{\partial t} = K_s \frac{\partial^2 T_s}{\partial z^2} \qquad (8)$$

$$\frac{\partial W}{\partial t} = -\rho_{sn} \frac{\partial h}{\partial t} - M_E + P - r \qquad (9)$$

where T_s is the soil temperature, K_s the conductivity of soil, W the available moisture which should be less than or equal to the available moisture capacity of soil W_{max}, otherwise, the runoff r occurs. h is the equivalent water depth of snow-cover and ρ_{sn} the equivalent water density of snow. Assume that the heat transfer vanishes at the bottom of shblayer of soil, i. e.,

$$\frac{\partial T_s}{\partial z} = 0 \qquad z = -D \qquad (10)$$

The heat balance equation at the surface can be written in the following form

$$S(\varphi, \iota)(1 - \alpha(h, W)) + C_p \rho K \frac{\partial T}{\partial z} - C_{ps} \rho_s K_s \frac{\partial T_s}{\partial z}$$

$$+ A - B - LM_E - M_s = 0 \qquad (11)$$

where $S(\varphi, t)$ is the coming insolation at the top of the atmosphere and allows annual variation

$$S(\varphi, t) = S_0(\varphi) + S_1(\varphi) \sin (2\pi t/\Omega + \theta) \qquad (12)$$

where $S_0(\varphi)$ are the annual mean solar radiation and amplitude varying with latitudes. M is the period of one year and θ is an initial phase angle. Albedo α is a function of snowdepth and soil moisture. When the ground is covered by snow, albedo varies with snow depth, whereas it varies with soil moisture during a snow-free period, i. e.,

$$\alpha = 0.60 \qquad\qquad h \geqslant 1 \qquad\qquad\qquad\qquad (13)$$

$$\alpha = 0.20 + 0.4\,h \qquad 1 > h > 0 \qquad (\text{Washington et al. 1977})$$

$$\alpha = 0.25 - 0.1\ W/W_{max} \qquad (W_{max} = 15\ \text{cm}) \qquad h = 0$$

where C_{ps} and ρ_{sn} are the specific heat and density of soil respectively. In Eq. (11), M_s is the heat flux due to snow melting and can be expressed by

$$M_s = -L_f\,\rho_{sn}\,\frac{\partial h}{\partial t} = C^*(T_{so} - 273)/\Delta t \qquad\qquad (14)$$

where T_{so} is the surface soil moisture, C^* the constant 0.23 (Arakawa 1972) Δ_t time step and L_f the latent heat of fusion. The rate of evapotranspiration is calculated by the bulk scheme

$$M_E = -\beta\,\rho\,C_E V(q - q_s(T_{so}) \qquad\qquad (15)$$

where the soil moisture availability $\beta = W/W\text{max}$, C_E is the drag coefficient, V mean wind speed and $q_s\,(T_{so})$ saturated specific humidity at the surface.

The Eqs. (1), (2), (3), (4), (8), and (9) constitute a set of progressive equations of the air-land system.

Calculation Approaches

For the sake of convenience, let

$$F = B - A$$

then Eqs. (1) –(3) reduce to two equations

$$\frac{\partial^2 F}{\partial z^2} + \frac{1}{h}\,\frac{\partial F}{\partial z} + (\alpha_1 \rho_w)^2\,F = 8\sigma \overline{T}^{\,3}\,\alpha_1 \rho_w\,\frac{\partial T}{\partial z} \qquad\qquad (16)$$

$$\frac{\partial T}{\partial t} = \frac{\partial}{\partial z}\left(K\frac{\partial T}{\partial z}\right) + \frac{1}{C_p \rho}\,\frac{\partial F}{\partial z} \qquad\qquad (17)$$

and the corresponding boundary conditions (5), (6) become

$$\frac{\partial F}{\partial z} + \alpha_1 \rho_w F = 2\sigma T^4 \qquad\qquad z = \infty \qquad\qquad (18)$$

$$\frac{\partial F}{\partial z} - \alpha_1 \rho_w F = 0 \qquad\qquad\quad z = 0 \qquad\qquad (19)$$

Integrating the soil temperature equation (8) from $-D$ to the surface and using (10), we have the equation of temperature of the soil sublayer

$$\frac{\partial T_s}{\partial t} = \frac{1}{C_{ps}\rho_s D}[S(1-\alpha)-F+KC_p\rho\frac{\partial T}{\partial z}-LM_E-M_s] \qquad (20)$$

The integrating domain of the atmosphere extents from the ground up to 20 km with grid size 500 m and the depth of soil sublayer D is 5 meters. The time step is 2 hours for both the atmospheric and soil models. The model starts running at the spring equinox with initial values of climatic means of March. After one model year, the experiments of snow melting with anomalous snow depth begin.

Results

First, we simulated annual variations of temperature and moisture in the atmosphere and soil, as well as the components of heat and water balance at the surface. The integration with climatic mean initial snow depth is treated as "reference", while those with different initial snowdepth are the "anomalies". Comparing the anomalies to reference, the time-lag climatic effect can be investigated. The results shown later were obtained from the experiments at 65° latitudinal belt and the initial snowdepth of reference is taken as 5cm.

Basic Climate Means

Figure 1 shows the annual variation of surface temperature at the 65° latitudinal belt. The dashed line represents the observed temperature of Tura (64°10′N,

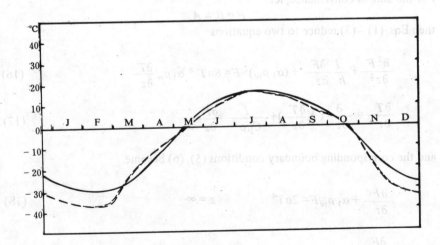

Fig. 1. Annual variation of soil temperature. Units are in °C. Solid line: computed, and dashed line; observed at Tura (64°10′N, 100°04′E)

100°04′E) which is located at the center of north Asia with snowcover during the entire cold seasons. The calculated temperature (solid line) is fairly close to observation. The maximum temperatures for both occure in July and the minimun in February. Snow is melting in May as the temperature rises across 0°C. The deficiency is that the calculated winter temperature is about 4° higher than that observed.

The vertical distributions of temperature in the atmosphere are calculated as shown in Fig. 2. The calculation of the temperature at lower and higher levels before snow melting (March, the solid line in Fig. 2A) is better, but the inverse layer near the ground and the tropopause do not appear. For the profile of July, after snow melting the computed temperature is close to that observed. From March to July, The temperature in the entire troposphere increases, especially at the lower levels; and the tropopause rises to about 12 km.

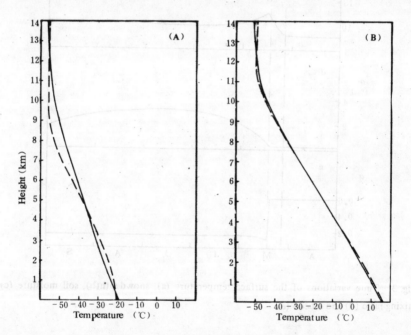

Fig. 2. Profiles of air temperature (A) Before snowmelting (March) and (B) After snowmelting (July). Units are °C

Snow Melting Process

From spring to autumn, a typical snow melting process and its time-lag effect are presented in Figs. 3 and 4, i.e., the variations of temperature and moisture in air and soil, and the components of the heat balance at the surface. We can see that each curve displays three stages of development.

Before snow melting. Before 5 May, surface temperature is lower than 0°C. Over snow surface albedo keeps up a high value 0.6. With the increase of insolation, the

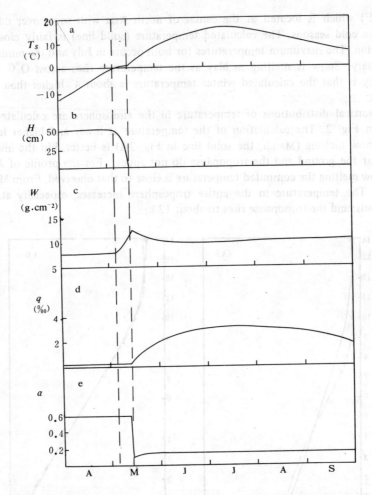

Fig. 3. Time variations of the surface temperature (a), snowdepth(b), soil moisture (c), air mixing ratio (d) and albedo (e)

temperature in air and soil rise gradually. At the same time, due to the small lapse rate of temperature near the ground, the sensible heat flux is weak. The sublimation in daytime almost compensates the freezing at night, thus, net evapotranspiration can be negligible. The moisture in air and soil remains constant.

During snow melting. Since 5 May, surface temperature exceeds 0°C, snow begins to melt. The depth of snow is diminishing and surface temperature remains about 0°C. As the depth reduces to a specific value (10 cm), the surface albedo decreases. Water due to snow melt permeates soil to increase its moisture content. The sensible heat flux enhances gradually. This process will last for about one week and its length depends on the initial depth of snow.

After snow melting. All snow melted by 12 May, soon after which, soil moisture

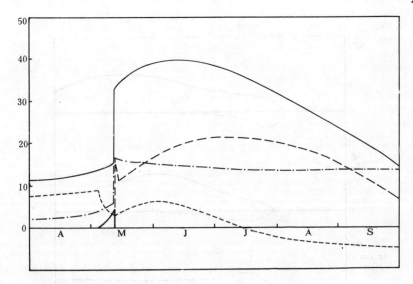

Fig. 4. Time variations of the heat balance components at the surface. Units are W.m ^{2}. —— net radiation − − − latent heat −.−.− sensible heat —— snowmelt . . . . . . heat coduction in soil

reaches its maximum and then decreases by evapotranspiration. Albedo reduces steeply to 0.13 of wet snow-free ground. Net radiation increases rapidly to be over 30 Wm^{-2} As a result, surface temperature rises and then air temperature is increased through turbulent heat transport. During the entire summer the intensity of latent heat flux is larger than that of sensible heat flux.

Following seasons turning, the distributions of temperature and moisture in air-land system tend to reach a new equilibrium until August even September.

Effect of Anomalous Snow-depth

To examine the short-term climatic effect of anomalous snowdepth, two experiments were performed, one with initial snow-depth 7.5 cm which is fifty percent more than "reference" (hereafter refered to as MS) and another with 2.5cm being fifty percent less than "reference" (LS).

The top panel of Fig. 5 shows the surface temperature differences between the anomalous and reference. The departure from reference for LS is positive with maximum of about 3°C occured in Aug., say, surface temperature is higher than normal. The departure for MS is negative, and its maximum of about −1°C appears in May. In LS years, a comparatively small amount of snow melted, soil is drier than that in reference case (Fig. 5c). But the lower atmosphere is more humid corresponding to higher air temperature (Fig. 5b). In contract to LS, the storage of water in soil after snowmelting for MS is relatively abundant (Fig. 5c). After August, all differences except for soil moisture reduce gradually and the climatic effect of anomalous snowdepth tend to disappear.

Such evolution of water budget must interact with the heat balance between air and soil. Because of the dependence of albedo on soil moisture on snow-free

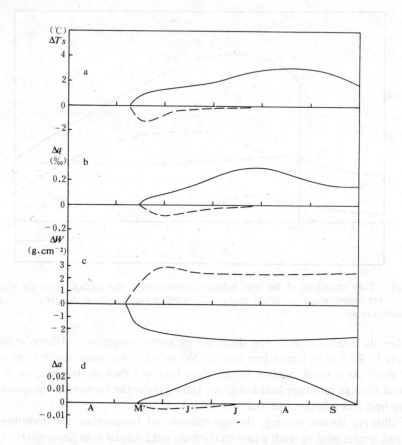

Fig. 5. Time variations of departures from reference of surface temperature (a), air mixing ratio (b) soil moisture (c) and albedo (d). Solid lines represent the departures for LS experiment and dashed lines for MS.

ground [Eq. (13)], drier soil in LS years has higher albedo (Fig. 5d), thus the absorption of insolation is less than that in a normal year (Fig. 6a). On the other hand, the rate of evapotranspiration which related to moisture in air and soil weakens considerably (Fig. 6c). As a consequence, the decreasing of soil moisture leads to increase in soil and air temperature. Usually, soil is rather wet under a snowcover surface. In MS years, thus, even though much water has been added to soil, the change in evapotranspiration is still small. A slight decrease in albedo and the corresponding absorbed insolation are present (Fig. 6a). Sensible heat flux weakens following the decrease in temperature near the ground (Fig. 6b).

Conclusions

In summary, the effect of snow melt with different snow depth on thermal and moisture conditions of the atmosphere and soil in later seasons is primarily realized

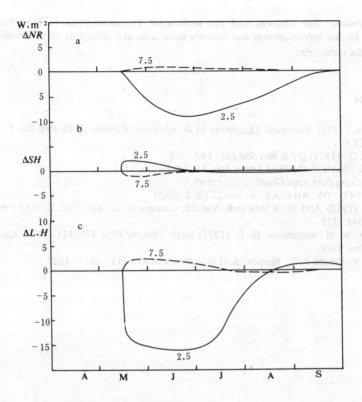

Fig. 6. As in Fig. 5, except for components of heat balance at the surface: net radiation (top), sensible heat flux (middle) and latent heat flux (bottom).

through the change in soil moisture, which alters the rate of evapotranspiration and albedo at the surface, resulting in a succession of processes in the atmosphere and soil. Two feedback processes probably develop in snow melt simulations. If the initial snowdepth is, as an example, larger than normal, much melted snow is added directly to the soil moisture content, causing two results: reduction of albedo and increase in evapotranspiration. The former results in increase of absorption of insolation and of the temperature in soil and air, furthermore, enhancement of sensible heat transfer. The latter, intensified evapotranspiration, reduces the soil temperature, and then air temperature by sensible heat exchange, thus the air moisture content decrease and the rate of precipitation enhances (because of conservation of relative humidity). This in turn maintains the wet soil and intensifies evapotranspiration. For these simulations, it seems that the latter feedback mechanism dominates the evolution of snow melting. Therefore, the temperature would be higher than normal in LS years and lower in MS years. These effects will continue till at least August. As for the anomalous circulation associated with anomalous temperature field, it is an interesting problem which needs to be investigated.

Acknowledgment. Qian Xiaowei took part in the work. The author wishes to thank Professor T. C. Yeh for his encouragement and valuable discussions and comments on the preliminary version of the manuscript.

References

Arakawa, A. (1972) Numerical Simulation of Weather and Climate. Tech Rep No 7 . Dept Met UCLA

Charney, J. G. (1975) *Q J R Met Soc 101: 193—202*

*H*ahn, D. G. Shukla, J. (1976) *J Atmos Soc* 33: 2461—1

Ji Jinjun (1984) *Pure Appl Geophys* (in press)

Kibel A (1943) OK. AH CCCP 39: 18—22 (in Russian)

Namias, J. (1963) Arid Zone Research, Vol 20. Changes of Climate. Proc. Rome Symp UN-ESCO, pp 345—359

Washington, W. M. Williamson, D. L. (1977) *Meth Comput Phys* 17: 111—172 (Academic Press, New York)

Yeh, T. C., Wetherald R. T: Manabe, S. (1983) *Mon Wea Rev* 111: 1013—1024

The Mean Transfer of Sensible Heat in the Atmosphere over China and the Distribution of Its Sources and Sinks

Gao Guodong[1] and Lu Yurong[1]

Abstract — Using 10-year average data for the period of 1960 – 1969 from about 100 radiosonding stations, we calculated the meridional and zonal mean transfer of sensible heat in the atmosphere (from the ground surface to the height of 100 hPa comprising 11 layers) and the magnitude and direction of its result over China for each month and the whole year. The distribution characteristics of the convergence and divergence (sinks and sources) of sensible heat were also analyzed.

Introduction

It is known that solar radiation is the energy source of the earth. When the earth and its surrounding atmosphere receive the energy of solar radiation and increase their own temperatures, on one hand they re-emit longwave radiation and on the other hand heat exchanges between different parts of the atmosphere and between the atmosphere and the ground surface take place by sensible heat transfer, latent heat transfer and by other ways. In different regions on the earth, owing to different humidity and temperature conditions, the thermodynamic state, the heat transfer, and heat exchange are quite different.

The sensible heat transfer is one of the main components of the heat balance in the atmosphere and on the ground surface. Calculating and analyzing the characteristics of sensible heat transfer and the sink-source distribution are not only of practical importance in understanding the cold and warm status, the origin and change of heat in atmosphere, and in studying climatic formation and climatic change, but also valuable in studying energy transfer in the atmosphere and energy balance and in climatic model and atmospheric modeling.

In our previous papers on the heat balance of the earth's surface in China, the sensible heat transfer between the ground surface and atmosphere was calculated and discussed[*]. In this paper we report the result of the investigation on the horizontal sensible heat transfer between different parts in the atmosphere.

Calculation Method

The formula to calculate the mean sensible heat transfer in the atmosphere is

$$\overline{Q_s} = -\frac{C_p}{g} \int_{p_o}^{p_z} \overline{T}(p) \cdot \overline{V}_{dp}$$

---

1. Department of Meteorology, Nanjing University, Nanjing, China
* See References under Gao Guodong and Lu Yurong.

where g is the gravitational acceleration; C_p—specific heat at constant pressure; $\overline{T}$—the average absolute temperature in a certain time interval; $\overline{V}$—the average wind velocity in a certain time interval and p_0 and p_z are respectively the atmospheric pressure at ground surface and height z.

Assuming $\overline{u}$ to be the zonal mean wind velocity and $\overline{v}$ the meridional mean wind velocity, the zonal sensible heat transfer $\overline{Q}_{su}$ is

$$\overline{Q}_{su} = \frac{-C_p}{g} \int_{P_0}^{P_z} \overline{T}(p) \cdot \overline{U}_{dp}$$

and the meridional sensible heat transfer $\overline{Q}_{sv}$ is

$$\overline{Q}_{sv} = \frac{-C_p}{g} \int_{P_0}^{P_z} \overline{T}(p)\overline{v} \; dp$$

the convergence (sink) and divergence (source) of sensible heat can be calculated by the following formula

$$\overline{F}_s = \frac{\partial \overline{Q}_{su}}{\partial x} + \frac{\partial \overline{Q}_{sv}}{\partial y}$$

We use the grid method of calculation with grid size of 260 km.

Calculated Results

The sensible heat transfer in the atmosphere is dominated by the structure of the temperature field in the atmosphere and that of the wind fields. It is caused by the combined effect of temperature and wind. The distribution of temperature in the atmosphere relates to the radiation balance in the atmosphere and the heat balance. The wind direction and speed are determined by the atmospheric circulation at the given time and at the given place. Therefore, the characteristics of the sensible heat transfer and the distribution of convergence and divergence in China are closely related and conformed to the state of radiation balance, the stratifications of air temperature and the characteristics of atmospheric circulation in China.

The maximum of the yearly average sensible heat transfer in the whole atmospheric layer in China is located in the region on the south of middle and lower Changjiang river with its value reaching 8×10^7 cal/cm·sec. It decreases toward both higher and lower latitudes and in the western areas. In the west and northeast of China the sensible heat transfer is about 5×10^7 cal/cm·sec and in Guangdong area about 3×10^7 cal/cm·sec. The zonal sensible heat transfer, mainly by westerly wind, has its maximum in the plain to the south of Changjiang with the value over 8×10^7 cal/cm·sec. In Sichuan Basin it is 6×10^7 cal/cm·sec and in Northeast China, South China and in the whole western area, it is about $4 - 5 \times 10^7$ cal/cm·sec. The confluent line of southward and northward meridional sensible heat transfer (the zero-value line) lies approximately along the latitude of $30°N$. The northerly wind transfer occurs on its northern side and the southerly wind transfer on its southern side. The amount of the sensible heat transfer is not large, -0.5×10^7 cal/cm·sec around South China and 2×10^7 cal/cm·sec in Northeast

China.

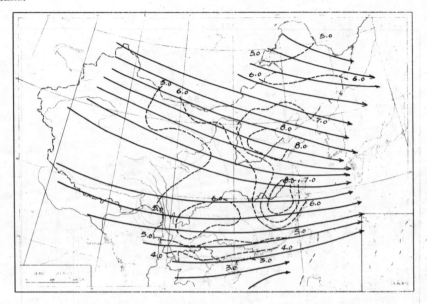

Fig. 1. The yearly average sensible heat transfer and its direction in the whole atmospheric layer over China ($\times 10^7$ cal/cm·sec)

In winter, controlled by the cold air from Siberia, the sensible heat transfer in the atmosphere over the mainland of China is principally carried by the north-west flow. in the east coastal areas and in the plain on the south of Changjiang the maximum sensible heat transfer in January may reach 13×10^7 cal/cm·sec, while in the north of Northeast China, the sensible heat transfer is only 5×10^7 cal/cm·sec. In the western area it decreases from Tibet towards Xinjiang with its value from $7-8 \times 10^7$ cal/cm·sec to $4-5 \times 10^7$ cal/cm·sec. After winter the sensible heat transfer decreases month by month and the gradient of contour lines also decreases gradually. The direction of sensible heat transfer deflects gradually clockwise due to the influence of the winter monsoon. Approaching June, the southeast coastal area begins to be affected by subtropical high pressure and the westward and northward sensible heat transfers occur.

In July and August, in the vast areas of the whole of South China, Southwest China and in the South of the Tibetan Plateau a change occurs, resulting in westward and northward transfers. The maximum westward sensible heat transfer is still less than 1×10^7 cal/cm·sec. In June, it only appears in the coastal areas of Guangdong and Fujian provinces, while in July the easterly sensible heat transfer increases to over 2×10^7 cal/cm·sec and spreads toward the whole areas to the south of Changjiang. In August again it decreases to 1×10^7 cal/cm·sec. After September the westward sensible transfer retreats from China. To the north of the zero-value line the sensible heat transfer is still eastward and southward, and the maximum sensible heat transfer shifts with the advance and retreat of the summer monsoon. In June the center of maximum eastward sensible heat transfer lies in Chang-

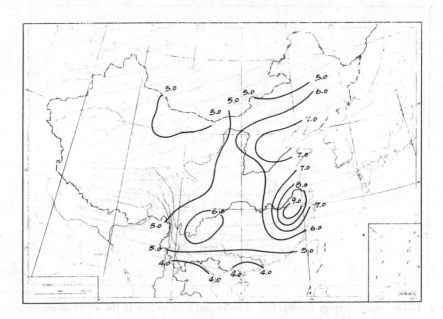

Fig. 2. The yearly average zonal sensible heat transfer in the atmosphere over China (× 10⁷ cal/cm·sec) (negative is westward, positive is eastward)

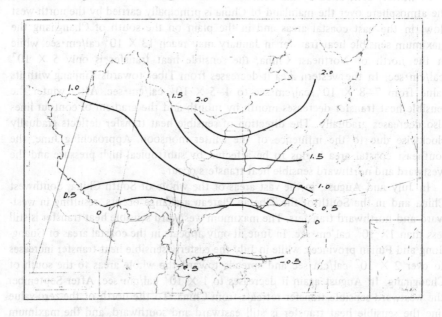

Fig. 3. The yearly average meridional sensible heat transfer in the atmosphere over China (×10⁷ cal/cm·sec) (negative is northward, positive is southward)

jiang-Huaihe areas. In July it shifts northward to North China, in August, to Northeast China and in September it begins to retreat southward. The eastward sensible transfer does not change too much from month to month. The maximum values are all around 5×10^7 cal/cm·sec. (Figs. 4, 5, 6, 7).

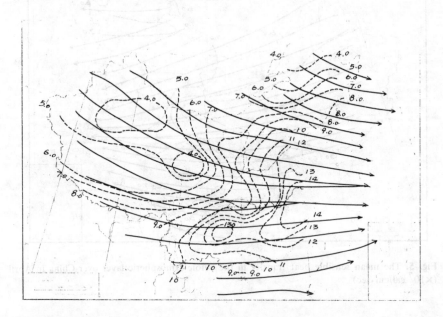

Fig. 4. The mean sensible heat transfer in the whole atmospheric layer over China in January ($\times 10^7$ cal/cm·sec)

The dividing line of convergence and divergence of sensible heat in the atmosphere for the whole year in China (zero line) lies approximately along the Huanghe valley. On its north there is a sensible heat divergent region with the maximum centre around Hetao area (in the area around the Great Bend of the Huanghe River), about -2×10^6 cal/cm²·year. The center of maximum convergence of sensible heat lies in the southeast area of China with the value of 5×10^6 cal/cm²·year. The western part of China is the sensible heat convergent region but the convergence is not large, about $0-2 \times 10^6$ cal/cm²·year.

In each winter month, the western part of China and the southeast coastal area are sensible heat convergent regions while the whole interior of the main land and the Northeast are all divergent regions. The magnitude of sensible heat convergence and of divergence are about the same. In the coldest month (January) they are both below 1.5×10^6 cal/cm²·month. In spring months, the sensible heat divergent region shrinks month by month and shifts northwards with its value getting smaller and smaller. In April and May the whole east part is almost a convergent region with the value below 1×10^6 cal/cm²·month. After July, the whole of China is a sensible heat convergent region with the maximum located in the area between Changjiang and the Huanghe River. In July its value is 0.5×10^6 cal/cm² Jmonth and in September increases to 1×10^6 cal/cm²·month. From October in autumn,

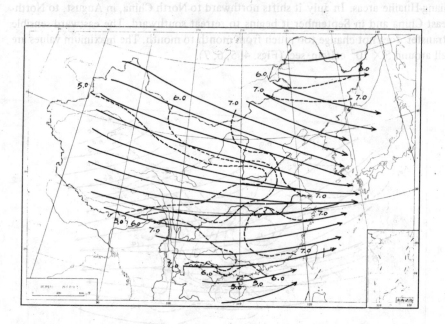

Fig. 5. The mean sensible heat transfer in the whole atmospheric layer over China in April ($\times 10^7$ cal/cm·sec)

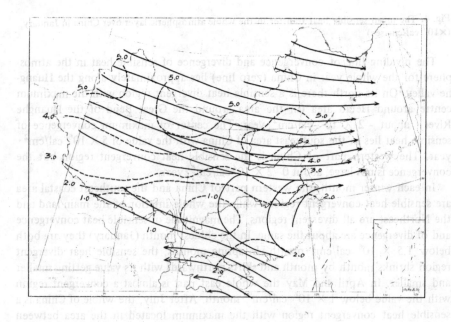

Fig. 6. The mean sensible heat transfer in the whole atmospheric layer over China in July ($\times 10^7$ cal/cm·sec)

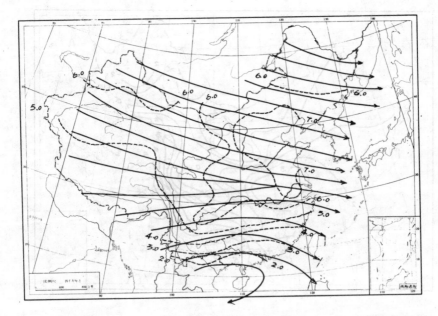

Fig. 7. The mean sensible heat transfer in the whole atmospheric layer over China in October
($\times 10^7$ cal/cm·sec)

the sensible heat convergent region begins to shift southwards. The value in
the convergence center is always below 0.5×10^6 cal/cm^2·month. The north of
Huanghe River is the sensible heat divergent region where the center of divergence
lies in Inner Mongolia and Shensi-Gansu area with the maximum value reaching
around -0.75×10^6 cal/cm^2·month. Later, the region expands month by month
and resumes the winter situation. (Figs. 8,9,10)

Comparing with the turbulent heat exchange between ground surface and the
atmosphere (Lu Yurong and Gao Guodong 1981a) calculated by us, it may be
found that whether in the whole year or in winter and summer, the sensible heat
convergence in the atmosphere is larger than the turbulent heat exchange by 2
orders of magnitude, but on the whole there is some similarity between the distributions of the contour lines of both quantities. The area with small turbulent
heat exchange between ground surface and atmosphere is just the area with large
sensible heat convergence in the atmosphere and the region with large turbulent
heat exchange between ground surface and atmosphere coincides with the region of
large sensible heat divergence in the atmosphere.

We also made a comparative study of the meridional sensible heat transfer in
the atmosphere in the area of 20°–50°N calculated by Sellers (1965) and ours.
The value given by Sellers is -1.5 to -2.0×10^{22} cal/year· latitudinal circle,
equivalent to -0.13 to -0.27×10^6 cal/cm·sec and the sensible heat transfer in the
atmosphere in China obtained by us in the region between 20 –40°N is -5×10^6
to $+ 15 \times 10^6$ cal/cm·sec. Since China only occupies a small part of the latitudinal

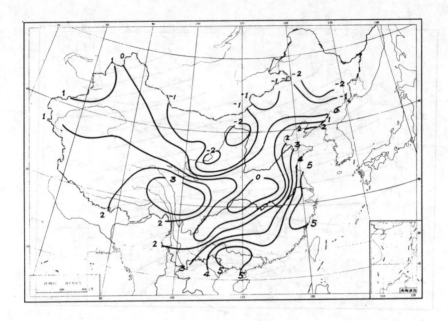

Fig. 8. The yearly sensible heat divergence (convergence) in the atmosphere over China ($\times 10^6$ cal/cm^2·year) (negative is divergence, positive is convergence)

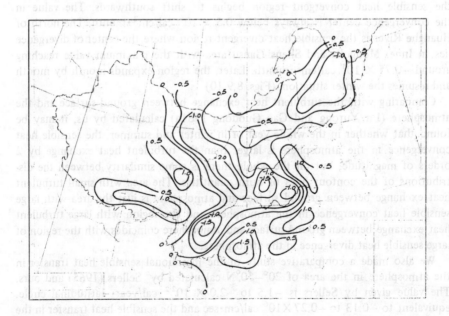

Fig. 9. The sensible heat divergence (convergence) in the atmosphere over China in January ($\times 10^6$ cal/cm^2·month).

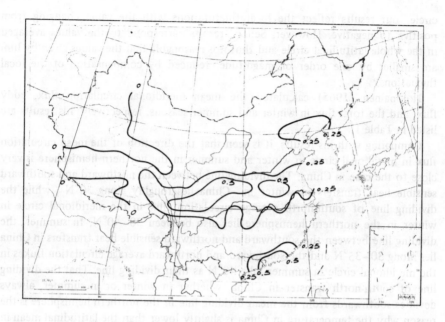

Fig. 10. The sensible heat divergence (convergence) in the atmosphere over China in July ($\times 10^6$ cal/cm$^2 \cdot$ month)

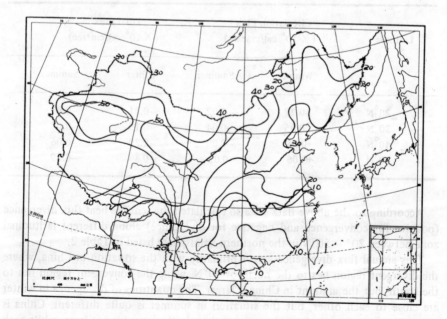

Fig. 11. The yearly turbulent heat exchange between the ground surface and atmosphere over China ($\times 10^3$ cal/cm$^2 \cdot$year)

306

circle, our results reflect the local fluctutations with the value changing from positive to negative. However, Sellers' results correspond to the values averaged in the whole latitudinal circle and thus it is reasonable that the values given by him are smaller by one order of magnitude, reduced by compensation of the local flucuations.

Holopainen (1965) calculated the mean meridional circulation flux, eddy flux, and the total flux in winter and summer seasons, respectively; his results are listed in Table 1.

Comparing with his results, it is seen that the direction of the mean circulation flux in meridional circle in winter and summer in the northern hemisphere is very close to the case in China. The dividing line between the northward and southward sensible heat transfers in winter in China lies roughly along 30°N , while the dividing line of south-north average circulation flux in the meridional circle in winter in the northern hemisphere lies also between 30—40°N. In summer, the dividing line between the southward and northward sensible heat transfers in China lies along 30—35°N and the southward and northward average circulation fluxes in the meridional circle in summer take 40°N as their dividing line. That the dividing line of south-north transfer in China, whether in winter or in summer, always deviates southwards from the mean dividing line in the northern hemisphere is the reason why the temperature in China is slightly lower than the latitudinal mean in the northern hemisphere.

Table 1. The Circulation Flux and Total Flux Calculatied by Holopainen

Latitude	Circulation Flux ($\times 10^6$ cal/cm·sec)		Total Flux ($\times 10^6$ cal/cm·sec)	
	Winter	Summer	Winter	Summer
20°N	−0.27	−0.02	−0.31	−0.01
30°N	−0.05	−0.04	−0.19	−0.05
40°N	+0.05	−0.00	−0.30	−0.06
50°N	+0.06	+0.01	−0.39	−0.09

According to the above data we also calculated the sensible heat flux divergence (positive for convergence and negative for divergence) along different latitudinal zone between 20° − 50°N in the northern hemisphere listed in Table 2.

This annual flux divergence agrees primarily with the situation in China, where the divergent region lies to the north of 30°N, while the convergent region lies to the south, but the amount in China is larger. The quantities of divergence in winter are close to each other, but the situation in summer is quite different. China is primarily a convergent region in summer though its amount is not large, while each latitudinal zone in the northern hemisphere is the divergent region on average. This is mainly caused by the fact that there exists strong high pressure over the ocean

Table 2. The Calculatied Sensible Heat Flux Divergence According to the Data of Holopainen

Latitudinal zone	Whole year cal/year	Winter cal/month	Summer cal/month
20–30°N	+0.008 × 10⁵	+0.03 × 10⁵	−0.009 × 10⁵
30–40	−0.12 × 10⁵	−0.02 × 10⁵	−0.001 × 10⁵
40–50	−0.07 × 10⁵	−0.007 × 10⁵	−0.003 × 10⁵

while in mainland of China hot low pressure occurs in summer.

References

Lu Yurong, Gao Guodong (1976/1978) *J Nanjing Univ (NatSci Ed)* **(I)** (1976) 2: 90–108; **(II)** (1978) 2: 83–89

Gao Guodong, Lu Yurong (1979) *Sci Atmos Sin* 3.1: 12–20

Lu Yurong, Gao Guodong (1981a), *Sci Atmos Sin* 5, 1: 78–84

Lu Yurong, Gao Guodong (1981b) A study of the water balance in China. *Proc Symp Variat Global Water Budget*, 10 – 15 August 1981. Oxford UK, pp 73–75

Gao Guodong, Lu Yurong (1982, *The radiation balance and the heat balance on the surface of China*. Scientific Press, Beijing, p 165

Lu Yurong, Gao Guodong (1983) *Sci Sin* (Ser B) 26 2: 186–195

Gao Guodong, Lu Yurong (1983) *Sci Sin* (Ser B) 26 4: 386–395

Sellers, W. D. (1965) *Physical Climatology*. The Univ Chicago Press, pp 272

Holopainen E O. (1965),*Tellus* 17: 285–294

The Mean Transfer of Latent Heat in the Atmosphere over China and the Distribution of Its Sources and Sinks

Lu Yurong[1] and Gao Guodong[1]

Abstract — Utilizing the 10-year average data (1960–1969) from about 100 radiosonding station, we calculated the longitudinal and latitudinal mean transfer of latent heat in the atmosphere (from the ground to the height of 100 hPa comprising 11 layers) and the magnitude and direction of its resultant over China for each month and the whole year. The distribution characteristics of the convergence and divergence (sinks and sources) of latent heat were also analyzed.

Foreword

This paper presents the results of a part of our research on "The Study of the Radiation Balance, Heat Balance and Water Balance in China."(see References)

The basic factor affecting the formation and change of climate is the state of heat budget in the atmosphere. The transfer of latent heat is the main component of heat balance in the atmosphere.

The study of latent heat transfer (LHT) is to investigate the contribution of water transformation in the atmosphere to the heat budget. Calculating and analyzing the characteristics of latent heat transfer in the atmosphere and the sink-source distribution over China is of practical importance in understanding the cold and warm conditions of China and in understanding the origin and change of heat-water in the atmosphere. It is also a fundamental work for studying the formation and the change of the climate of China. In addition, it provides valuable parameters for establishing the climatic model and atmospheric modle for China.

Methods of Calculation

Latent heat transfer can be divided into two aspects. One is the vertical latent heat transfer between the ground and the atmosphere, that is the transfer of evaporation latent heat. We have finished the research on this part and published our result (Gao Guodong et al. 1978 1980; Lu Yurong and Gao Guodong 1981). In this paper, we report the investigation on the horizontal latent heat transfer in the atmosphere (Lu Yurong and Gao Guodong 1983 b).

The formula for calculating the latent heat transfer in the atmosphere is

$$\overline{Q}_L = -\frac{L}{g} \int_{p_0}^{p_z} \overline{q}(p)\overline{V}\,dp$$

---

1. Department of Meteorology, Nanjing University, Nanjing, China

where g is the gravitational acceleration; L is the latent heat; $\bar{q}$ is the mean specific humidity in a certain time interval; $\bar{V}$ is the mean velocity of wind in a certain time interval and p_0, p_z are respectively the atmospheric pressure on ground surface and height z. Let $\bar{u}$ be the mean latitudinal velocity of wind and $\bar{V}$ the mean longitudinal velocity of wind, then the zonal transfer $\overline{Q}_{Lu}$ and meridional transfer $\overline{Q}_{Lv}$ of latent heat are expressed by

$$\overline{Q}_{Lu} = -\frac{-L}{g} \int_{p_0}^{p_z} \bar{q}(p)\,\bar{u}\,dp$$

$$\overline{Q}_{Lv} = \frac{-L}{g} \int_{p_0}^{p_z} \bar{q}(p)\,\bar{v}\,dp$$

and the convergence/divergence (sink/source) of latent heat can be calculated by

$$\overline{F}_L = \frac{\partial \overline{Q}_{Lu}}{\partial x} + \frac{\partial \overline{Q}_{Lv}}{\partial y}$$

We used grid method for calculation; the grid distance is 260 km.

Main Results

Firstly, the origins of latent heat transfer in the atmosphere over China are mainly from three directions.

1) The latent heat released from the vapor is brought in by the southwest air stream from the Indian Ocean and the Bay of Bengal. It influences the south-western part of China and the vast area to the south of Changjiang River. It is the largest latent heat transfer. In the southwestern part of China and the region along the Nanling mountains, the magnitude can be more than 0.7×10^6 cal/cm·sec. In other regions it is about 0.6×10^6 cal/cm·sec. (Fig. 8)

2) The latent heat released is brought in by the northwest air stream from the Atlantic Ocean and the Arctic Ocean. It influences the vast area to the north of the Changjiang River and the Huai River. In the regions along the Huanghe River and the Huai River and in Northeast China, it is about 0.5×10^6 cal/cm·sec. Towards the west in the Xinjiang area, it decreases to less than 0.2×10^6 cal/cm·sec.

3) The latent heat released is brought in by the southerly flow from the South China Sea (including the East China Sea). It only affects the southeast coastal areas where the latent heat transfer reaches about 0.5×10^6 cal/cm·sec. The latent heat transfers from these three directions behave differently in their amounts and directions at different levels. At 850 hPa the transfers in these three directions are clearly distinguishable. At 700 hPa the transfer by southly current is no longer apparent. At the level up to 500 hPa the southwest and the northwest streams join together as an eastward current of latent heat transfer. Since the latent heat transfer is determined by the joint effect of moisture content and wind velo-

city, the height of maximum latent heat transfer is at about the 700 hPa level; it moves slightly higher in summer and slightly lower in winter (Figs. 1,2,3)

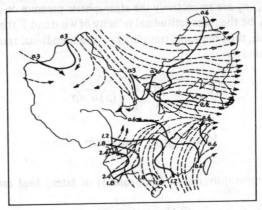

Fig. 1. The direction and magnitude of annual mean latent heat transfer at 850 hPa over China ($\times 10^3$ cal/cm·sec. hPa)

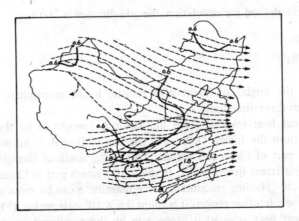

Fig. 2. The direction and magnitude of annual mean latent heat transfer at 700 hPa over China. ($\times 10^3$ cal/cm·sec. hPa)

Secondly, the annual longitudinal moisture transfer of the whole atmospheric layer appears in the south of the Changjiang River as a strip distribution. The maximum latent heat transfer eastward is formed in a belt around 25°–27°N with the value of 0.6–0.7 $\times 10^3$ cal/cm·sec. It decreases towards higher or lower latitudes.

In winter, except in the southeast coastal area below 900 hPa and in some areas in Xinjiang (owing to the topographical influence) below 850 hPa, where there is a westward latent heat transfer, in all other areas from the ground to high altitude only an eastward latent heat transfer exists with the maximum eastward transfer at the layer of 600–700 hPa. In summer the westward latent heat transfer becomes

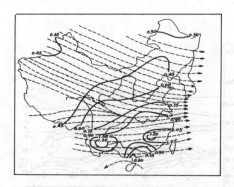

Fig. 3. The direction and magnitude of annual mean latent heat transfer at 500 hPa over China. ($\times 10^3$ cal/cm·sec hPa)

strong. In the South from Fujian above 800–700 hPa to the east border of the Tibet Plateau above 400 hPa the latent heat transfers westward.

The north-south boundary line of the annual latitudinal latent heat transfer is roughly along 30°N, merging into the Changjiang River valley in the east and reaching the north bank of the Yarlung Zangbu River valley in the west. To its north southward latent heat transfer exists and northward latent heat transfer is to its south. The northward latent heat transfer, parallel to the latitude, has its maximum value of -0.4×10^3 cal/cm·sec, while the maximum value of southward latent heat transfer is about 0.1×10^3 cal/cm·sec in North China.

In winter, except that on the ground surface in the west the weak northward transfer occurs owing to topographical effect, in all other areas and at high altitude only southward transfer exists with the maximum latent heat transfer at the layer of 600 hPa. In summer, the high latitude areas are still controlled by northerly winds, while to the south of the Changjiang River, in the southeast coastal areas and the whole south China southerly wind predominates. (Figs. 4,5,6,7)

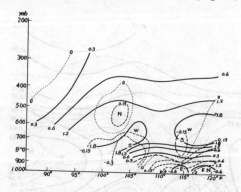

Fig. 4. The zonal and meridional latent heat transfer at various altitudes in the atmosphere along 25°N over China in January. ($\times 10^3$ cal/cm·sec. hPa) (*solid lines:* zonal transfer, *dotted lines:* meridional transfer, *dashed lines:* zero lines. West and north winds are positive; east and south winds are negative)

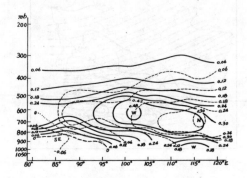

Fig. 5. As in Fig. 4, but for 45°N in January

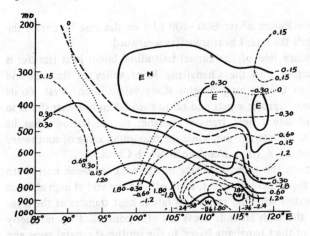

Fig. 6. As in Fig. 4, but for 25°N in July

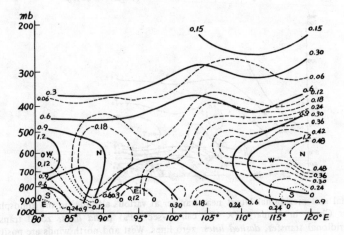

Fig. 7. As in Fig. 4, but for 45°N in July

Thirdly, the annual maximum resultant latent heat transfer in the whole atmospheric layer in China lies in the plains of two lakes to the south of the Changjiang River and in regions along the Southeast Transversal mountain range and Yunnan province with the latitudes between $24°-27°N$. Its value is larger than 0.7×10^6 cal/cm·sec. The next is the Liaodong Peninsula in Northeast China and the minimum latent heat trasnfer is in Northwest China; it is less than 0.1×10^6 cal/cm·sec. In Northwest China higher value occurs in the region at the northern side of Tianshan mountain. The Tibetan Plateau has an obvious shielding and shunt. effect on latent heat transfer. The level of maximum latent heat above the plateau is nearly 100 hPa higher than in the surrounding region. The lower layer of the southwest air stream is obstructed from flowing northwards. Therefore only in the southeast part of the plateau, in the Yarlung-zangbu River valley, where the terrain is relatively low, is the latent heat transfer comparatively high. Going deep into the interior of the plateau the value decreases. The plateau obstructs the transfer from flowing toward the northwest and thus the northwest inland region in China is the area with minimum latent heat transfer. (Fig. 8)

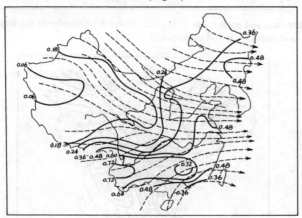

Fig. 8. The direction and magnitude of annual resultant heat transfer in the whole atmospheric layer over China ($\times 10^6$ cal/cm·sec)

Fourthly, the latent heat transfer in China is closely related to the atmospheric circulation and the advance and retreat of the monsoon. In winter the whole of China is under the control of cold high pressure from Siberia, the transfers by west wind and north wind predominate, which results in a zonal distribution. The vapor content is not high and thus the latent heat transfer is not significant. In South China, it is more than 0.6×10^6 cal/cm·sec. The latent heat transfer decreases with increasing latitude and in Northeast China it becomes about 0.15×10^6 cal/cm·sec. (Fig. 9).

In summer the subtropical high pressure extends westwards and northwards, strengthening the moisture transfer by the east flow and south flow and therefore the latent heat transfer becomes much larger than in winter. In July, the center of the maximum transfer lies in the basin of two lakes with the value over 1.8×10^6 cal/cm·sec, and decreases with increasing distance from the coast. In the Xinjiang area it drops down to 0.3×10^6 cal/cm·sec. (Fig. 10)

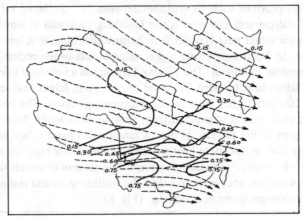

Fig. 9. The direction and magnitude of resultant latent heat transfer in the whole atmospheric layer over China in January ($\times 10^6$ cal/cm·sec)

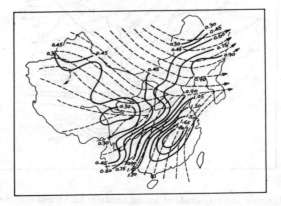

Fig. 10. The direction and magnitude of resultant latent heat transfer in the whole atmospheric layer over China in July ($\times 10^6$ cal/cm·sec)

Fifthly, the confluent lines of latent heat transfers brought by different air streams coincide well with the lines of climate demarcation. For example, the annual meanposition of the confluent line of southwest latent heat transfer and northwest latent heat transfer lies approximately along the Changjiang-Huai River valley to the south of the Qinling mountains. This is the demarcation line of wet climate and dry climate in China. However, the intersection line of southwest latent heat transfer with the southeast one lies roughtly along the Nanling mountains. The latent heat transfers originating from the Bay of Bengal and South China Sea are plentiful. The annual mean transfers in these two flows do not differ remarkably, but the difference of seasonal variation is comparatively uniform within a year, while in southwest areas the annual amplitude is notable due to the clear distinction between dry and wet seasons therein.

Sixthly, the zero line of yearly latent heat convergence is roughly along $39° - 34°N$, that is between the Huai River and Qinling mountain range, which coincides with the intersection line of southwest and northwest latent heat transfer. The area to its south is the latent heat convergent region and the divergent region lies to its north. The zero line shifts slightly southwards to the Changjiang valley in winter and moves northwards to the north of the Huanghe River in summer. The maximum convergence of latent heat is higher than 2.16×10^5 cal/year in Guangxi, Yunnan province, and the Sichuan Basin. The values in all divergent regions are not large. The maximum in Inner Mongolia is less than -0.12×10^5 cal/year. (Figs. 11, 12, 13)

The distribution of latent heat sinks and sources coincides well with that of precipitation in China. The place with rich precipitation is just the area with large

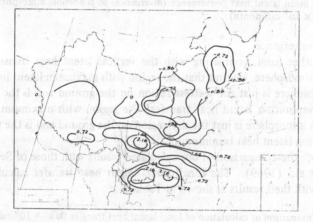

Fig. 11. The annual mean latent heat convergence (divergence) in the whole atmospheric layer over China. ($\times 10^5$ cal/year)

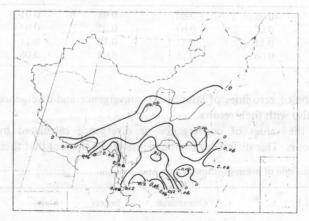

Fig. 12. The mean latent heat convergence (divergence) in the whole atmospheric layer over China in January ($\times 10^5$ cal/month)

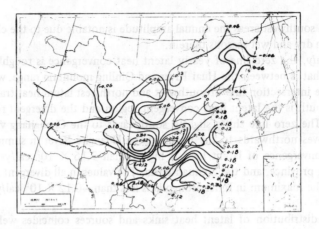

Fig. 13. The mean latent heat convergence (divergence) in the whole atmospheric layer over china in July (× 10^5 cal/month)

latent heat convergence.

On the other hand, comparing with the vertical latent heat transfer between ground and atmosphere, we find that the region with maximum latent heat convergence in atmosphere is just the wettest region on the ground and is the region with maximum evaporation latent heat transfer. The region with maximum latent heat divergence in atmosphere is just the dry region on the ground and is the place where the evaporation latent heat is minimum.

Seventhly, the comparison of our calculated results with those of Sellers (1945) and Starr et al. (1969). The longitudinal latent heat transfer calculated by us agrees well with their results as shown in Table 1.

Table 1. Comparison of calculation of longi latent heat transfer flux (× 10^6 cal/cm·sec)

Latitude	Our resules	Sellers	Starr	Peixoto Mckean
20	−0.01	−0.02	−0.01	−0.01
30	0.07	0.07	0.07	0.07
40	0.11	0.11	0.13	0.11
50	0.06	0.06	0.04	0.06

The location of zero lines of latent heat convergence and divergence calculated by us agrees also with their results.

However, the values of convergence or divergence calculated by them are smaller than ours. This difference can be explained by the fact that their calculated

Table 2. Comparison of mean-position of zero lines of latent heat convergence and divergence

Claulated by	Our resules	Sellers	Starr	Peixoto Crisi
The mean-position of zero lines	32–34°N	23°N	39°N	38°N

Table 3. Comparison of calculations of latent heat divergence/convergence (cal/year)

Latitude	Our resules		Sellers
20–30°N	2.16×10^5	0.28×10^5	
30–40°N	0.72×10^5	0.01×10^5	Convergence
40–50°N	-0.36×10^5	-0.15×10^5	Divergence

values are averaged over the whole latitude circle, and thus the positive values are compensated with the negative ones in the circle while our calculated values are averaged over only a partion of the latitude circle across China.

References

Gao Guodong, Lu Yurong, (1979) *Sci Atmos Sin* 3, 1: 12–20

Gao Guodong, Lu Yurong (1982) *The radiation balance and the heat balance on the surface in China.* Science Press, Beijing, 165pp

Gao Guodong, Lu Yurong (1983) *Sci Sin* (Ser B) 26, 4: 386 –395

Gao Guodong, Lu Yurong, Li Huaijin (1978) *Acta Geogr* 33, 2: 102 – 108

Gao Guodong, Lu Yurong, Li Huaijing (1980) *Acta Meteorol Sini* 38, 2: 165–176

Lu Yurong, Gao Guodong (1976/1978) *J Nanjing Univ (Nat Sci Ed)* (I) 1976 2: 90–108:;(II) 1978 2: 83–89

Lu Yurong, Gao Guodong, (1981) A study of the water balance in China. Proc Symp Variat Global Water Budget, August 1981. Oxford U K, pp 73–75

Lu Yurong, Gao Guodong (1981) A study of the dry and wet condition in China, idem, pp 76–78

Lu Yurong, Gao Guodong (1982) *Agric Meteorol* (Quarterly) 3: 14–18

Lu Yurong, Gao Guodong (1983 a) *Sci Sin* (Ser B) 24, 2: 186–195

Lu Yurong, Gao Guodong (1983 b) *Plateau Meteorol* 2, 4: 34–48

Lu Yurong, Gao Guodong, Lu Huaijin (1979) *J Nanjing Univ* 1: 125–138

Sellers WD (1965) *Physical Climatology.* Univ Chicago Press, 272 pp

Starr, V.P, Peixoto J. P. Mckean R.C. (1969) *Pure Appl* Geophys 75:300–331

The Thermal Forcing of Sea-Land Heating Difference and the Summertime Anticyclone over Southern Asia

Zhu Zhengxin[1]

Abstract — In addition to the polar vortex, the south Asian high is the most strong and stable system of upper troposphere in the Northern summer. It has great influence on climate and weather distribution in vast areas of southern Asia including China. Its climatic features and activity regularities, as well as the causes of its formation, have been studied extensively by many authors, but a satisfactory theoretical explanation of the dynamic mechanism is still due. In this paper, a highly truncated spectral model is used to clarify the effects of thermal forcing on the formation and maintenance of the south Asian high. Furthermore, the essential causes of its seasonal variations, the controlling effects of thermal forcing, and the sudden changes of the circulation patterns in June and October are discussed.

The Model

A two-level, quasigeostrophic spectral model is used. The vorticity and thermodynamic equations can be written as

$$[\frac{\partial}{\partial t} + \bar{u}(y,p)\frac{\partial}{\partial x}]\nabla^2\psi + (\beta - \frac{\partial^2\bar{u}}{\partial y^2})\frac{\partial\psi}{\partial x} = f_0\frac{\partial\omega}{\partial p}, \tag{1}$$

$$[\frac{\partial}{\partial t} + \bar{u}(y,p)\frac{\partial}{\partial x}]\frac{\partial\psi}{\partial p} + J(\psi, \frac{\partial\psi}{\partial p}) - \frac{\partial\bar{u}}{\partial p}\frac{\partial\psi}{\partial x} + \frac{\sigma}{f_0}\omega = -\frac{RQ}{c_p p f_0},$$

where $\bar{u}(y, p)$ is the prescribed basic zonal velocity, ψ the eddy streamfunction, Q diabatic heating rate. In (1) the nonlinear term of vorticity advection $J(\psi, \nabla^2\psi)$ has been omitted since it is one order smaller than the term of temperature advection in potential vorticity equation in the case of ultra-long wave. These equations are similar to those used in a previous work by Zhu Zhengxin and Zhu Baozheng (1982) which deals with blocking dynamics, except that the basic zonal wind $\bar{u}$ is variable with latitudes. This is because the south Asian high is situated at the dividing zone of easterly and westerly, therefore the basic zonal wind cannot be treated as a constant with latitudes.

By using the conventional method of two-level model with the mid-level at 400 hPa, and introducing the eddy friction in terms of vertical motion at the lower boundary caused by Ekman pumping, the equations of the two-level model can be written as

$$\frac{\partial}{\partial t}\nabla^2\psi = -\bar{u}(y)\frac{\partial}{\partial x}\nabla^2\psi - \bar{u}_T(y)\frac{\partial}{\partial x}\nabla^2\epsilon - (\beta - \frac{\partial^2\bar{u}}{\partial y^2})\frac{\partial\psi}{\partial x} - r\nabla^2(\psi - 2\epsilon) \tag{3}$$

---

1. Nanjing Institute of Meteorology, Nanjing, China

$$\frac{\partial}{\partial t}(\nabla^2 - \lambda)\,\epsilon = \lambda J(\psi,\epsilon) + \bar{u}(y)\lambda\frac{\partial\epsilon}{\partial x} - \bar{u}_T(y)\lambda\frac{\partial\psi}{\partial x} - \bar{u}(y)\frac{\partial}{\partial x}\nabla^2\epsilon -$$

$$\bar{u}_T(y)\frac{\partial}{\partial x}\nabla^2\psi - (\beta - \frac{\partial^2\bar{u}_T}{\partial y^2})\frac{\partial\epsilon}{\partial x} + r\nabla^2(\psi - 2\epsilon) - SQ\,, \qquad (4)$$

where $\quad \psi = (\psi_1 + \psi_3)/2\,, \quad \epsilon = (\psi_1 - \psi_3)/2\,, \quad \bar{u}(y) = (\bar{u}_1 + \bar{u}_3)/2\,,$

$$\bar{u}_T(y) = (\bar{u}_1 - \bar{u}_3)/2\,, \quad S = \frac{Rf_0}{c_p\sigma p_2^2}\,,$$

$$\lambda = \frac{2f_0^2}{\sigma p_2^2}\,, \qquad r = \frac{f_0 g}{RT_4}\sqrt{\frac{\gamma}{2f_0}}\,,$$

where ψ, ϵ are streamfunction and thermal wind streamfunction at 400 hPa respectively, γ the eddy viscosity coefficient, and $\bar{u}(y)$ and $\bar{u}_T(y)$ are assumed quadratic forms.

$$\bar{u}(y) = a_1 + b_1 y + c_1 y^2\,,$$
$$\bar{u}_T(y) = a_2 + b_2 y + c_2 y^2. \qquad (5)$$

Choosing the proper values of $a_1 \cdots\cdots c_2$, the actual zonal wind distribution in the subtropics with easterly in southern region and westerly in northern region and a jet near the northern boundary can approximately be fit in general. The channel region is of a width πD and length of $2\pi L$ on a β-plane centered at 30° N, which corresponds to a subtropical belt ranging a half latitude circle and spanning 28° latitudes with $L = 3000$ km and $D = 900$ km.

A highly truncated spectral method is adopted. The basic orthogonal functions are

$$F_A = \sqrt{2}\,\cos\frac{y}{D}\,, \qquad F_k = 2\cos\frac{x}{L}\sin\frac{y}{D}\,, \qquad F_L = 2\sin\frac{x}{L}\sin\frac{y}{D}$$

The diabatic heating field is specified in the form

$$Q = Q_A F_A + Q_K F_K\,, \qquad (6)$$

where the two terms on the right side indicate the meridional and sea-land heating differences respectively. The distribution of the idealized heating field shown in Fig. 1 is somewhat similar to the actual distribution with a source over the land of northern Africa and southern Asia and a sink over the Pacific ocean. The highly truncated spectral equations can then be written as:

$$-\dot{\psi}_A = r(\psi_A - 2\epsilon_A)\,, \qquad (7)$$

$$-L\dot{\psi}_k = -(\beta_1 - \bar{u})\psi_L + \bar{u}_T\epsilon_L + rL(\psi_k - 2\epsilon_k)\,, \qquad (8)$$

$$-L\dot{\psi}_L = (\beta_1 - \bar{u})\psi_k - \bar{u}_T\epsilon_k + rL(\psi_L - 2\epsilon_L)\,, \qquad (9)$$

$$-(D^{-2} + \lambda)Lk\dot{\epsilon}_A = \alpha_1(\psi_k\epsilon_L - \epsilon_L\psi_k) - rLkD^{-2}(\psi_A - 2\epsilon_A) - S_1 Q_A\,, \qquad (10)$$

$$-(k^{-1} + \lambda)LK\dot{\epsilon}_k = \alpha_1(\psi_A\epsilon_L - \psi_L\epsilon_A) + [\bar{u}(1 + \lambda k) - \beta_2]\epsilon_L \qquad (11)$$
$$+ \bar{u}_T(1 - \lambda k)\psi_L - rL(\psi_k - 2\epsilon_K) - S_1 Q_k\,,$$

$$-(k^{-1}+\lambda)Lk\dot{\bar{\epsilon}}_L = \alpha_1(\psi_k\epsilon_A - \psi_A\epsilon_k) - [\bar{\mu}(1+\lambda k) - \beta_2]\epsilon_K \qquad (12)'$$

$$-\bar{\mu}_T(1-\lambda k)\psi_K - rL(\psi_L - 2\epsilon_L),$$

$$\bar{\mu} = a_1 + \frac{1}{2}b_1\pi D + \frac{1}{3}c_1\pi^2 D^2 - \frac{1}{2}c_1 D^2,$$

$$\bar{\mu}_T = a_2 + \frac{1}{2}b_2\pi D + \frac{1}{3}c_2\pi^2 D^2 - \frac{1}{2}c_2 D^2,$$

$$k = (D^{-2} + K^{-2})^{-1}, \qquad \alpha_1 = \frac{8\sqrt{2}}{3\pi}\frac{\lambda k}{D}, \qquad S_1 = sLk,$$

$$\beta_1 = (\beta - 2c_1)k, \qquad \beta_2 = (\beta - 2c_2)k.$$

The Equilibrium Solutions and Their Stabilities

The steady solutions of equations (7)–(12) can be obtained analytically by algebraic operations.

$$\bar{\epsilon}_K = [1 + (\frac{2-d_1}{d_2})^2]^{-1}\left\{\frac{2-d_1}{d_2}\eta\frac{Q_A}{Q_K} \pm \sqrt{[1+(\frac{2-d_1}{d_2})^2]\frac{S_1 Q_A}{\alpha_1 d_2} - \eta^2\frac{Q_A^2}{Q_K^2}}\right\}, \qquad (13)$$

$$\bar{\epsilon}_A = -\bar{\epsilon}_k\frac{Q_k}{Q_A} + \frac{e_1}{\alpha_1 d_2}, \qquad (14)$$

$$\bar{\epsilon}_L = -(\frac{2-d_1}{d_2})\bar{\epsilon}_k + \eta\frac{Q_A}{Q_k}, \qquad (15)$$

$$\bar{\psi}_A = 2\bar{\epsilon}_A, \qquad (16)$$

$$\bar{\psi}_K = d_1\bar{\epsilon}_k - d_2\bar{\epsilon}_L, \qquad (17)$$

$$\bar{\psi}_L = d_2\bar{\epsilon}_k + d_1\bar{\epsilon}_L, \qquad (18)$$

where
$$d_1 = [2r^2 L^2 + \bar{\mu}_T(\beta_1 - \bar{\mu})] / [(\beta_1 - \bar{\mu})^2 + r^2 L^2],$$
$$d_2 = rL[\bar{\mu}_T - 2(\beta_1 - \bar{\mu})] / [(\beta_1 - \bar{\mu})^2 + r^2 L^2],$$
$$e_1 = \bar{\mu}_T d_2(1-\lambda k) + rL(2-d_1),$$
$$e_2 = \bar{\mu}\lambda k + \bar{\mu}_T d_1(1-\lambda k) - (\beta_2 - \bar{\mu}) + rL d_2,$$
$$\eta = \frac{e_1(2-d_1) + e_2 d_2}{\alpha_1 d_2^2}.$$

From (13)–(18) it is noted that there are two equilibrium states under the same forcing. Their features of the flow fields and the stabilities are very different, which will be illustrated later. Superposing perturbances on the equilibrium state, then linearizing equations (7)–(12), the coefficient matrix of the linearized equations is derived. When the maximum real parts of the eigenvalues of the matrix Max σ_r is negative, the state is stable.

Taking the vertical temperature gradient to be $0.6°C/100$ m, $\gamma = 1.647$ m^2/s, $s = 0.9699 \cdot 10^{-5}$ g/s.cal, and the zonal wind is given as

$$\bar{u} = -1.5 + \frac{11}{\pi D}y + \frac{21}{\pi^2 D^2}y^2, \qquad \bar{u}_T = \frac{3}{\pi D}y + \frac{9}{\pi^2 D^2}y^2.$$

Several calculated examples of these two kinds of equilibrium states and their stabilities are given in Table 1.

Table 1. The calculated results of two kinds of equilibrium state. Definitions of all variables can be found in the text.

Q_A Q_K 10^{-5} cal/g.s	Equilibrium state	$\bar\psi_A$	$\bar\psi_K$	$\bar\psi_L$ 10^7 m^2/s	$\bar\epsilon_A$	$\bar\epsilon_K$	$\bar\epsilon_L$	Max σ_r 10^{-7}s^{-1}	Stability
0.06 0.16	I	0.1269	−0.3589	0.8305	0.0635	−0.0347	0.3889	−0.2628	yes
	II	0.8461	−0.6286	0.6507	0.4231	−0.1696	0.3517	0.2604	no
0.06 0.17	I	−0.3190	−0.1804	0.8865	−0.1595	0.0460	0.3877	−0.5220	yes
	II	1.2919	−0.7489	0.5075	0.6459	−0.2383	0.3093	0.5380	no
0.06 0.18	I	−0.6088	−0.0738	0.9017	−0.3044	0.0917	0.3795	−0.6366	yes
	II	1.5817	−0.8039	0.4149	0.7909	−0.2734	0.2788	0.6700	no
0.06 0.19	I	−0.8484	0.0058	0.9046	−0.4242	0.1247	0.3699	−0.3198	yes
	II	1.8214	−0.8373	0.3426	0.9107	−0.2968	0.2537	0.7505	no
0.06 0.20	I	−1.0613	0.0694	0.9020	−0.5307	0.1504	0.3603	0.0278	no
	II	2.0343	−0.8593	0.2829	1.0172	−0.3139	0.2322	3.4270	no
0.05 0.17	I	−0.7619	−0.0201	0.8256	−0.3809	0.1035	0.3411	−0.3449	yes
	II	1.7349	−0.7544	0.3361	0.8674	−0.2637	0.2398	0.6411	no

From Table 1 it can be seen that state 2 corresponding to the minus sign in (13) is usually unstable so that it will not be discussed any more. However, state 1 corresponding to the positive sign in (13) is stable in most cases. Weak instability occurs only when the thermal forcing is too strong.

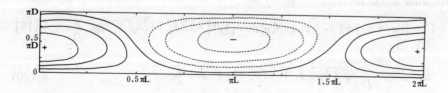

Fig. 1. Distribution of diabatic heating, the isopleth interval is 10^{-6} cal/g·s

322

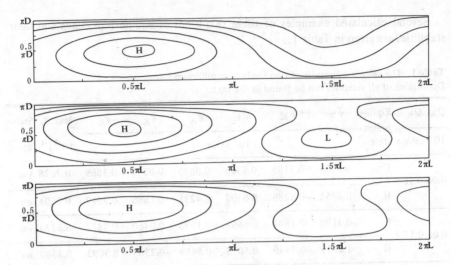

Fig. 2. Flow fields of equilibrium state 1 at 200 hPa. The isopleth interval is 10 m/s, corresponding to 72.9 m height difference. (a) $Q = 0.06.10^{-5}$, $Q = 0.17.10^{-5}$ cal/g.s; (b) $Q = 0.06.10^{-5}$, $Q = 0.19.10^{-5}$ cal/g.s; (c) $Q = 0.05.10^{-5}$, $Q = 0.1710^{-5}$ cal/g.s

Figure 2 (a)–(c) shows the flow fields of state 1 at upper level, (b) and (c) correspond to the cases when Q_K is larger or Q_A is smaller than the value in case (a) respectively. It is obvious that the high at 200 hPa is flat in Fig. 2 and occupies most of the area. The high is centered in the east of the heating source about 0.5 πL away. Its features are similar to that of the south Asian high not only in the configuration but also in the position since the actual location of the south Asian high is also in the east of the continent. The trough situated in the east of the cooling source is somewhat analogous to the Pacific trough in the upper troposphere. Hence, the features of the upper level flow fields of equilibrium state 1 and its stability may reflect the main characteristics of the actual south Asian high. However, it should also be pointed out that the main discrepancy is that the location of the high is somewhat in the east of its actual position. This may be due to the quasigeostrophic approximation and the omission of topography in the model.

The zonal mean velocity U_1 at the upper level and the zonal wind u_1 at the longitude where the high is centered are

$$U_1 = [(a_1 + a_2) + (b_1 + b_2)y + (c_1 + c_2)y^2] + \frac{(\psi_A + \epsilon_A)}{D}\sqrt{2}\, \sin \frac{y}{D}, \quad (19)$$

$$u_1 = U_1 - \frac{2}{D}\sqrt{(\psi_K + \epsilon_K)^2 + (\psi_L + \epsilon_L)^2}\, \cos \frac{y}{D}. \quad (20)$$

Figure 3 shows the profile of U_1 with curves a, b and c corresponding to the cases (a), (b) and (c) in Fig. 2 respectively, and the dashed line is the specified basic zonal wind. The distribution of zonal mean wind is similar to the actual one in

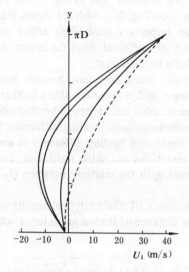

Fig. 3. The zonal mean wind profile at
the 200 hPa level

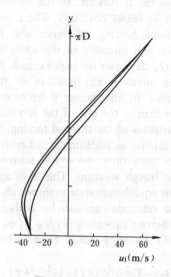

Fig. 4. Zonal wind at the longitude
where the high is centered

the subtropics with the easterly in its southern part and the westerly in the northern part. Figure 4 shows the profile of u_1. The profile is similar to that in Fig. 3 except that the speeds of both the easterly and westerly are much larger than those of zonal mean wind. Especially, a notable easterly jet is found in the southern area in case b and case c. This is due to the zonal mean wind superposed by the velocity field related to the south Asian anticyclone, which results in the much stronger easterly.

Based on the above analysis, it can be concluded that the flow fields at 200 hPa for the equilibrium state 1 account for the main characteristics of subtropical planetary wave systems in the upper troposphere such as the south Asian high and the Pacific trough. The similarity is not a coincidence, bue rather reveals that the south Asian high and the Pacific trough is a sort of stable, nonlinear equilibrium state induced by the ocean-continent heating difference and topography. In this aspect, it does have the dynamic mechanism similar to that of the blocking situation as discussed in (Zhu Zhengxin and Zhu Baozhen 1982). However, there is another important factor involved in the dynamics of south Asian high, i.e., the dynamic effect of the Hadley cell which causes the distribution of the basic zonal wind.

The Controlling Effects of Thermal Forcing and the Seasonal Variation of the South Asian High

Comparing Fig. 2 (a) with (b), it is noticed that when the ocean-continent heating

324

difference increases, the high will intensify and expand notably, and its center will shift to the northwest. In the meantime, the intensity and extent of the Pacific trough decreases obviously. The comparison of Fig. 2(a) with (c) shows that the meridional heating difference also has an important controlling effect on the position and intensity of the south Asian high, different from the effect of Q_K, when Q_A decreases the high expands and shifts to northwest.

The intensity and position of the south Asian high have notable seasonal variations. In midsummer it becomes stronger and its position shifts further northwest than in the rest of the summer. The seasonal variations can be attributed to the variation of the thermal forcing. The ocean-continent heating difference is the most intense in midsummer, whereas the meridional heating difference is weaker. At the same time, the south Asian high intensifies and shifts northwest, and the Pacific trough weakens. This is in agreement with the relations between Q_K, Q_A and the equilibrium state as shown above.

The relations can also be analyzed in terms of the analytical results of the state. Setting $\phi_0 = \arctan(\bar{\epsilon}_L/\bar{\epsilon}_K)$, $L\phi_0$ is the distance of the warm center at 400 hPa east of the heating source. Then we have

$$\text{tg}\phi_0 = -(\frac{2-d_1}{d_2}) + [1+(\frac{2-d_1}{d_2})^2] / [(\frac{2-d_1}{d_2}) + \sqrt{[1+(\frac{2-d_1}{d_2})^2]\frac{S_1 Q_K^2}{\alpha_1 \eta^2 Q_A} - 1}] \ . \ (21)$$

According to (21), when Q_K increases or Q_A decreases, ϕ_0 will decrease, so the center will move westwards. This is why the position of the south Asian high and Pacific trough shift westwards in these cases.

The controlling effects of thermal forcing on the high and the trough can be examined from the effects of Q_K, Q_A on the wave component and the zonal flow component of the equilibrium state. The amplitude of the wave component can be expressed as:

$$\sqrt{\bar{\epsilon}_K^2 + \bar{\epsilon}_L^2} = \sqrt{\frac{S_1 Q_A}{\alpha_1 d_2}} \tag{22}$$

As shown by (22) the wave amplitude is proportional to the square root of Q_A but independent of Q_K. This seems to contradict the relations between Q_K, Q_A and the intensity of the high as shown in the former examples. However, in addition to the contribution of the wave component, the intensity and position of the high and the trough are also determined by the zonal flow component of the state.

The zonal mean streamfunction $\bar{\psi}_1$ on 200 hpa level can be derived from (19):

$$\bar{\psi}_1 = -[(a_1+a_2)y + \frac{1}{2}(b_1+b_2)y^2 + \frac{1}{3}(c_1+c_2)y^3] + (\bar{\psi}_A + \bar{\epsilon}_A)\sqrt{2}\cos\frac{y}{D} + const \ , \tag{23}$$

where the first term on the right hand corresponds to the prescribed basic zonal flow, the second term corresponds to the zonal flow component of the equilibrium state. From Table 1 it is found that when Q_K increases or Q_A decreases the negative value of $(\bar{\psi}_A + \bar{\epsilon}_A)$ will increase; consequently, the zonal mean high pressure zone will intensify and move northwards. This can be seen in Fig. 5 by comparing the curves a, b, and c, where case (b) and case (c) correspond to a larger

Q_K or a smaller Q_A in comparison with case (a). The zonal mean high pressure zone in case (b) and case (c) is much more strong and northward than in case (a).

Hence, although the amplitude of the wave component of the equilibrium state does not change when Q_K increases according to (22), the high of the wave component superposed on the intensified zonal mean high pressure zone does intensify, expand, and shift northwards, whereas the low of the wave component superposed on the high pressure zone will weaken as a trough. This may be the intrinsic cause of the seasonal variations of the south Asian high and the Pacific trough.

On the other hand, when Q_A decreases, the zonal mean high pressure zone will also intensify and shift northwards, but according to (22), the amplitude of the wave component becomes smaller, so that it cannot determine whether the high consisting of the weakened wave component and the intensified zonal mean high pressure zone will intensify or weaken. However, two conclusions can be drowned in this case, i.e., the high will shift northwards and the zonal mean subtropical high pressure zone will intensify.

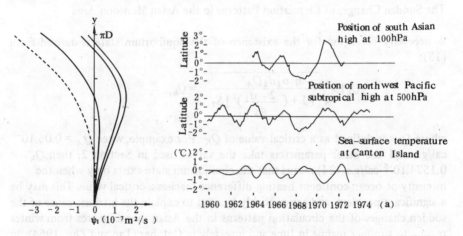

Fig. 5. The distribution of zonal mean stream-function ψ_1 at 200 hPa level

Fig. 6. SST Variation oat Cantou Island, the latitudinal locations of the South Asian high at 100 hPa and the subtropical high over the Northwest Pacific at 500 hPa

The controlling effects of difference in meridional heating on the position and intensity of the south Asian high and other subtropical systems can also be verified by some observational evidence. Figure 6 shows the sea-surface temperature at Canton Island (2°48′ S, 171°43′ W), the latitudinal locations of the south Asian high at 100 hPa and the subtropical high over the northwest Pacific at 500 hPa respectively (Zhu, 1980). The diagrams show an obvious relation of opposite phase between the sea-surface temperature and the positions of the highs. Canton Island is located in the west of the cool water belt in the equatorial Pacific. The positive departure of the temperature indicates the warming of the cool water zone, then the meridional heating difference becomes larger due to the increase of equatorial

326

sea-surface temperature. It is also evident in Fig. 6 that in the mean time the positions of the south Asian high and the northwest Pacific subtropical high deviate southwards from their mean positions. This observational evidence supports the results of the analysis of the controlling effects of Q_A on the south Asian high as discussed above.

The flow field of the equilibrium state at the lower level will be addressed briefly as follows. There is a zonal mean high pressure zone at 600 hPa level, which weakens as well when Q_A increases. Therefore when the meridional heating difference increases, the zonal mean subtropical high pressure zone will weaken and shift southwards. This may cause the weakening of the trade wind and then the equatorial sea-surface temperature will increase further. Hence, it seems that the relationship between the meridional heating difference and the zonal mean subtropical high pressure zone consists of a sort of positive feedback loop, which might play an important role in the mechanism of El-Nino events.

The Sudden Changes of Circulation Patterns in the Asian Monsoon Area

A necessary condition for the existence of the equilibrium state is derived from (13):

$$|Q_k| \geqslant \sqrt{\frac{\eta^2 \alpha_1 d_2 Q_A}{[1+(\frac{2-d_1}{d_2})^2]S_1}} \equiv Q_{kc} \quad ,$$

where Q_{kc} is defined as a critical value of Q_K. For example, when $Q_A = 0.06.10^{-5}$ cal/g.s and the other parameters take the values used in Section 2, then $Q_{kc} = 0.1574.10^{-5}$ cal/g.s. (24) means that the equilibrium state exists only when the intensity of ocean-continent heating difference reaches a critical value. This may be a significant problem, and can probably be used to explain the intrinsic cause of the sudden changes of the circulation patterns in the Asian monsoon area from winter regime to summer regime in June and inversely in October (Tao and Zhu, 1964). In fact, the sudden changes of the circulation patterns can be regarded as the processes of onset or decay of the equilibrium type of south Asian anticyclone. The intrinsic cause of the sudden changes may be due to the fact that when the intensity of the thermal forcing, mainly caused by ocean-continent heating difference, reaches or falls below a certain critical value, the equilibrium state will be set up or can no longer be maintained.

References

Tao Shiyan Zhu Fukang, (1964) *Acta Meteorol Sin* 34: 385–395
Zhu Zhengxin, Zhu Baozhen, (1982) *Sci Sin* (Ser B) 25: 1201–1212
Zhu Fukang, et al. (1980) *The South Asian High* Science Press, Beijing, p21

Simulation Capability and Senstitivity of the Regional Circulation to Orography in the Low Resolution Spectral Model: The Summer Asian Monsoon Circulation

Ni Yunqi[1] , Bette L. Otto-Bliesner[2] and David D. Houghton[2]

Abstract — In order to investigate simulation capability and sensitivity of the regional circulation to orography in a low resolution spectral model, the effect of orography on the summer monsoon circulation in the Asian continent is examined by comparing general circulation model simulations with and without orography included. The model is a 5-level global low-resolution spectral model of atmospheric circulation which incorporates the primitive equations augmented by physical parameterization and mountains. The statistical significance of effects caused by the change in topography specification is determined by comparison of these changes with natural model interannual variability in a 5-year simulation of the model with orography. Many of these effects were shown to be statistically significant to the 5% level in terms of the natural model interannual variability. The control simulation of the model with mountains captures the large-scale features observed in the south Asian monsoon. They include a warm anticyclonic circulation and easterly jet in the upper troposphere located over Tibet and southeast of Tibet, a cyclonic system and westerly jet in the lower troposphere located over Tibet and the vicinity of South Asia.

Results clearly show how orograpohy serves to modify the position and intensity of monsoon circulation. Comparison between simulation with mountains and without mountains reveals that the effect of orography enhances and modifies the cyclonic system over all the south Asian continent and weakens the southwesterly jet in the lower troposphere. In the upper troposphere, it moves the center of the south Asian anticyclone northward from a latitude south of the Himalayas to the highest peaks of Tibet, intensifies the easterly jet and forms an apparent warm core. In contrast, the experiment of Hahn and Manabe (1975) suggested that the effect of orography was to produce the cyclonic circulation over south Asia, i.e. without mountains they had only a continental midlatitude low centered at 50°N. The experiments also show that orography not only contributes to latent heating due to uplift moisture but also causes important sensible heating west of Tibet in July. This latter effect was not shown in the Hahn and Manabe study. This study demonstrates a regional simulation capability of a low resolution global spectral model including topographyic influences, which supports other applications of such models.

Introduction

In the past two decades, general circulation models (GCM's) have been applied to the study of the summer monsoon over eastern Africa, southern Asia, and the the nearby oceans for a better understanding of the large-scale features of the monsoon (Washington 1970; Murakami et al. 1970; Godbole 1973; Alyea 1972;

1. Department of Meteorology, Nanjing University, Nanjing, China
2. Department of Meteorology, University of Wisconsin, Madison, WI537-6, USA

328

Abbott 1973, Manabe et al. 1974; Washington and Daggupaty, 1975; Hahn
and Manabe 1975; Gilchrist 1974, 1976). Despite differences in the models,
all these models qualitatively simulated the upper-troposphere tropical easterly jet,
the Somali jet and the low-level westerly flow in the vicinity of India. However, the
precipitation amounts in the models varied greatly from each other and from those
observed. Research results have not fully clarified how the thermal effects of the
Tibetan Plateau influence the development and maintenance of the south Asian
monsoon (Reiter and Gao 1982), and debate exists whether the temperature
structure is caused by sensible heat from the Earth's surface or by latent heat from
precipitation over the continent. It is evident that mountains play an important role
in this situation.

Hahn and Manabe (1975) studied the effects of mountains on Asian monsoon
circulation by comparing the simulation in their model with mountains with that
without mountains using a GCM with 11 layers and a horizontal resolution of
approximately 270 km. Their experiments reveal that the presence of mountains is
instrumental in positioning the south Asian low pressure system. In the experiment
without mountains, the low-pressure system moved far to the north and east of the
Himalayas. In the model with mountains, higher temperatures are maintained in the
middle and upper troposphere, and upward motion and latent heating dominate
over the Tibetan Plateau. Conversely in the simulation without mountains,
downward motion and sensible heating by the earth's surface dominate in this
region, and high temperature over the Tibetan region extends the low pressure
southward over the plains of south Asia. They concluded that the monsoon
circulation extended further into the Asian continent in the control experiments
with mountains. The onset of the south Asian monsoon is also influenced by
the effects of mountains. In the simulation with mountains, the onset occurred
rather abruptly with the jet stream quickly moving north of the Himalayas into its
summertime position. In the model without mountains, the subtropical jet gradually moves northward over a period of about two months, finally reaching a
summertime position approximately $10°$ further south than that in the model
with mountains. They attributed these differences to the mechanical and thermodynamical effects of the Tibetan Plateau. Although their model is capable of simulating the mean July circulation features which are associated with the south Asian
monsoon, there are deficiencies in the simulated results, such as the lowest pressure
of the simulated south Asian low being deep and located too far toward the east, and
the center of the simulated anticyclonic circulation in the upper troposphere being
located about $5°$ too far south. It has been demonstrated by Hahn and Manabe
(1975) that their high resolution global circulation model successfully simulates
some of the important features of the south Asian monsoon. Their studies also
isolate effects of mountains on the south Asian monsoon.

The important question here is whether or not a similar regional simulation
and sensitivity capability can be obtained with a low-resolution global spectral
model. This will clarify the scope of application of such low-resolution models
which offer a unique capabilitiy to provide statistical ensembles for climate
ahalysis. The results from the low-resolution spectral model with and without
mountains are compared. The important features of the south Asian monsoon are

examined and the role of mountains in the south Asian summer monsoon identified. Comparison of results to observation and hgh resolution GCM results provides verification of the regional simulation capability of a low-resolution global spectral model.

An important consideration in this study is the determination of the significance of the differences between the simulations made with and without topography present. The problem is to determine how much of these differences is actually due to the mountains and how much is due to the inevitable differences in the model's solutions caused by natural model variability associated with the essentially unpredictable synoptic-scale motions an other effects. A statistical test-student test is used to examine the climatic signal in the solutions. Simulation differences (the without mountain case minus the mountain case) are compared to the standard deviation for model variability determined from and ensemble of simulations made with the mountain case, the effects of topography can be considered statistically sigificant at the 1% level if the difference equals 5 standard deviations, at the 5% level if the differenc equals 3 standard deviations.

In this study, a brief description of the model is provided in Section 2, a discussion of the simulated south Asian monsoon circulation in the model highlighting the role of mountains, especially the thermodynamical effects of Tibetan Plateau, is presented in Section 3. Discussion and conclusions follow in Section 4.

Brief Decription of the Model

The model with mountains (the M model) is the same as that used by Otto-Bliesener et al. (1982). A comprehensive description of the M model is contained in the paper by Otto-Bliesner et al. (1982). Therefore, the following description is brief.

The model has a global spherical coordinate domain and incorporates the primitive euqations augmented by physical paramenterization. Model variables are discretized in the vertical and carried at five equally spaced sigma levels in the free atmosphere, plus a surface level. In addition, vertical velocity is calculated at intermediate levels.

These equations are solved by using the Galerkin procedure, where the horizontal distribution of model variables are represented as linear combinations of surface spherical harmonics. The horizontal resolution is defined by triangular truncation at wavenumber 10. The transform method of solution is used with a $11.6°$ by $11.25°$ latitude-longitude transform grid. The model makes use of the semi-implicit time integration scheme has a timestep of 90 minutes.

In the model, all physical parameterizations are incorporated into the model in grid-point space. Those processes, important for long-term integration, are as follows: radiation including the shortwave radiation absorbed by the model atmosphere and the surface, the longwave radiation cooling; condensation and convection; horizontal and vertical diffusion. More detailed description can be found in Otto-Bliesner et al. (1982).

This paper includes a study of the effects of orgography on the monsoonal circulation by comparison of the model runs with and without the mountains, denoted M model and NM model respectively. For the model without mountains, the dynamical and thermodynamical equations and physical processes are those in the M model except for eliminating all land elevations, and initial conditions are taken from a computer data tape of the M model. But in order to reduce to magnitude of surface pressure adjustment responding to the large change in land elevations, surface pressure is initially set to 1013.25 hPa everywhere rather than using values from the M model.

Monsoon Simulated in the Model and the Role of Mountains in Monsoon Circulation

Lower and Middle Troposphere

Geopotential height fields. The July time-mean 900 hPa geopotential height fields as computed by the M model is compared with its oberved 850 hPa geopotential height fields in Fig.1a–b.

The low pressure belt extending from Arabia to the east coast of Asia associated with the south Asian monsoon is simulated although its intensity is overestimated and the position of low center is 10° longitude too far west. The location of the subtropical high over the western Pacific is similar to observation although its intensity is underestimated.

The difference map between the 900 hPa geopotential heights in the NM model and the M model is represented in Fig. 1c. The difference between the 900 hPa geopotential heights in both models exceeds 50 m over east of Tibet, east coast of Asia and the western Pacific which is 3 standard deviations in terms of natural model variability. Comparing the 900 hPa geopotential height in the M model with that in the NM model, we find that the continental low is located about 35°N, 100°N in the NM model, far to the east of the low center in the M model which is centered about 35°N, 80°E near the highest point of the Himalayas. In addition the intensity of the tropical anticyclone and gradients from the subtropical anticyclone toward the Asian low belt are reduced in the NM model. According to the statistical test described above, these differences between results from both models are significant at the 5% level. Results show that the effect of Tibet is not only favorable to the maintenance of a low center near the highest point of the Himalayas, but also to the large-scale South Asian monsoon circulation by intensifying the subtropical anticyclone over the western Pacific and height gradients from the subtropical anticyclone toward the south Asian low. The latter result differs from Hahn and Manabe (1975), who characterized the primary effect of mountains on the Asian monsoon as shifting the low center to much nearer the Equator from its position at about 50°N, 125°E in the model without mountains.

Wind field. The July time-mean 900 hPa flow field as computed by the M model is compared with its observed 850 hPa wind field in Fig. 2b and 2a. The observed flow field in the lower troposphere associated with the south Asian monsoon is characterize by southeast flow from the Southern Hemisphere, becoming a south-

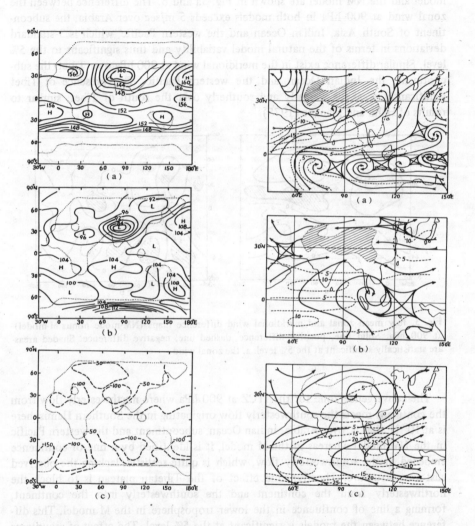

Fig.1. Horizontal distribution of July mean geopotential height (10 m) for the observed 850 hPa surface (a) and simulated 900 hPa surface (M-model) (b); and July mean geopotential height difference map (NM-model minus M-model, units: m) (c)

Fig. 2. July mean flow field at the 850 hPa level (observed) and 900 hPa level, M-model (b) and NM model (c). Dashed lines are isotaches (m/sec)

westerly as it crosses the Equator, and converging into the vicinity of the south Asian low pressure belt. The axis of the southwesterly jet is located at about 10°N. The observed characteristics of the south Asian monsoon are simulated by the M model except that the intensity of the southwesterly jet located at 60°E and south of the Equator is underestimated.

The difference maps of the zonal and the meridional wind between the M model and the NM model are shown in Fig. 3a and b. The difference between the zonal wind at 900 hPa in both models exceeds 5 m/sec over Arabia, the subcontinent of South Asia, Indian Ocean and the western Pacific, which is 3 standard deviations in terms of the natural model variability and thus significant to the 5% level. Similar difference exist in the meridional wind at 900 hPa over Tibet? the subcontinent, the Indian Ocean and the western Pacific. The presence of Tibet significantly weakens westerly and southerly over the south of Tibet, similar to results of Hahn and Manabe (1975).

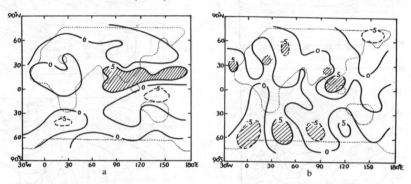

Fig. 3. July mean zonal and meridional wind difference maps (NM-mode minus M-model) at 900 hPa. Solid line, positive difference; dashed line, negative difference. Shaded areas are statistically significant at the 5% level. a. the zonal wind

The convergence zone for the ITCZ at 900 hPa where northwesterly flow from the continent meets the southwesterly flow originating in the Southern Hemisphere is at 10° N, that is, over Arabia, Indian Ocean, subcontinent and the western Pacific in the NM model whereas in the M model, it is identified by a line of confluence embeded in the southwesterly flow, which is quite realistic, lacking the observed east-west structure. The dynamic effect of the Tibetan plateau is to block the northwesterly from the continent and the southwesterly into the continent, forming a line of confluence in the lower troposphere in the M model. This difference between the models is significant at the 5% level. The effect of mountains on the wind field in the lower troposphere verified by our experiments is in agreement with Hahn and Manabe (1975). It is noteworthy that the cyclonic circulation at 500 hPa over south Asia is not simulated in the NM model; whereas in the M model and observations the low-level cyclonic circulation over the south of Tibet reaches 500 hPa level.

Temperature field. The July time-mean 900 hPa temperature for the M model and 850 hPa temperature for observation are shown in Fig. 4b and 4a. In the model, maximum 900 hPa temperatures are found over the Arabian Peninsula and northwestern Africa in good agreement with the observed 850 hPa temperature field. In the NM model, the highest temperatures at 900 hPa are shifted to the south Asian continent (see Fig. 4c).

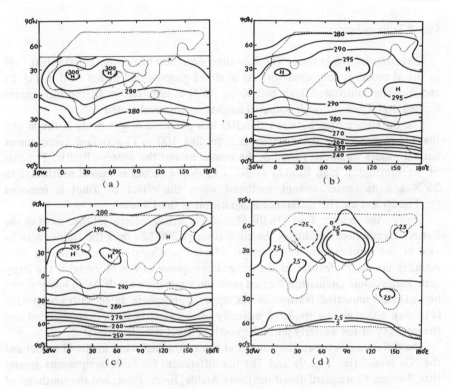

Fig. 4. July mean temperature field of the 800 hPa surface (observation, units: °K) (a) and 900 hPa surface ((M-model) (b) and NM model (c) units: °K) and the temperature difference (NM-model minus M-model minus M-model) (d). Solid, dashed lines and shaded areas in the difference map are the same as in Fig. 3

A difference map between surface temperature in the NM model and the M model is presented in Fig. 4d. The difference between the surface temperature in both models exceeds 5° C (3 standard deviations of natural model variability) in the Tibetan area. The presence of Tibet decreases temperature of surface and lower troposphere in Tibet and the vicinity in agreement with Hahn and Manabe (1975).

Moisture. By comparing qualitatively the simulated 900 hPa and 500 hPa moisture with the observations, it is evident that the centers of 500 hPa and 900 hPa moisture maximum near 30°N are simulated, but the center of 500 hPa maximum moisture is located too far north.

In the NM model, the 900 hPa moisture has a high moisture area in a long east-towest region from the western Pacific to the west coast of Africa and the south-to-north moisture gradient is uniform with the center of maximum moisture located at about 30°N, 90°E. In the M model the mixing ratios at 500 hPa are larger and the maxima are located further east over eastern Asia and the western

Pacific.

Upper Troposphere

Geopotential height field. The July time-mean 300 hPa geopotential height field in the M model is compared with its observed geopotential hiehgt field in Fig. 5a and 5b. The subtropical high belt along 30°N is simulated with the high center located over Tibet although potential heights are underestimated.

The difference map between the 300 hPa potential height fields in the M and the NM model shows the differences exceeding 100 m (5 standard deviations of natural model variability) over all the continent and the western Pacific. The axis of the subtropical high located at 30°N in the M model is moved southward to 25°N and its center toward southeast when the effect of Tibet is removed (cf. Fig. 5b and c). This difference is significant at the 1% level.

Upper tropospheric flow. In the M model, a strong easterly jet over most of the Eastern Hemisphere dominates the circulation at 300 hPa over the tropics with its axis in the southeastern Tibet region oriented northeast-to-southwest (Fig. 6b). Another important feature in the upper troposphere of the M model is the large-scale anticyclonic circulation located over the warm-core south Asian low pressure belt. These simulated features in the upper troposphere agree with observation (Fig. 6a), although the modeled intensity of the easterly jet is overestimated and the intensity of the westerly jet in the north of Tibet is underestimated.

The difference between the zonal and the meridional wind in the M model and the NM model (see Fig. 7a and 7b) has differences for both components greater than 5 m/sec (3 standard deviations) over Arabia, India, Tibet, and the southeast of Asia. The zonal and meridional wind in the upper troposphere are significantly intensified at the 5% level when mountains are included. Another effect of mountains is to move the axis of westerly and easterly jets from 70°N and near the Equator in the NM model to 55°N and 20°N in the M model respectively, and the center of anticyclone from a latitude south of the Himalayas in the NM model (see Fig. 6c) to the highest peak of Tibet in the M model, (see Fig. 6b). These results differ from those of Hahn and Manabe (1975), who showed that the main effects of mountains were to move the center of the south Asian anticyclone and to shift the westerly and easterly jets. In both our and Hahn and Manabe experiments, the upper tropospheric anticyclone lies almost directly over the south Asian low at the surface and the highest mountains in the M model; however, the upper tropospheric anticyclone still lies over the south Asia low at the surface, although the low located a little far to southeast of Tibet is weaker in our NM model, whereas in Hahn and Manabes NM experiment, the anticyclone lies to the south of Tibet and the surface low lies far to the northeast of Tibet (about 30°latitude). In fact, the structure of Asian monsoon circulation did not exist in the NM model of Hahn and Manabe's experiment. According to our experiment, the inclusion of Tibet has an important influence on the location and intensity of Asian Monsoon circulation but not on its existence (or formation). This problem will be further discussed in Section 4.

Temperature field. The simulation position and intensity of the warm tem-

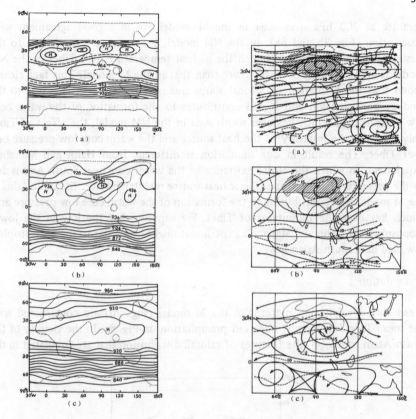

perature at 300 hPa over Asia in the NM-model are very close with
observed ones. With what shown in the NM-model, it is due to the
warm source. And the global isotherm of air temperature in the NM-
model is higher than that at 300 hPa. So that, it is clearly
shows ... ... local warm area is warm than others to the
... ... contributes to rise thermal low, otherwise, core
low ... ... South Asia in the M-model. It ... that either
only ... ... the heat source and the warm core low pressure belt
over Tibet. The result of air circulation is different from that in the M-model's
experiment ... ... existence of the warm core low area is due
to the ... ... heat source in it. ... ... results in
the ... ... the formation of the warm core low pressure area
which ... ... over Tibet. We suppose ... ... lower
tropospheric ... ... experiment may ... ... tropical
low ... ...

Fig. 5 Horizontal distribution of observed (a) and simulated July mean geopotential heights (10 m) of the 300 hPa surface for M-model (b) and NM-model (c)

Fig. 6. Observed (a) simulated July mean flow field at the 300 hPa surface for M-model (b) and NM-model (c). Dashed line, istoaches (m/sec)

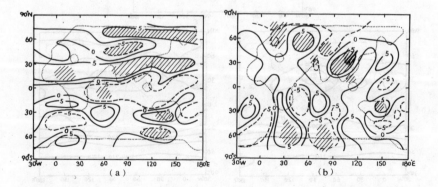

Fig. 7. As in Fig. 3 except for 300 hPa

perature at 300 hPa over Asia in the M model are in a good agreement with observation (Fig. 8b and 8a). In the NM model, a warm area from Arabia to the western Pacific still exists although the highest temperature over Tibet in the NM model is approximately 9°–10° lower than that in the M model. This fact clearly shows that the formation of a local warm area over the south Asia is due to the land-sea configuration and it still contributes to the formation of the warm core low pressure area which lies over south Asia in the NM model; the effect of Tibet only increases the intensity of the heat source and the warm core low pressure belt over Tibet. The result in our simulation is different from Hahn and Manabe's experiment, in which the entire existence of the warm core over South Asia is due to the presence of a midtropospheric heat source maintained over the mountains in the M model, and contributing to the formation of the warm core low pressure area which lies over the mountains of Tibet. We suppose that the low in the lower troposphere in Hahn and Manabe's experiment might be similar to the extratropical low, and no a south Asian low.

Precipitation

Mean July precipitation patterns in the M model (Fig. 9b) are constructed with the mean June-July-August observed precipitation in Fig. 9a. In the vicinity of the south Asian monsoon, the features of rainfall distribution that are simulated in the

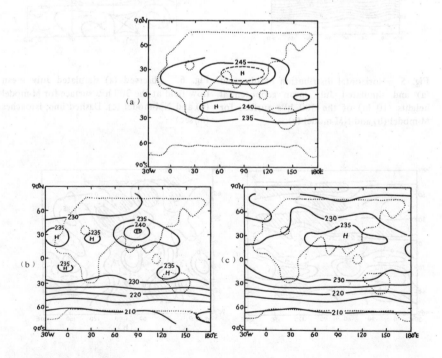

Fig. 8. Observed (a) for M-model (b) and NM-model (c) and simulated July mean temperature field of the 300 hPa surface (°K)

M model are the regions of maximum rainfall in the Western Pacific at about 10°N, 145°E, and the areas of minimum rainfall in the Arabian peninsula and Arabian Ocean. The most striking discrepancies of rainfall distribution are underestimates of rainfall over the Indian peninsula and the extension of large precipitation amounts too far north over Asia.

Figure 9d shows the difference map between the precipitation in the M model and the NM model. The difference between the precipitation in both models exceeds 0.2 cm/day (3 standard deviations of natural model varibility) in Tibet, the southeast coast of Asia and the western Pacific. Figure 9d shows that precipitation increases the southeast coast of Asia and the western Pacific, decreases in the Tibet area, with the maximum center of precipitation located in the western Pacific. The model precipitation distribution is different from Hahn and Manabe's experiment in which a desert-like climate is simulated in south Asia in the model simulation without mountains. The model results clearly show that Tibet increases rainfall and maintains the maximum center of precipitation in Tibet; however, the southeast of Tibet receives less rainfall than in the NM model (see Fig. 9c).

The Vertical and Meridional Motion

In the M model, upward motion exists over most of Tibet. This is in close

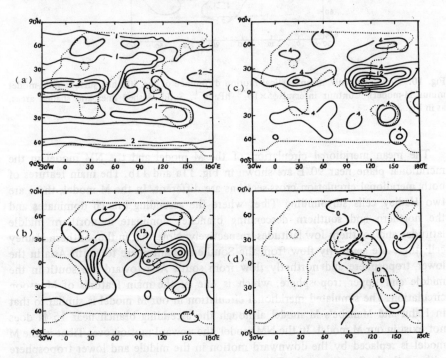

Fig. 9 Distribution of observed (a) and simulated July mean precipitation rate for M-model (b) and NM-model (c) and simulated precipitation difference (NM-model minus M-model) (d). Units: cm/day. Solid, dashed lines and shaded areas are the same as in Fig. 3.

338

agreement with observations.

The difference map between the vertical motion at 500 hPa in the M model and in the NM model is presented in Fig. 10. The difference exceeds 0.05×10^{-4} hPa sec^{-1} and (3 standard deviations) north over south of Tibet, the southeast coast of Asia. In the NM model, raising motion exits in the southeastern Tibet, the southeast of Asia and the western Pacific and sinking motion in most of Tibet and the subcontinent. This is different from the result of Hahn and Manabe's experiment where broad bands of descending motion are both to the north and south of a narrow band of intense upward motion oriented roughly east-west and centered about 10°N.

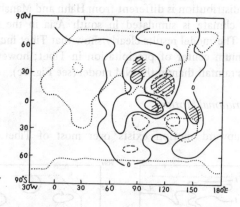

Fig. 10. Simulated July mean vertical motion difference of the 500 hPa surface (NM-model minus M-model), Contour intervals 5×10^{-6} hPa s^{-1}. Solid, dashed lines and shaded areas, as in Fig. 3

The mean meridional circulation of the M model and the NM model in the meridional plane near 90°E are shown in Fig. 11a and 11b. The main features of both meridional circulation cross-sections are different. In the M model, there are two Hadley cells surrounding Tibet where the ascending branch dominates and the northern and southern descending branches dominate in northern middle latitudes and southern low latitudes, respectively. Note that in the southern Hadley cell, there are southerly flow from the Southern Hemisphere to south Asia in the lower troposphere and northerly flow from south Asia toward the south in the middle and upper troposphere, which is one of the main features of Monsoon circulation. The simulated meridional circulation in our M model is similar to that in Hahn and Manabe's M model although the ascending branch near 15°N does not exist in our M model. In the NM model, the upward motion over Tibet in the M model is replaced by the downward motion in the middle and lower troposphere with northerly flow across the Tibet region, and there are two Hadley cells over the Tibet region in the upper troposphere. The main features of this meridional circulation are not typical of monsoon circulation since the monsoonal circulation system moves to the southeast of Tibet in the NM model. The main features of meridional circulation along 90°E in the NM model are similar to those in Hahn and Manabe's experiment (1975).

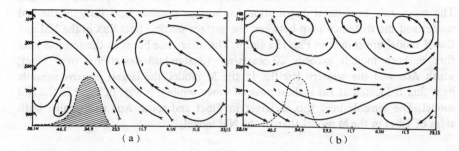

Fig. 11. Simulated July mean meridional circulation profiles along 90°E for M-modle (a) and NM-modle (c)

Figure 12 shows the mean meridional circulation of the NM model in the meridional plane near 112.5°E. In Fig. 12, main features of the meridional circulation are similar to those in the M model. The monsoonal circulation still exists when Tibetan Plateau is removed although it is shifted to the southest, which is much different from Hahn and Manabe's simulation without mountains that the main features of the monsoon meridional circulation disappear in the NM model.

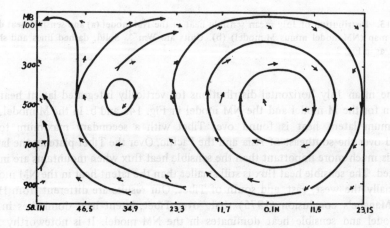

Fig. 12. Simulated July mean meridional circulation profile along 112.5°E in the NM-modle

The Thermal Effects of Orography

Two mechanisms for maintaining high temperature over the moutnains are sensible heating from the earth's surface and latent heating released by uplifting moist air. The mean July horizontal distribution for surface sensible heat flux for the NM model and the differnce map between the surface sensible heat flux in the NM and the M model are shown in Fig. 13a and b. The difference between the sensible heat flux in both models exceeds 20 watts/m^2 (3 standard deviations) over Tibet, south Asia and the western Pacific. In the M model, the largest positive sensible heat flux is located at the highest peak of Tibet, while it is west of Tibet in the NM model. The sensible heat flux is larger in Tibet and south Asia and smaller both sides of Tibet in the M model than in the NM model.

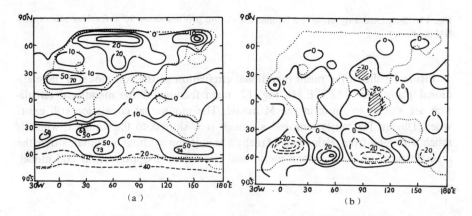

(a)

(b)

Fig. 13. ·Distribution of July mean sensible heat in the NM model (a) and sensible heat difference map (NM-model minus M-model) (b). Units are Wm^{-2} Solid, dashed lines and shaded areas, as in Fig. 3

The mean July horizontal distributions for vertically integrated latent heat are shown for the M model and the NM model in Fig. 14a and b. In the M model, the maximum latent heat is found over Tibet with a secondary maximum to be found over the southeast of Asia and the Pacific. Over the Tibet plateau, the latent heat is much more important than the sensible heat flux when mountains are incorporated. The sensible heat flux is still smaller than the latent heat in the NM model, especially the west, east, and south of Tibet. Our results are different from Hahn and Manabe's conclusion (1975) that over Tibet, latent heat dominates in the M model and sensible heat dominates in the NM model. It is noteworthy that the extent of the 300 hPa high temperature agrees quite well with the area where the latent heat dominates in both models. This shows that the latent heat is the most important heat source for the formation and maintenance of the high temperature areas in both models and it contributes to the formation of the warm core low pressure area which lies over the Tibet in the M model and southeast of the Tibet in the NM model.

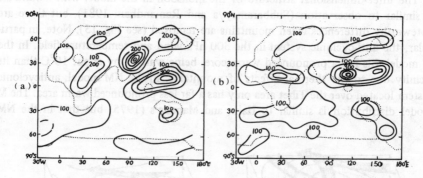

Fig. 14. Distribution of July mean latent heat (watts/m^2) in the M modle (a) and NM model (b)

On the whole, the thermal effects of Tibetan Plateau are to maintain the south Asia low located over the Tibet due to latent heating induced by mountain effects.

Discussion and Conclusions

This study demonstrated that the low resolution global spectral model with mountains is capable of simulating the large-scale features of the summer Asia monsoon circulation. In the M model, a large warm-anticyclonic circulation and the easterly jet in the upper troposphere are located over Tibet and southeastern Tibet, and a cyclonic system — south Asia low and the westerly jet in the lower tropophere are located over Tibet and the vicinity of south Asia, respectively. The intensity of the easterly jet in the upper troposphere is overestimated and that of the westerly jet in the lower tropopshere is underestimated.

The broad-scale horizontal distribution of flow associated with the summer monsoon in the NM model and that in the M model seem to be similar. A warm anticyclonic system exists in the upper troposphere and a cyclonic system (South Asia low) in the lower troposphere in both models. The depth of the cyclonic system is under 500 hPa level in the NM model, while its depth reaches 500 hPa in the M model. A strong easterly jet is present in both models in the upper troposphere, but it is located about 10° too far south in the NM model. A strong westerly jet, which crosses the Equator, exists in both models in the lower troposphere. The jet is much stronger in the NM model than in the M model. These circulation features suggest that the main effect of mountains is to shift the South Asia anticyclone northward and South Asia low northwestward to a location closely corresponding to the highest peaks of Tibet and to weaken the flow in the lower troposphere and to increase the depth of the cyclonic system. In contrast, the experiment of Hahn and Manabe (1975) suggested that the effect of orography was to produce the South Asia low, i.e., without mountains they had only a continental mid-latitude low centered at 50°N.

The three-dimensional structure of the monsoon in the model with mountains is similar to observation (Subbaramayya and Ramandham 1981), but there are noteworthy differences when mountains are removed (see Fig. 15). Note, in particular, the anti-cyclonic system in the 300 hPa wind and temperature field. In the M model this has a pronounced warm core, being warmer by about 10°C than its vicinity due to the thermal effects of mountains. In the NM model, anticyclonic system located over the Tibet area only has a far less pronounced warm area. The M model distribution is similar to Hahn and Manabe's (1975) but that in the NM

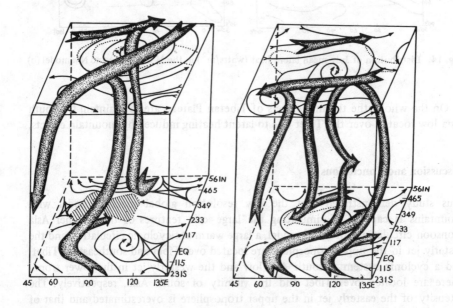

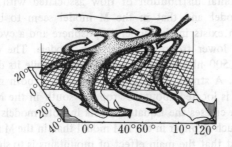

Fig. 15. Observed and simulated three dimensional structure of South Asian monsoon

343

model is different from their results, which did not have a warm area implying that the main feature of temperature field in the upper troposphere associated with South Asian Monsoon disappears when mountains are removed.

The differences in the distribution of vertical motion reflect the characteristics of two different kinds of monsoon. In the M model, upward motion is present throughout the tropopshere in the monsoon region resulting in a release of latent heat and vertical transport of sensible heat from the earth's surface. The low-level westerly jet supplies the moisture from the Indian ocean and strong upward motion transports the moisture into the middle level of the troposhere to produce large amounts of precipitation in Tibet. The strong upward motion is associated with the uplift effect of orography. Conversely, in the NM model, downward motions exist over Tibet except for the south and southwest of Tibet. The main precipitation areas are not in the Tibet area and the Indian peninsula but in the Indo-China peninsula and the western Pacific Ocean because of the distribution of vertical motion in the NM model. Our experiment suggests that the existence of Tibet increases rainfall and maintains the maximum center of precipitation in the Tibet area. However, in Hahn and Manabe's experiment (1975), downward motion and dry northwest flow dominate over South Asia. Consequently, a more continental climate extends southward over South Asia in the NM model. Therefore, the monsoon climate extending farther northward into Asia is due to mountains in their simulation.

The thermal effects of the Tibetan Plateau are to maintain the South Asia low located over Tibet due to latent heating. We must emphasize that the south Asia low still exists and its maintenace is still dependent on the latent heating even if the Tibetan Plateau were removed. This conclusion is much different from Hahn and Manabe's (1975), in which the latent heating dominates over Tibet in the M model and sensible heating dominates over Tibet in the NM model.

The analysis described above and the agreement between the simulations and with the observation (cf. Figs. 15b and 15c) indicates that the thermal effect of continentality is the dominating factor for simulation of the formation of monsoon, but the presence of mountains is instrumental in determining the position and structure of monsoonal circulation with both dynamical and thermal effects of orography, being of similar importance for the simulated monsoon circulation in the M model. This is the most important difference between our simulation and Hahn and Manabe's (1975), in which the formation of the South Asian monsoon is dependent on the existence of Tibet.

Acknowledgement The authors wish to thank Dr. Akira Kasahara at NCAR for his review and constructive comments of this research and Dr. Robert Gallimore for his comments on the topography smoothing procedures. This research was sponsored by the Climate Dynamics Research Division, National Science Foundation through NSF Grant ATM 81-12464. Computing support by the National Center for Atmospheric Research is also gratefully acknowledged. NCAR is sponsored by the National Science Foundation.

References

Abbott, D.A. (1973): *Tech Rep* No 73-2. *Dep Meteorol, Florida State Univ*, 190 pp

344

Alyea, F. (1972): *Atmos Sci Pap* No 193. *Dep Atmos Sci, Colorado State Univ*, 120pp

Gilchrist, A., (1974): *Tech Note* II/29. *Meteorol Office, Bracknell*, 35 pp

Gilchrist, A., (1976): *Tech Note* II/46. *Meteorol Office, Bracknell*

Godbole, R.V., (1973): *Ind J Meteorol. Geophys* 24: 1-14

Hahn, D.G Manabe S. (1975): *J Atmos Sci* 32: 1515-1540

Manabe, S., Hahn D.G. Holloway J.L. Jr, (1974): *J Atmos Sci* 31:43-83

Murakami, T., Godbole, R.V.., Kelkar, R.R., (1970): *Numerical simulation of the monsoon along 80° E. Proc Conf Summer monsoon southeast Asia. Navy Weather Research Facility, Norfolk Virginia*, pp 39-51

Otto-Bliesner, B.L., Branstator G.W. Houghton, D.D. (1982): *J Atmos. Sci,* 39: 929-947

Reiter, E.R., Deng-Yi Gao (1982): *Mon Wea Rev* 110: 1694-1711

Subbarmayya, I Ramanham R., (1981): *On the onset of the Indian southwest monsoon and the monsoon general circulation.* In: Lighthill S.J, Pearce P.R.P. (eds) Monsoon dynamics. Cambridge Univ Press, 1981

Washington, W.M., (1970): *On the simulation of the Indian monsoon and tropical easterly jet stream with the NCAR general circulation model. Proc Symp Trop Meteorol,* 2-11 June 1970. *University of Hawaii, Honolulu, Hawaii, sponsored by the American meteorological society,* WMO OVI-1-6

Washington, W.M., Daggupaty, S. M., (1975): *Mon Wea Rev* 103: 105-14

The Influence of Qinghai-Xizang Plateau on the Climate over China

Zou Jinshang[1]

Comprehensive Abstract

The Qinghai-Xizang Plateau is the largest plateau in the world. Its major axis is over 3,000 km long and its minor axis over 1,400 km. Its average height above sea level is over 4 km, occupying almost 1/3 of the troposphere. Such a huge protrusion in the alternating belt of the westerlies and easterlies will no doubt have a profound dynamic and thermal influence on the general circulation and climate.

In this paper, based on meteorological data for 1960–1969 and 1973–1982, some aspects of the influence of Qinghai-Xizang Plateau on the climate over China are investigated. A part of the results obtained from climatological analysis can be sketched as follows:

The Existence of Arid and Semi-arid Areas to the North of Qinghai-Xizang Plateau in Connection with the Shelter Effect of the Plateau

Computation results shown in Table 1 indicate that the annual mean value of net vapor transfer in the atmosphere over the arid and semi-arid areas is smaller by far than that over the wet region. It is well known that maximum of vapor transfer values occurs in the layer of 850–700 hPa. Of course, the huge Qinghai-Xizang Plateau is the principal barrier of moisture inflow in the southern border of arid areas.

The maximum of net water vapor transfer over the arid and semi-arid area occurs in July. In winter months, the amounts of net water vapor transfer are all negative. To compensate the outflow of water vapor, there must be strong evaporation from the earth's surface of snowpack on the mountains. We conclude that dry process in arid and semi-arid area occurs in winter months.

The Influence of Qinghai-Xizang Plateau on Jet Streams

The most striking phenomenon on daily weather maps in winter is the split and confluence of the westerlies. In half of the troposphere there are two branches of the westerlies, encircling the plateau, one flowing to the north and the other to the south of the plateau.

In summer, we can also see two jet streams in the high troposphere: the one to the north of plateau is termed the subtropical jet with the core at 200 hPa level,

---

1. Department of Meteorology, Nanjing University, Nanjing, China

Table 1. Comparison of vapor transfer values over arid and semi-arid areas with that over wet areas

	Arid and semi-arid areas Transfer values (km^3)			Wet areas		
Border	Total import	Total export	Net	Total import	Total export	Net
E	888.7	5036.4	-4147.7	1445.5	5147.6	-3702.1
S	3473.5	2258.6	1214.9	6569.8	2107.0	4462.8
W	3512.4	887.8	2624.6	1965.7	261.0	1074.7
N	2622.7	1t 76.7	864.0	1462.7	2090.1	-627.4
Sum	10497.3	9959.5	537.8	11443.7	9605.7	1838.0

located near $42°$ N; the other to the south of plateau is called the easterly jet with the core at 100–150 hPa level, located near $15°– 17°$ N. The intensification of these two jets is attributed to the effect of warm source over plateau in summer.

Relations Between Qinghai-Xizang High at 100 hPa and Rainfall Belt in East China

Numerous meteorologists have noted that the movement of ridge line of "Qinghai-Xizang high" is related to the seasonal variation of the rainfall belt over eastern China. When the center of the "Qinghai-Xizang high" is located over the west of plateau the vertical monsoon cell intensifies in meridional section during the flood season. This leads to the development of rain-belt in connection with the ascending current in front of the monsoon cell. Thus we can say that the Qinghai-Xizang high is an active center in the high troposphere, controlling the climate over China.

Influence of the Development of the Severe Moving Systems over East China and, Canversely, on the Climate over Dlateau

Two selected cases are investigated. During the period of 21–24 August 1966, a well-developed typhoon, invading the region of middle and lower reaches of the Changjiang (Yangtze) River, moved north-westward, while a cyclonic vortex over the plateau stagnated for two days. It led to the occurrence of heavy rain and a flash flood in the Tou-Tou river.

In addition, typhoon activities over the South China Sea are frequently, while the cyclonic activity is very few over the Qinghai-Xizang Plateau, and persistent drought occurs.

A Numerical Simulation of the Cross-Equatorial Propagation of Orographic Disturbances

Chen Xiongshan[1]

Abstract

It is meaningful to study the dynamic links between the weather systems in the Northern Hemisphere and those in the Southern Hemisphere. Considerable observational evidence appears to exist. In a linear stationary model the circular mountain at $30°N$ perturbs the zonal flow in the Northern Hemisphere, but does not perturb the zonal flow in the Southern Hemisphere due to the existence of the equatorial easterlies, as if equatorial easterlies provide an effective barrier for stationary waves in a linear model.

In the present study we use the nonlinear shallow water equations including orography to study the cross-equatorial propagation of orographic disturbances. It is a global spectral model. We use the rhomboidal truncation of wavenumber 15 and a 11.25 degree interval along a latitudinal circle for real space variables. This means that we have used the spectral model in which the aliasing terms are retained. We have chosen the center of the circular mountain to be at the position of the Tibetan Plateau. Equations are integrated using a semi-implicit time difference scheme with time step of one hour.

The 500 hPa zonal average values of height and wind on January 1, 1979 adopted from FGGE data are used as the initial values for the weak equatorial easterly case. In this case the zonal average value of the equatorial easterly is 1.6 m s^{-1}, and the width of the easterly is 1500 km. Over the orography there is the strong westerly of 18 m s^{-1}. In the numerical experiment we find that the orographic disturbances develop and propagate substantially into the middle latitudes of the Southern Hemisphere. The amplitude of disturbances of the Southern Hemisphere is 50 m at 5 days. At the same time, the equatorial westerly occurs at the longitude of the mountain, as if the westerly wind equatorial "duct" allows the penetration of orographic disturbances towards the Southern Hemisphere.

For the strong equatorial easterly case we use the zonal average values of July 1, 1979 data. In this case the zonal average value of equatorial easterly is 4.9 m s^{-1}, and the width of the easterly is 3000 km. Over the orography there is the weak westerly of 8 m s^{-1} which may excite weak orographic disturbances. We find that the orographic disturbances can penetrate through the strong equatorial easterlies and disturb the flow pattern of the Southern Hemisphere with the amplitude of 50 m at 20 days. In this case the equatorial westerly does not occur. In a nonlinear model the equatorial easterly cannot prevent the propagation of distur-

---

1. Institute of Atmospheric Physics, Academia Sinica, Beijing, China

348

bances from the Northern Hemisphere to the Southern Hemisphere. This result is quite different from the result of the linear model.

Section IV

Impact of Human Activity and
Some Other Natural Factors on Climate

Relationship Between Seasonal Climate Variation and Lower Stratospheric Circulation and Its Application to Prediction

Huang Shisong[1] and Tang Mingmin[1]

Abstract — The transition modes of lower stratospheric circulation from winter to summer are described. It is demonstrated that the seasonal climate variation such as the amount of precipitation and the dates of beginning and ending of the first rainy season in South China, the beginning date and duration of the plum rain season in the middle and lower Changjiang River reaches, the occurrence of disastrous cold summers in North China and so on are closely related with the stratospheric circulation in winter. One can apply these connection phenomena to make seasonal climate prediction.

Seasonal Transition of Stratospheric Circulation from Winter to Summer

The analysis shows that the main feature of the summer circulation on the 10 hPa, 30 hPa, and 50 hPa surface is characterized by the development of a large circum-polar anticyclone dominating the entire Northern Hemisphere with easterly winds prevailing everywhere except the equatorial region where may prevail the westerly winds (Institut für Meteorologie 1958). During the winter months two different types of circulation pattern may be found; either a large circum-polar cyclone dominates the entire Hemisphere with westerly winds prevailing everywhere, or the cyclone dominates only the region of high-and mid-latitudes, and a high pressure belt exists in the subtropics with easterly winds prevailing over the low latitudes. Figures 1 and 2 illustrate the examples at 10 hPa level. It should be noted that the strong subtropical high cell is situated over the Western Pacific and East Asia area.

Accordingly, there are two modes of circulation transition from the winter pattern to the summer pattern. The first transition mode is usually associated with a sudden warming process. If the sudden warming process takes place in the winter months, the associated high pressure system entering the polar region will weaken and die out, and the circum-polar cyclone will be reestablished. But if the sudden warming process takes place in late March or April, the high pressure system associated with warmings will be maintained in the polar region and will intensify and extend southward, merging with the subtropical high and developing into a great circum-polar high system of hemispheric scale. See Fig. 3.

In the second transition mode (Fig. 4), there is no sudden warming process and no associated high pressure system entering the polar region during late March or April. The original circum-polar cyclone itself will weaken and fill up gradually due to progressive rising of pressure over the polar region, and then a high pressure system will appear, which strengthens, expands, and merges with the high pressure

---

1. Department of Meteorology, Nanjing University, China

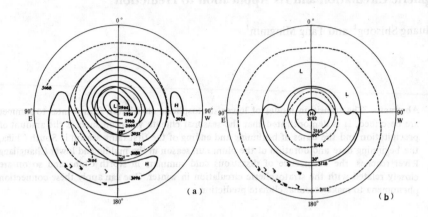

Fig. 1. Monthly mean 10 hPa height maps for February (a) and May (b) of 1968

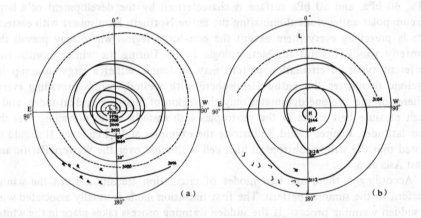

Fig. 2. Monthly mean 10 hPa height maps for February (a) and May (b) of 1971

system in subtropic latitudes, which usually start to develop in the month of April or May. Finally, a well-developed large circum-polar anticyclone dominates the entire Hemisphere.

The first transition mode happens usually when subtropical high pressure belt exists a in the cold months. The second transition mode happens usually when there is no subtropical high belt in the cold months. As shown in Figs. 1 and 2, the winter circum-polar cyclone will be deeper but less extensive in the first case than in the second case. The date of primary establishment of the summer circulation pattern which is defined as the first day of easterly winds prevailing almost

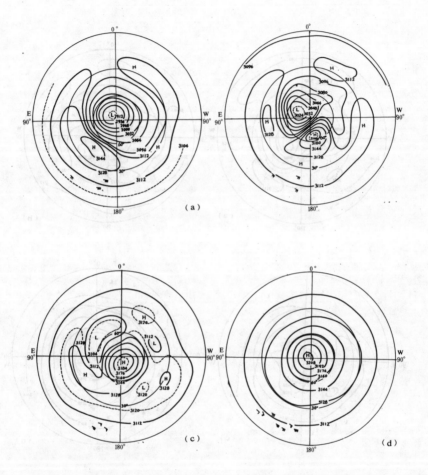

Fig. 3. Daily 10 hPa height maps for (a) 1st, April; (b) 18th, April; (c) 8th, May and (d) 29th, May of 1968

over the entire hemisphere will be earlier in the first case. As for the 10 hPa level circulation, it will be established generally on or before 5th of May in the first case and after 5th of May in the second case.

The seasonal transition modes on 30 hPa and 50 hPa surfaces are substantially similar to those on 10 hPa surface.

Evidently, the long-term temporal variation of the stratospheric circulation features and the seasonal transition modes can be simply expressed by the variation of upper zonal winds of any single station at low latitudes. Figure. 5a shows the variation of monthly mean zonal winds at 30 hPa and 50 hPa levels at the station Majuro Island during the years from 1964 to 1983. Easterly winds prevailing in the

354

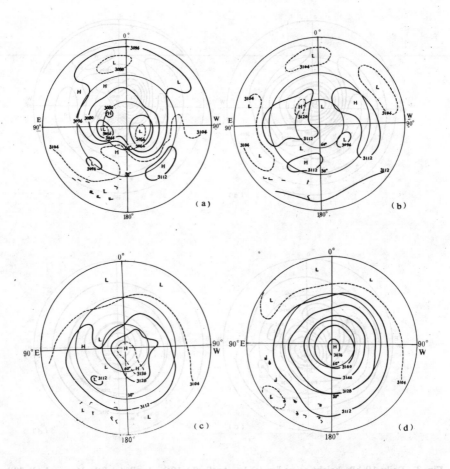

Fig. 4. Daily 10 hPa height maps for (a) 2nd, April; (b) 30th, April; (c) 14th, May and (d) 28th, May of 1971

cold months imply that a subtropical high over the West Pacific and East Asia area existed during these months. Prevailing westerly winds imply no subtropical high in winter and early spring months.

Figure. 5a exhibits more or less the characteristics of so-called quasi-biennial Oscillation (Reed and Rogers 1962; Belmont et al. 1975) over the equatorial region which is typically illustrated by the variation of zonal winds at Singapore (Fig. 5b) But Fig. 5a and b are not quite the same; it is easy to see that the westerly winds above Majuro Island appear only during the cold seasons with a duration of less than about seven months, much shorter than the one over Singapore.

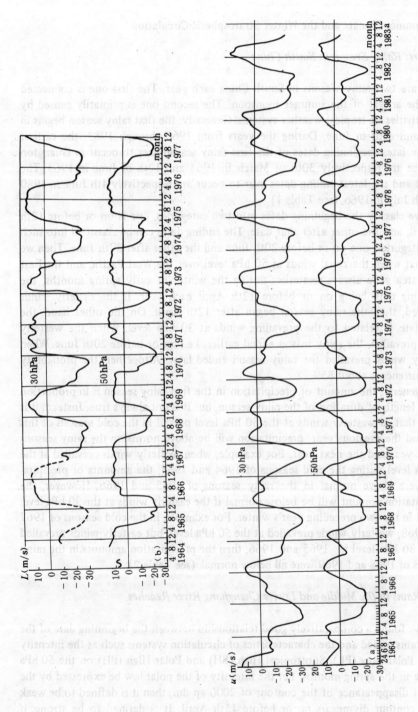

Fig. 5. Long-term variation of monthly mean zonal wind at 30 hPa and 50 hPa levels of the stations (a) Majuro (7°05'N, 171° 23'E) and (b) Singapore (1°21'N, 103°54'E)

The Summer Climate and the Winter Stratospheric Circulation

The First Rainy Season in South China

There are two rainy seasons in South China each year. The first one is connected with the arrival of the summer monsoon. The second one is primarily caused by the activities of tropical weather systems. Generally, the first rainy season begins in April and ends in June. During the years from 1960 through 1983, the earliest and the latest beginning dates of the first rainy season ever to occur in Guangton province are respectively 30th of March in 1964 and 12th of June in 1963. The earliest and the latest ending dates ever to occur are respectively 4th June in 1980 and 4th July in 1966. (see Table 1)

If we classify the beginning dates into two categories, one is on or before 12th of April, and the other after that date. The ending dates can be classified into such two categories, one on or before 20th June and the other after 20th June. Then we find that when the zonal winds at 50 hPa level over the West Pacific and the East Asian area were always westerlies during the winter or early spring months, the beginning date being on or before 12th April each year. If the easterly winds prevailed, then the rainy season began after 12th April. On the other hand, the final date is related to the prevailing winds at 30 hPa level. When the westerly winds prevailed, the rainy season ended earlier, i.e., on or before 20th June. When easterly winds prevailed the rainy season ended later. Altogether, the probability of occurrence is about 85%.

In general, the amount of precipitation in the first rainy season is in proportion to the length of duration of the rainy season, but it is not always true. Instead, it is found that, if westerly winds at the 50 hPa level prevail in the cold seasons of this year and the previous year, precipitation will be above normal in the rainy seasons of this year and the next year. For example, when westerly winds prevailed at the 50 hPa level during the cold seasons of 1964 and 1965, the amounts of precipitation were above normal in the rainy seasons of 1965 and 1966. However, the precipitation amount will be below normal if the easterly winds at the 30 hPa level prevail in the two preceding year's winter. For example, in the cold seasons of 1967 and 1968, westerly winds prevailed at the 50 hPa level, but easterly winds prevailed at the 30 hPa level in 1965 and 1966, then the precipitation amounts in the rainy seasons of 1968 and 1969 were all below normal. (see Table 2)

Plum Rains in the Middle and Lower Changjiang River Reaches.

We have found a comparatively good relationship between the beginning date of the plum rains period and the characteristics of circulation systems such as the intensity of the Polar Low (PL), Subtropical High (SH) and Polar High (PH) on the 50 hPa surface in the spring months. Let the intensity of the polar low be expressed by the date of disappearance of the contour of 2000 gp dm, then it is defined to be weak if the contour disappears on or before 15th April. It is defined to be strong, if

Table 1. Beginning and ending dates of the first raing season of Guangdong province in the years of 1960–1983

Year	60	61	62	63	64	65	66	67	68	69	70	71	72	73	74	75	76	77	78	79	80	81	82	83
Date of beginning	27/4	5/5	12/6	30/3	4/4	28/4	31/3	1/4	13/4	10/4	1/4	5/4	2/4	7/4	13/4	8/4	27/5	6/4	8/9	11/4	12/4	22/4	3/4	
Date of ending	15/6	11/6	30/6	17/6	20/6	28/6	4/7	8/6	24/6	18/6	28/6	27/6	19/6	12/6	26/6	16/6	2/7	4/6	1/7	9/6	18/6			

Table 2. Percentage of departure of April-June total precipitation from normal (%) at ten stations in Guangdong Province

Station	\ Year	64	65	66	67	68	69	70	71	72	73	74	75	76	77
Santao		−21	5	38	−53	−5	−21	−44	−13	10	68	32	3	−10	−4
Hoyuan		−11	12	62	−26	66	−19	−33	−25	−2	49	7	43	−12	−27
Hweyang		9	15	41	−33	2	−19	−8	−9	14	45	−16	43	8	−15
Guangzhou		−21	30	10	−33	−30	−7	−22	2	27	25	−14	69	−15	−22
Zhenjiang		−64	48	−17	54	−30	4	5	44	5	42	29	52	−2	−19
Meixian		2	11	0	0	11	−30	−35	−22	17	32	17	29	7	−8
Lianping		26	14	1	−33	30	−18	−22	−33	−1	53	26	48	13	−15
Fugang		−15	20	21	−20	24	−90	7	−11	−6	−19	10	23	−25	−4
Qingyuan		−15	32	10	−26	−21	−37	7	46	18	32	−3	6	−11	−21
Nanyung		27	1	4	−36	−14	−23	11	−33	−5	54	−5	17	29	−8
Ten stations mean		−8.3	18.8	17.0	−20.6	4.7	26.0	−13.4	−5.4	7.7	38.1	8.3	33.3	−1.8	−14.3

the contour disappears after 15th April. In a similar way, the intensity of the Western Pacific subtropical high is expressed by the date of the appearance of the contour of 2072 gp dm, and it is defined to be weak or strong when the contour appears after the 15th or on or before the 15th May respectively. The intensity of the Polar High is expressed by the geopotential height reached by the end of May. Then, as shown in Table 3, when PL, SH, and PH were all weak, the plum rains period began very early, i.e., before 10th June. If PL and SH were weak, but PH was rather strong, it began in the second ten days of June, this means noraml. When PL and SH were strong, no matter how was the PH, the rain period began late, i.e., in the last ten days of June.

No similar definite relationship has been found for the ending date of the plum rains period. However, by comparing Table 4 with Fig. 5a, one can find that in the year with subtropical highs appearing on the 30 hPa surface during the preceding winter months, the duration of the plum rains period was longer than 20 days. So we can roughly estimate the ending date of the plum rains period. If the plum rains period begins late and the duration is short, the total amount of precipitation is usually much below normal.

Disastrous Cold Summers in North East China

In North East China, a persistant low temperature occurring during May through August or during a growing peiod in June and/or August may cause a disastrous harvest of crops. The loss may amount to about 20–30% of the normal, such as in the years 1964, 1969, 1971, 1972, and 1976. It is found that no such disastrous cold summer ever happened in the year when easterly winds prevailed at 30 hPa level over the tropical region during the preceding winter and spring months. Therefore, if there exists a large subtropical high on the 30 hPa surface over the Western Pacific and the East Asian area in the preceding cold months, no disastrous cold summer will happen in North East China.

Conclusions

Up to now, much research work has been done on the interrelation between the troposphere and stratosphere (Neyama 1966; Huang Shi-song el al. 1965; Ebdon 1975; Murgatrod and O'Niell 1980; Newell 1981), but none has touched the problem raised in this paper. However, from the above demonstration it can be concluded that there seems to be a response of seasonal climate variation to the feature of stratospheric circulation. One can apply such a connection phenomenon to make seasonal climate prediction, though the physical processes and mechanism involved are not yet clear and need to be investigated.

359

Table 3. Spring circulation systems on 50 hPa surface and beginning date of plum rains period (Nanjing) E: early, L: late, S: strong, W: weak

Year	1971	1974	1970	1972	1975	1976	1958	1966	1967	1968	1965	1969
Intensity of polar low (date of disappearance of the contour 2000 gp dm)	5/4 E W	1/4 E W	11/4 E W	15/4 E W	1/4 E W	6/4 E W	21/4 L S	5/4 E W	30/4 L S	25/4 L S	18/4 L S	21/4 L S
Intensity of western Pacific subtropical high (date of apperance of the contour 2072 gp dm)	25/5 L W	30/5 L W	30/4 E S	19/5 L W	19/5 L W	24/5 L W	20/4 E S	12/5 E S	7/5 E S	5/5 E S	13/5 E S	13/5 E S
Intensity of polar high (geopotential height at High center during the end of May in pg dm)	2082 — 2088	2082 — 2088	2088 — 2090	2092 — 2096	2090 — 2096	2090 — 2100	2096 — 2096	2096	2096 — 2104	2084 — 2088	2090	2090
Date of beginning of plum rains period	9/6	9/6	17/6	20/6	16/6	16/6	26/6	24/6	24/6	23/6	30/6	30/6

Table 4. Beginning date, ending date and duration (days) of the plum rains period of Nanjing in the years of 1964–1980 and the wind at 30 hPa level above Majuro during the preceding cold months, (E: easterly wind, W: westerly wind)

Year		1964	1965	1966	1967	1968	1969	1970	1971	1972	1973	1974	1975	1976	1977	1978	1979	1980
Wind at 30 hpa		W	E	E	W	E	W	E	W	W	W	E	E	E	E	W	E	E
Plum Rains period	Beginning date	24/6	30/6	24/6	24/6	23/6	30/6	17/6	9/6	20/6	16/6	9/6	16/6	16/6	28/6	23/6	23/6	9/6
	Ending date	1/7	22/7	13/7	5/7	20/7	18/7	20/7	20/6	5/7	29/6	18/7	17/7	15/7	21/7	25/6	25/7	21/7
	Duration	8	23	20	12	28	19	34	18	16	14	40	32	30	19	3	33	43

360

References

Belmont, A.D. et al. (1975) *J. Appl Meteor* **14**: 585–594
Ebdon, R.A. (1975) *Meteool Mag.* **104**: 282–297
Huang Shi-song et al. (1965) *Acta Sci Natu Univ Nankinensis* 9, 1: 123–132
Institut für Meteorologie Univ Berlie (1958, 1965–1976) Daily and Monthly Northern Hemisphere 10, 30 and 50 hPa Synoptic Weather Maps. *Meteorol Abh*
Murgatroyd, R. J. O'Niell A. (1980) *Phil os Trans R Soc,* Lond, A 296: 87–102
Newell, R.E. (1981) *J Atmos Sci* 28:2789–2796
Neyama, Y. (1966) *J Meteorol Soc Jpn* 44: 159–166
Reed, R.J., Rogers D.G. (1962) *J Atmos Sci* 19: 127–135

The Interaction Between the Subtropical High and the Polar Ice in the Northern Hemisphere

Fang Zhifang[1]

Abstract − The interaction between the subtropical high and the polar ice in the Northern Hemisphere are studied by the lag cross-correlation matrix method. The statistical results show that the interaction between them is characterized by the seasonality.

Introduction

Polar sea ice plays a striking role in the air-sea interaction in high latitudes. Recently, a number of meteorologists (e.g., Walsh 1979; Fu, 1981) have already performed some investigations on the influence of polar ice on the atmospheric circulation and the climate in some regions. Their results have shown that polar ice not only has evident local action, but also affects the atmospheric circulation and the climate in mid-latitudes and even in another Hemisphere.

However, the variation of polar ice itself results from the variation of the atmospheric circulation and the air-sea interaction has time-lag effect. In this paper we mainly discuss the interaction between the subtropical atmospheric circulation and the polar sea ice in the Northern Hemisphere.

Data and Method

We use the monthly mean sea ice cover data in the Arctic regions during the 25-year period from 1953−1977 given by John Walsh. The Arctic sea ice cover are refers to the regions to the north of 50°N in Arctic sea, North Atlantic and North Pacific. The whole Arctic region is divided into four parts with a center at the North pole, according to the longitude range, i.e., 160°−110°W, 110°W−20°W, 20°W−70°E, and 70°E−160°E denoted by A, B, C and D respectively.

Differences in polar-ice-covered area over two successive months are used as the polar ice melting-freezing index of the later month of the two to reflect the month-to-month variation of polar ice cover.

Data of the intensity of the subtropical high for both Northern Hemisphere and Northwest Pacific are provided by the National Meteorological Administration of China.

By using the cross-correlation function, the correlation coefficient $R(t,\tau)$ as a funtion of t,τ, where t is the interaction month, and τ is the time lag in months between two parameters, constructs the lagging correlation matrix $R(t, \tau)$.

1. Chengdu Institute of Meteorology, Chengdu, China

According to the method described above, therefore, lag correlation matrices are obtained by relating the extent of polar ice for the whole region to the intensity of the subtropical high in the Northern Hemisphere, the polar ice melting-freezing index for whole region to the intensity of the subtropical high in the Northern Hemisphere and the polar ice extent for each region to the intensity of the subtropical high in Northwest Pacific.

By means of these matrices we may study the interaction between Arctic sea ice and atmospheric circulation.

Effect of Arctic Sea Ice on Atmospheric Circulation

Figure 1a presents the effect of the polar ice extent on the subtropical circulation index. According to the statistical test, the area with $R \geqslant 0.4$ reaches the significant level $\alpha = 0.05$.

It is shown that there is a sensitive period when the polar ice has the strong effect on the Northern Hemisphere subtropical high. Polar ice in early winter (November and December) apparently affects the intensity of the subtropical high in the same month and one month later. With the correlation coefficient −0.48, polar ice during the whole winter (November−February) is negatively correlated with the subtropical high in August, September, which reflects that the more the winter polar ice expands, the weaker the summer subtropical high will be. On the other hand, March−October polar ice has little influence on subtropical atmospheric circulation.

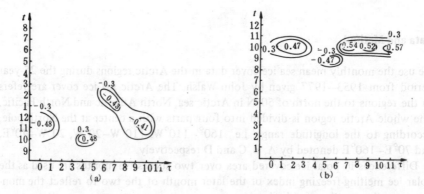

Fig. 1. The lag correlation map. (a) the lag correlation between the polar-ice-covered area and the intensity of the subtropical high; (b) the lag correlation between the melting-freezing index of the polar ice and the subtropical high

Figure 1b shows that the effect of the polar ice melting-freezing index on the subtropical circulation has a "key month", that is, October. The influence of the polar ice melting-freezing index in October on the winter subtropical circulation with lag of 1−3 month is described by a positive relationship with R=+0.47. Its influence on the summer subtropical circulation is more obviously with R = + 0.57 at time

The melting-freezing index in October is negative because it is the freeze-up period. Therefore the positive correlation coefficient indicates that the more the polar ice freezes in October, the weaker the subtropical high in May—September of the following year.

It is seen by the comparison of Fig.1a with Fig.1b that there is a general agreement between the correlation regions in these two figures. The effect of the polar ice on the atmospheric circulation is characterized by the seasonality. The winter polar ice extent is established by the amount of increase in polar ice in October. However, there is an apparent correlation between the winter polar ice area and the summer subtropical high. For any other month the effect of polar ice on the atmospheric circulation is weaker.

Figure 2a reflects the influence of the polar ice area in each region on the intensity of the subtropical high in the Northwest pacific. In Fig. 2a, it can be seen that the polar-ice-covered area in region A affects the Northwest Pacific subtropical high. The orientation of the isocorrelation parallels to the main diagonal of the corresponding matrix. This fact shows that the polar ice extent in region A in January—May has an influence on the intensity of the Northwest Pacific subtropical high in July, correlation coefficient −0.61. This result indicates that the greater the polar-ice-covered area in region A in winter-spring, the weaker the Northwest Pacific subtropical high in the following July.

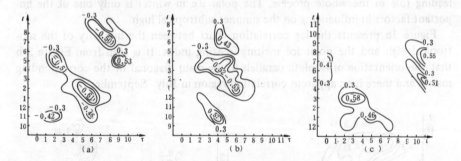

Fig. 2. The lag correlation between the polar ice extent for each region and the intensity of the subtropical high in the Northwest Pacific, (a) regionA; (b) region B; (c) region c puls region D

Figure 2b shows that the polar ice extent in region B has an influence on the Northwest Pacific subtropical high with a negative correlation coefficient −0.53 between the polar ice in January—April and intensity of the subtropical high in July. Figure 2c shows that the polar-ice-covered area in Quasi-Eastern Hemisphere (region C plus region D) has an influence on the Northwest Pacific subtropical high with a positive correlation coefficient 0.58 between the polar ice in January—March and the intensity of the subtropical high in June. Therefore, it is clear that the effects of the polar ice extent in the Eastern Hemisphere on the Northwest Pacific

364

subtropical high in summer are quite different from those in Western Hemisphere. The sea ice extent during the spring-winter in Quasi-Western Hemisphere has an effect upon the subtropical high in July, which is the negative correlation. While the polar ice extent during the spring-winter in Quasi-Eastern Hemisphere has an effect upon the subtropical high in June, which is the positive correlation.

The Effect of the Atmospheric Circulation on the Polar Ice

Figure 3a presents the lag correlation chart between the subtropical high intensity and the polar-ice-covered area. There are three good correlation regions. One has the scope of a greater correlation. The orientation of its isopleth parallels the main diagonal of the corresponding matrix. This correlation region shows that the subtropical high in summer (May–September) has an influence on the polar ice extent for the whole region in the winter (November–February). The maximum correlation coefficient is −0.69. This fact indicates that the polar ice cover in winter is controlled mainly by the subtropical high in summer. Therefore, when the subtropical high in summer is strong, the polar ice cover area in winter is smaller.

Comparing Fig.1a with Fig.3a, it can be seen that the effect of atmospheric circulation in summer on the polar ice cover area in winter is much stronger than the reverse. The anomaly of the polar ice affects the anomaly of the atmospheric circulation and vice versa. Therefore, there is a feedback process between polar ice and atmospheric circulation. However, the subtropical high in summer plays the leading role in the whole process. The polar ice in winter is only one of the important factors in influencing on the summer subtropical high.

Figure 3b presents the lag correlation chart between the intensity of the subtropical high and the polar ice melting-freezing index. It is seen from Figure 3 b that the orientation of isopleth parallels to the main diagonal of the corresponding matrix and there is an apparent correlation region in July–September.

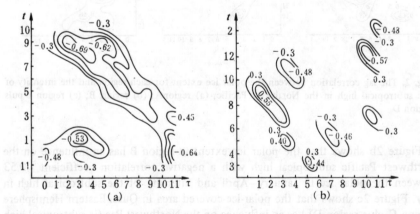

Fig. 3. (a) The lag correlation between the subtropical high and the polar ice-covered area. (b) The lag correlation between the intensity of the subtropical high and the polar ice melting-freezing index

In July—September, the intensity of the subtropical high has an influence on the increasing amount of the polar ice in October and their correlation coefficient reaches to 0.55. The positive correlation shows that the stronger subtropical high in July—September results in delaying the freezing of the polar ice and the less increasing amount of the polar ice in October.

As pointed out by Untersteine (1975), the heat is stored in the Arctic Ocean during the summer, a part of which delays the freezing of the polar ice in fall. If the subtropical high in summer is extensive and strong, then the earth-air system is favorable to the receipt of radiative energy and storage of heat. The energy which stores in mid-lower latitudes is transported to higher latitudes by the ocean current and the atmospheric circulation. Therefore, this is favorable to the storage of energy in Arctic Ocean. As a result, the increase of the polar ice in fall is delayed and the polar ice extent in winter reduced.

Summary

According to the foregoing analysis and discussion, it may be noted that
(1) There is a lag correlation and a seasonal relationship between the subtropical high and the polar ice in the Northern Hemisphere. The lag time is half a year or even a year. The polar ice is the dominating factor in their interaction, while the subtropical high dominates in summer.
(2) The effect of the spring-winter polar ice in the Eastern Hemisphere on the subtropical high in the Northwest Pacific is different from that of the polar ice in the Western Hemisphere.
(3) The winter polar ice cover in the Northern Hemisphere affects the summer subtropical high and vice versa. There is a positive feedback between the polar ice cover and the subtropical high.

References

Fu Congbin (1981) *Kexue Tongbao* 27: 71–75
Untersteiner, N. (1975) The Physical Basis of Climate and Climate Modelling. GARP Publ Ser No 16, World Meteorol Organiz, Int Council Sci Unions, pp 364–370
Walsh, J. E. (1979) *J Geophys Res* 84: 6915–6928

On the Influence of Deforestation of the Tropical Forest on the Local Climate in Xishuangbanna, China

Zhang Keying[1]

Abstract

Xishuangbanna (21°13'—22°34' N, 99°58'—101°50' E, located in the southern part of Yunnan Province) is famous for its lush tropical forest, but the forest cover percentage has been reduced to less than half in the past 31 years (during 1949—1980), i.e., it was reduced from 69 % in 1949 to 30 percent in 1980 or by 1.3 % per year. In the last six years (during 1974—1980) the reduction rate has been 5% per year.

This article is based on long-term data of three stations, i.e., Jinghong (21° 52' N, 101°04' E, Alt. 553m) as investigation station of the forest area, Lancang (22°34' N, 99°56' E, Alt. 1055m) as the station for comparison, and Simao (22°40' N, 101°24' E, Alt. 1302m) as the control station. By using the appropriate test-elimination method, we ascertained the effect of deforestation of the tropical forest in Xishuangbanna on the local climate as follows:

1. Annual rainfall reduces by about 4 % (50—60 mm). In the dry season (from November to next April), the rainfall reduces about 15 %, i.e., about 25 mm. In the rainy season (from May to October), it reduces about 4 %, i.e., about 30mm.
2. Annual (or the dry season) mean relative humidity lowers by 2 or 3 %. In May, it lowers by about 5 %. Annual mean, dry season mean, and rainy season mean absolute humidity lowers by 0.3, 0.5, and 0.8 hPa, respectively.
3. In the cold season (from December to the following February) the mean air temperature falls 0.3°C, and mean minimum temperature and extreme minimum temperature fell 1°C and 1—1.5°C. In the dry-hot season (March to April) mean air temperature increases 0.4°C—0.8°C (in late 1970's). The annual range of air temperature increases by 0.5° C or more.
4. Annual number of fog days in foggy season (from November to February of next year) decreases by 7—15 (in late 1970s). Annual sunshine hour increases 330—370 hours; the sunshine hours of the foggy season increases by 80—100 hours owing to the foglayer becoming thinner, and the time of dissipation of the foglayer becoming nearly two hours earlier.

In summary, the local climate becomes drier and colder in the cold season and drier and hotter in the dry-hot season, and the sunshine hours increase remarkably owing to deforestation, although the micro-climate of this region remains unchanged (no appreciable change by significant test).

---

1. Laboratory of Ecology, Kunming Branch of Chinese Academy of Science, China

The Coupeld Oscillation of Solar Activity and the Strength of Subtropical Marine High in the Northern Hemisphere

Xu Qun[1] and Jin Long[2]

Comprehensive Abstract

There is high correlation ($P \leqslant 0.1\%$) between the series of monthly mean sunspot numbers and that of the monthly area index ($\geqslant$ 5880GPM) of subtropical Marine High at 500 hPa level in the N. H., the former leading the latter by 0–3 years based on data of the last 28 years (1954–81). The power spectrum analysis of the area index also reveals significant cycles of 11 and 22 years, coupling well with the same cycles of solar activity (the significance level of coherence estimation is 1%).

All parameters of the West Pacific High (the area index, the longitude of the western-most point of the ridge line, the latitude of central ridge line and the latitude of northern limit of 5880 G.P.M. line) have shown the prominent cycle of 11 years, but their phases responding to solar activity are different; when solar activity is strengthened (weakened), the area index of the West Pacific High increases (decreases) with a phase lag of 1–2 years, its ridge point extends westward (withdraws eastward) with a phase lag of 2 years, and the latitude of the central ridge line and the northern limit of the Subtropical high would both move northward (southward) with a lag of 3 years.

In the coupled fluctuation of lower frequencies, the response of area index of East Pacific High to monthly sunspots is most sensitive; the correlation reaches its maximum at the phase lag of 14–15 months: $R = + 0.41$ (N=323).

Although the response phase of area index of North Atlantic Subtropical High lags far behind, the maximum of correlations with a time lag of 29--30 months, its value is high: $R_{Max} = + 0.49$ (N=308). The correlations between monthly sunspots and monthly area index of Subtropical High for whole N. H. also reached statistical significance ($P \leqslant 0.1\%$) as the latter lagged at 0–42 months; its maximum occurred at 16 and 24–29 months ($R_{Max} = +0.44$).

Response periods to sunspots of area indexes of all Subtropical Highs in the N. H. may be divided into 3 classes:
1. In the East Pacific High, the most prominent is "11-year" cycle.
2. For the Subtropical High of the North Atlantic and the whole N. H., not only the cycle of 11 years is prominent, but both the fluctuation of 22 years and long-term trend are also outstanding.
3. For the West Pacific High, the "11-year" cycle is of great importance, with the cycle of 3.5 years second.

---

1. Meteorological Institute of Jiangsu, Nanjing, China
2. Nanjing Institute of Meteorology, Nanjing, China

The formation of 3.5-year cycle may arise from variations of sea surface temperature of the East Equatorial Pacific (120°−180° W), and its lagging effect on the Subtropical High is most outstanding in the West Pacific in the following 2−4 months; therefore, the fluctuation in strength of the West Pacific High may arise from the double effects of "11-year" cycle of solar activity and of "3.5-year" cycle for sea surface temperature variation in the Equatorial East Pacific.

The oscillation of tree-ring width in Taiwan in the past 500 years also reveals the significant "11-year" cycle connected with solar activity, which may also be controlled by the North Pacific Subtropical High. These results strongly support our work on solar Subtropical High relationship.

From long historical data, it is shown that in the middle and lower Changjiang Valley flood occurred at the peak of solar "11-year" cycle. Now it can be interpreted as follows: When solar activity increases, the Subtropical High of West Pacific will strengthen and its western ridge will stretch westward in the following 1−2 years; the southwest stream transporting abundant warm and moist air from the Bay of Bengal will be established steadily in early summer and converge frequently over the middle and lower Changjiang valley with cold air from the north, thus causing plenty plum rain and flood in that region.

Section V

Prediction Methods for Monthly and

Seasonal Climate Variation

An Atmosphere-Ocean/Land-Coupled Anomaly Model for Monthly and Seasonal Forecasts

Chao Jiping[1], Wang Xiaoxi[1], Chen Yingyi[1], and Wang Lizhi[1]

Abstract — An anomaly atmosphere-ocean/land-coupled model for monthly and seasonal prediction is developed. By using a three-layer version of the anomaly general circulation model (AGCM) and the anomaly filtered model (AFM) which filtered the transient Rossby waves, a prediction of Feb. 1977 was carried out. It is shown that the correlation coefficients between predicted and observed anomalous fields with both models are much better than those of the persistence prediction, and are also better than those of the early one-layer version of AFM. In comparison with AFM, the results of AGCM are in better agreement with the observations. But the required computer time in AGCM is 200 times more than that in AFM. The results also show that the atmospheric initial fields seem to be unimportant in comparison with the effects of boundary heating from ocean and land.

Introduction

An atmosphere-ocean/land-coupled model for long-range forecasts with the time period ranging from a month to a season has been developed (Chao et al. 1982). Two basic ideas in this model are different from the usual general ciculation model (GCM). First, since we are only interested in the temporal evolution of the anomalous components rather than the climatological ones, the climatological components can thus be removed from the full field equations by separating all the system variables into their climatic and anomalous components and only the time-dependent anomalous system is retained. The observational climate data are also used in the model. The system consists of two main equations. One is the non-adiabatic vorticity equation for the atmosphere, and the other is the thermal equation for the underlying ocean and land. The two equations are connected at the earth's surface through the interface conditions which are the heat balance and the vanished vertical velocity. The second important idea of this model is as follows. By analyzing the linear case of this atmosphere-ocean/land system, it can be shown that there exist two basic types of dynamical processes corresponding to different time scales: the fast one and the slow one. The fast process has a time-period of the order of one week which is, in essence, the non-adiabatic transient Rossby wave, and the latter has a period of the order of several months. It is easily understood that the slow one is produced by the heating of ocean and/or of land, in another words, by the interaction of atmospheric and oceanic/or terrestrial physical processes. The fact that the growth rate of the shorter time-scale wave is about one order of magnitude larger than that of the longer time-scale waves is one of the difficultie₃ for the long-range numerical forecasts, because the evolution of the long-range

---

1. Institute of Atmospheric Physics, Academia Sinica, Beijing, China

process of smaller amplitude will be distorted by the short-range process with larger amplitude. It is suggested that the most efficient numerical model for monthly and seasonal forecasts should filter out this high-frequency noise. According to this idea, two approaches can be used. One way is that the transient Rossby wave can be filtered by omitting the partial derivative with respect to time in the atmospheric vorticity equation, but the time derivative term is still kept in the thermal equation for underlying ocean and land. With this assumption, the vorticity equation becomes time-independent, i.e., it is only a balance relationship between the anomalous geopotential height field and the earth's surface heating field. Therefore this model may be called as anomaly filtered model (AFM). Another way is to integrate these equations step by step to obtain forecasts for one month range and then to take a monthly mean. This approach seems more straightforward and it has the advantage that it would provide a direct effects of the high-frequency variations on the low-frequency changes. In the pressent paper, we call this approach as the anomaly general circulation model (AGCM).

The comparison of the prediction results between these two methods is given in this chapter.

Model

In the (x, y, p, t) coordinate system, the anomalous vorticity equation may be written as

$$\frac{\partial}{\partial t}(\Delta\phi') + \frac{1}{f}J(\overline{\phi}+\phi', \Delta\phi') + J(\phi', f+\frac{1}{f}\Delta\overline{\phi}) = f^2\frac{\partial\omega'}{\partial p} + k'\nabla^4\phi', \quad (1)$$

where ϕ' is the geopotencial height, the symbols "$-$" and "$'$" denote the climatological monthly mean and its deviation respectively, the last term in Eq. (1) represents the friction effect, and

$$J(A, B) = \frac{\partial A}{\partial x}\frac{\partial B}{\partial y} - \frac{\partial A}{\partial y}\frac{\partial B}{\partial x}$$

is the Jacobi operator. The first law of thermodynamics in the atmosphere is

$$\frac{\partial}{\partial t}(\frac{\partial\phi'}{\partial p}) + \frac{1}{f}J(\overline{\phi}+\phi', \frac{\partial\phi'}{\partial p}) - \frac{R}{pf}J(\phi', \overline{T}) - \frac{\partial}{\partial p}(k+k_r)\frac{\partial^2\phi'}{\partial p^2}$$

$$+ \frac{1}{\tau_r}\frac{\partial\phi'}{\partial p} = -\tilde{\sigma}_p\omega' + \frac{\mathcal{E}'Q}{\rho C_p}, \quad (2)$$

where k_r and τ_r are parameters introduced by considering the infrared radiation. Q is the heat exchange of condensation. The last term of Eq. (2) may be expressed by

$$\frac{\mathcal{E}'Q}{\rho C_p} = -\frac{L}{C_p}w\frac{\partial q'_s}{\partial z} \approx -\frac{L}{C_p}H_b(\text{div}\vec{V})_b\frac{\partial q'_s}{\partial z}, \quad (3)$$

where H_b is the thickness of the planetary boundary of the atmosphere. On the other hand, using the following formula

$$q \doteq q_s = 0.622\frac{e_s(T)}{p}, \quad (4)$$

we obtain approximately

$$\frac{\partial q'_s}{\partial z} \approx \left(\frac{\partial q'_s}{\partial z}\right)_0 \approx -\left(\gamma \frac{\partial \ln \bar{e}_s}{\partial T}\right)_0 \cdot q'_{s,o} \ , \tag{5}$$

and

$$q'_{s,o} \approx \left(\frac{\partial \bar{q}_s}{\partial T}\right)_0 \cdot T'_s \ . \tag{6}$$

Then we have

$$\frac{\mathcal{E}'_Q}{\rho C_p} = \frac{L}{C_p} H_b \ (\text{div } \vec{V})_b \cdot \gamma \left(\frac{\partial \ln \bar{e}_s}{\partial T}\right)_0 \cdot \left(\frac{\partial \bar{q}_s}{\partial T}\right)_0 \cdot T'_s \ . \tag{7}$$

Eliminating ω' from Eqs. (1) and (2), we obtain the non-adiabatic anomalous vorticity equation

$$\frac{\partial}{\partial t} \left(\Delta \phi' + \frac{f^2}{\tilde{\sigma}_p} \frac{\partial^2 \phi'}{\partial p^2}\right) + \frac{1}{f} J(\bar{\phi} + \phi', \Delta \phi') + J(\phi', f + \frac{1}{f} \Delta \bar{\phi})$$

$$+ k \Delta \phi' - \delta_1 k^* T'_s = \frac{f^2}{\tilde{\sigma}_p} \frac{\partial G}{\partial p} \ , \tag{8}$$

where

$$G = -\frac{1}{f} J\left(\bar{\phi} + \phi', \frac{\partial \phi'}{\partial p}\right) + \frac{R}{pf} J(\phi', \bar{T}) + \frac{\partial}{\partial p} (k + k_r) \frac{\partial^2 \phi'}{\partial p^2} - \frac{1}{\tau_r} \frac{\partial \phi'}{\partial p} \ , \tag{9}$$

$$k^* = fRH_b \gamma \frac{L}{C_p} (\text{div } \vec{V})_b \left(\frac{\partial \ln \bar{e}_s}{\partial T}\right) \left(\frac{\partial \bar{q}_s}{\partial T}\right) \frac{\partial}{\partial p} \left(\frac{1}{p\tilde{\sigma}_p}\right) \ , \tag{10}$$

$$\delta_1 = \begin{matrix} 1 \\ 0 \end{matrix} \qquad \text{when} \qquad \begin{matrix} (\text{div } \vec{V})_b < 0, \\ (\text{div } \vec{V})_b > 0. \end{matrix}$$

The geostrophic approximation

$$u' = -\frac{1}{f} \frac{\partial \phi'}{\partial y} \ , \qquad v' = \frac{1}{f} \frac{\partial \phi'}{\partial x} \ , \qquad \zeta' = \frac{1}{f} \Delta \phi' \ , \tag{11}$$

the hydrostatic relation

$$T' = -\frac{p}{R} \frac{\partial \phi'}{\partial p} \ , \tag{12}$$

and the following relationship

$$k' \nabla^4 \phi' \approx -k \nabla^2 \phi' \tag{13}$$

have been used in Eq. (8).

The boundary conditions for solving Eq. (8) are taken as

$$p = p_s, \qquad \omega' = 0 , \qquad (14) \qquad T'_s = -\frac{1}{R} \frac{\partial \phi'}{\partial \ln p} \tag{15}$$

$$p \to 0, \qquad \frac{\partial \phi'}{\partial p} = 0, \qquad (16) \qquad \omega' = 0 . \tag{17}$$

It is assumed that the heat flux of condensation vanishes at both lower and upper boundaries, and the earth's surface temperature is taken as the reference one, the Eq. (2) at both boundaries may then be written as

374

$$p = p_s \qquad \frac{\partial}{\partial t} \left(\frac{\partial \phi'}{\partial p} \right) - G = -\frac{R}{p_s} T_s' \; , \qquad (18)$$

$$p \to 0, \qquad \frac{\partial}{\partial t} \left(\frac{\partial \phi'}{\partial p} \right) - G = 0, \qquad (19)$$

The thermal equation for the underlying ocean and land is as follows

$$\frac{\partial^2 T_s'}{\partial z^2} - \frac{1}{k_s} \frac{\partial T_s'}{\partial t} = \frac{\delta}{k_s} \left[J(\psi_{s,} \, T_s') + J(\psi_{s,}' \, \overline{T}_s + T_s') \right] \equiv H_1 \; , \qquad (20)$$

where $\delta = 1$ for ocean and $\delta = 0$ for land, and ψ_s is the stream function of ocean current. The anomalous part ψ_s' can be calculated using the Ekman wind-driven theory, i.e.,

$$u_s' = \frac{\sqrt{2}}{2} \cdot \frac{0.0126}{\sqrt{\sin \phi_0}} \, (u_1' + v_1'), \qquad u_s' = \frac{\sqrt{2}}{2} \frac{0.0126}{\sqrt{\sin \phi_0}} \, (v_1' - \dot{u}_1') . \qquad (21)$$

In addition, the equation of heat balance at the earth's surface is [1]

$$z = 0, \qquad \rho_s C_{ps} k_s \frac{\partial T_s'}{\cdot \partial z} - \rho C_p \, k_T \frac{\partial T'}{\partial z} + \delta L \rho k_T \gamma \frac{\partial \ln \overline{e}_s}{\partial T} \frac{\partial \overline{q}_s}{\partial T} T_s'$$

$$= -\frac{s_0}{w_0} l_b \zeta_{bg}'. \qquad (22)$$

It can be taken as one of the boundary conditions in which "b" denotes the variable on the top of boundary layer. Another one is

$$z = -h, \qquad \frac{\partial T_s'}{\partial z} = 0 , \qquad (23)$$

where h is the depth of mixed layer.

The Numerical Schemes

The Method of Predicting the Earth's Surface Temperature Anomaly

At first we need to calculatein $\partial T'/\partial z$ in Eq. (22). Assuming that the temperature in the atmospheric boundary layer is governed by the heat conductive equation and using the quasi-stationary approximation we have

$$\frac{\partial^2 T'}{\partial z^2} - \frac{1}{k \tau_\gamma} T' = \frac{1}{k} \left[\mathbf{J} \left(\overline{\psi} + \psi', \, T' \right) + J(\psi', \, \overline{T}) \right] \equiv H_2 . \qquad (24)$$

The boundary conditions are

$$z = 0, \qquad T' = T_s' \; , \qquad (25)$$

$$z \to \infty, \qquad T' = 0 . \qquad (26)$$

After solving T' from Eqs. (24)–(26), we can write the Eq. (20) as follows:

$$\frac{\partial T'_{si,j}}{\partial t} + k_s H_{1i,j} + \left(\frac{\delta k_s}{hD_Q} + \frac{k_s}{hD_s}\right) T'_s + \frac{k\tau_\gamma}{D_s}\frac{k_s}{h}H'_{2i,j}$$

$$= -\frac{s_0 l_b}{w_0 \rho_s C_{ps} k_s} \zeta'_{bg} , \tag{27}$$

where

$$D_Q = \frac{\rho_s C_{ps} k_s}{L\rho k_T \gamma \dfrac{\partial \ln \bar{e}_s}{\partial T} \dfrac{\partial \bar{q}_s}{\partial T}} , \qquad D_s = \frac{\rho_s C_{ps} k_s \sqrt{k\tau_r}}{\rho C_p k_T} \tag{28}$$

Its difference form with respect to time is

$$T'^{t+\delta t}_{si,j} = \left(1 - \frac{\delta k_s \delta t}{hD_Q} - \frac{k_s \delta t}{hD_s}\right) T'^{t}_{si,j} - k_s \delta t H_{1i,j}$$

$$-\frac{k\tau_r \delta t}{D_s h} H_{2i,j} - \frac{S_0 l_b \delta t}{w_0 \rho_s C_{ps} k_s}(\Delta\phi')_b . \tag{29}$$

The Forecast of Anomalous Geopotential Height

The anomalous geopotential height can be calculated from Eq. (8) after the temperature of ocean or land is obtained. In order to overcome the interferences caused by the transient Rossby wave two approaches will be used as follows.

(a) The Anomalous Filtered Model (AFM)

The transient Rossby wave can be filtered out by omitting the partial derivative with respect to time in the vorticity equation for the atmosphere, i. e.,

$$k\Delta\phi' + \frac{1}{f}J(\bar{\phi} + \phi', \Delta\phi') + J(\phi', f + \frac{1}{f}\Delta\bar{\phi}) - \delta_1 k^* T'_s - \frac{f^2}{\sigma_p}\frac{\partial G}{\partial p} = 0 , \tag{30}$$

while the time derivative term is still retained in the thermal equation for underlying ocean and land. We may call this approach an anomalous filtered model model (AFM).

We write the Eq. (30) at three levels illustrated in Fig. 1.

```
   0    hPa ──────────── 7
 200    hPa ──────────── 6
 500    hPa ──────────── 4
 700    hPa ──────────── 2
1000    hPa ──────────── 1
```

Fig. 1. The vertical structure for atmosphere

$$p = 0, \qquad \frac{\partial \phi'}{\partial p} = 0, \qquad G_7 = 0, \tag{31}$$

$$i = 2, 4, 6, \qquad \Delta\phi'_i + \frac{1}{fk_i} J(\bar{\phi} + \phi', \Delta\phi')_i + \frac{1}{k_i} J(\phi', f + \frac{1}{f}\Delta\bar{\phi})_i$$

$$-\delta_1 k^* T'_s - \frac{f^2}{\tilde{\sigma}_p k_i}(\frac{\partial G}{\partial p})_i = 0, \tag{32}$$

$$p = 1000\, hPa \qquad G_1 = \frac{R}{p_s} T'_s, \qquad \phi'_1 = \phi'_2 - \frac{R}{g} \ln \frac{p_1}{p_2} T'_s. \tag{33}$$

The iteration has been carried out by the method of over-relaxation. The horizontal mesh size is taken as 540 km and the polar stereographic projection is used. An energy and momentum conservative representation of Jacobi terms proposed by Arakawa (Haltiner 1971) has also been adopted in the following calculation.

Another way to filter high-frequency noise is to take the monthly average of the daily predictions. The difference form of Eq. (8) is as follows:

$$(\Delta + \frac{f^2}{\tilde{\sigma}_p}\frac{\partial^2}{\partial p^2})\phi'^{(t+\delta t)} = \left\{ [(1 - k\delta t)\Delta + \frac{f^2}{\sigma_p}\frac{f^2}{\partial p^2}]\phi' - \frac{\delta t}{f} J(\bar{\phi} + \phi'\Delta\phi') \right.$$

$$\left. -\delta t J(\phi', f + \frac{1}{f}\Delta\bar{\phi}) + \delta_1 k^* T_s' + \frac{f^2\delta t}{\tilde{\sigma}_p}\frac{\partial G}{\partial p} \right\}^{(t)} \tag{34}$$

The time step is taken to be one hour. Before integrating the next time step the iteration method is used, in which the initial iteration field is taken as the solution of the local Green function. Two initial conditions are used in solving Eq. (34). One is the anomalous height field in Jan. 1977 and the another is zero.

The Experiments of One-Month Prediction

A prediction example of Feb. 1977 is given in this section. The predicted anomalous fields of 700 hPa, 500 hPa, and 300 hPa heights for Feb. 1977 are illustrated in Figs. 2, 3, and 4 respectively. The middle chart (b) is the observational field, the left one (a) is the prediction by AFM· and the right one (c) is the prediction by AGCM. It is found that both results predicted by the two kinds of initial conditions in AGCM are the same except in the first ten days. This shows that the atmospheric initial field seems not important in comparison with the effects of boundary heating from ocean and land. Figure. 5 represents the predicted and observed anomalous fields of earth's surface temperature. The correlation coefficient between observation and prediction tops 0.71. For the sake of comparison all correlation coefficients including persistance is listed in Table 1.

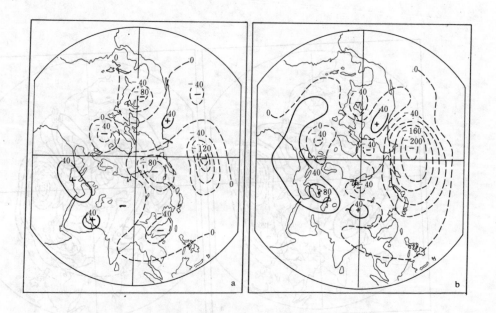

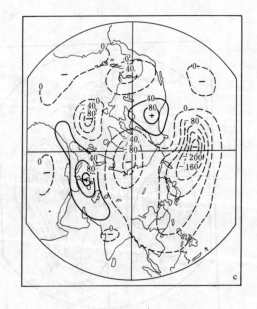

Fig. 2. Anomalous fields of 700 hPa height in Feb. 1977. Chart (b) is the observational anomaly and chart (a) and (c) are the predicted results by AFM and by AGCM respectively

378

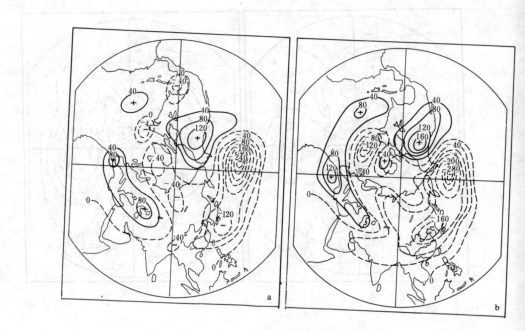

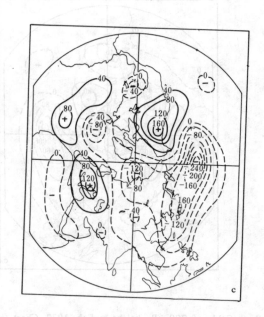

Fig. 3. As in Fig. 2 except for 500 hPa

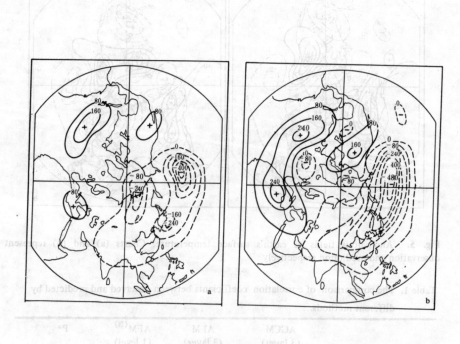

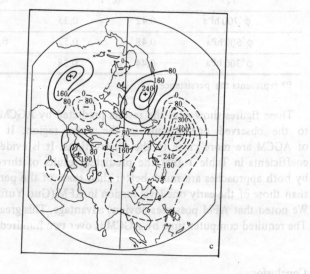

Fig. 4. As in Fig. 2 except for 300 hPa

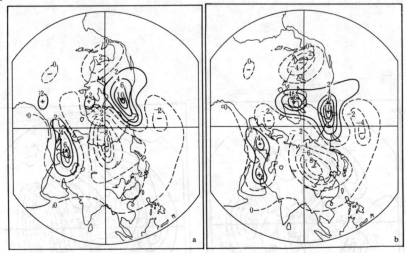

Fig. 5. Anomalous fields of earth's surface temperature. Charts (a) and (b) represent observation and prediction respectively

Table 1. The comparison of correlation coefficients between observed and predicted by different methods.

	AGCM (3 layer)	AFM (3 layer)	AFM[3] (1 level)	P*
T'_s	0.71		0.36	0.25
$\phi'700\,\text{hPa}$	0.42	0.33		0.28
$\phi'500\,\text{hPa}$	0.48	0.52	0.42	0.34
$\phi'300\,\text{hPa}$	0.44	0.38		0.33

P* represents the persistence.

These figures show that the results predicted by AGCM and by AFM are similar to the observed one except in the polar region. It seems that the results of AGCM are more close to the observations. It is evident from the correlation coefficients in Table 1 that the predicted results of three-layer version obtained by both approaches are much better than that of the persistance, and also better than those of the early one-level version in AFM (Guo Yufu and Chao Jiping 1984). We noted that AFM possesses obvious advantage. This greatly saves computer time. The required computer time by AGCM is over two hundred times that by AFM.

Conclusions

Using both approaches, the anomaly general circulation model (AGCM) and the anomaly filtered model (AFM), an example of Feb. 1977 is predicted by their three-layer versions. It is shown that both AGCM and AFM have the potential ability for long-range numerical forecasts. The correlation coefficients between the observation and the prediction of both models are much better than that of the persistance, and are also better than that of the early one-level version in AFM. By comparison with AFM, the results of AGCM are closer to the observations. But AFM is a much cheaper system, and a promising method for monthly and seasonal forecasts.

The results also show that the atmospheric initial field seems to be unimportant in comparison with the effects of boundary heating from ocean and land.

References

Chao Jiping, Guo Yufu, Xing Runan (1982) *J Meteorol Soc, Jpn,* 60: 282–291
Haltiner, G.J. (1971) Numerical weather prediction chap 12. Wiley, London
Guo Yu fu, Chao Jiping (1984) *Adv Atmos Sci* 1: 30–39

Statistical Prediction of the Frequency of Typhoons Based on Sea Surface Temperature

Takashi Aoki[1]

Abstract − The frequency of typhoon formation showed a minimum during El Nino events and maximum frequency is observed two years later. There is simultaneous correlation in the sense that as the sea surface temperature in the eastern Equatorial Pacific increases, the frequency of typhoon formation decreases, and vice versa.

There are significant correlations between the number of typhoons formed and sea surface temperature in the preceding year and two years before typhoon formation. High positive correlations are found for the sea surface temperature in the eastern Equatorial Pacific, while there are negative correlations for the northwestern part of the Pacific.

To predict the frequency of typhoon formation, multiple regression equation is developed by using sea surface temperatures as independent variables. The multiple correlation coefficient for typhoon formation is 0.85. This equation may be useful in long-range forecasting.

Introduction

In previous studies (Xie et al. 1983; Aoki and Yoshino 1984), it has been made clear that the frequency of typhoons has a close relationship to sea surface temperature (SST) of the preceding year and two years before. In the present study, the lagged correlations are examined and multiple regression analysis was performed to explore the possibility for long-range forecasting of the frequency of typhoon formation using SST in the North Pacific.

Data

Data concerning typhoon formation are extracted from the "Geophysical Review", issued by the Japan Meteorological Agency. The annual frequencies of typhoon formation during the period from 1953 to 1982 are examined. Data of SST are prepared for grid points at 5° latitude-longitude intervals in the North Pacific between 50° N and 10° S except the area south of 10° N and west of 180°.

The SST from 1949 to 1976 are available in a published data book of "Monthly Sea Surface Temperatures in the North Pacific" by the Institute of Geography, Academia Sinica and the Shanghai Central Meteorological Observatory. For additional data, "Fishing Information" and "Oceanographic Monthly Summary" compiled by NOAA are referred to for the period 1977−1980 and 1981− 1982, respectively.

In this study, three-month averaged SST, namely, winter (January to March),

---

1. Meteorological Research Institute, Tsukuba Ibaraki 305, Japan

spring (April to June), summer (July to September) and autumn (October to December) are used.

El Nino Events and the Frequency of Typhoon Formation

Warm episodes in equatorial Pacific SST are associated with large positive precipitation anomalies in the central Pacific and negative precipitation anomalies over much of Indonesia and India (Rasmusson and Carpenter 1982, 1983). Warm episodes also tend to be accompanied by Pacific/North American pattern of mid-tropospheric geopotential height which is characterized by below-normal height in the North Pacific and the southeastern United States, and above-normal height over western Canada (Wallace and Gutzler, 1981; Horel and Wallace, 1981). Tanaka (1982) pointed out that the El Nino years of 1965, 1972 and 1976 were years of weak monsoon.

Great concern for the relationship between the frequency of typhoon formation and El Nino events has arisen due to the results of such studies. In order to clarify this relationship, we examined a total of six events from 1953, 1957, 1965, 1969, 1972 and 1976 (Rasmusson and Carpenter 1982). The warm episode years and the preceding and following years will be referred to as WEY (0), WEY (−1) and WEY (1), respectively.

To obtain a picture of fluctuations during the five years before and after El Nino events, we constructed composites of the yearly averaged Southern Oscillation Index based on the Tahiti Darwin pressure difference (top of Fig. 1), the yearly

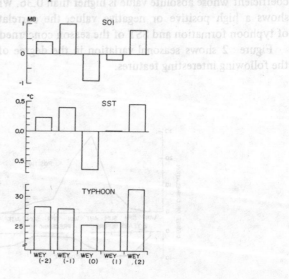

Fig. 1. Fluctuations of Southern Oscillation Index (top), equatorial sea surface temperature (middle) and frequency of typhoon formation (bottom) during the five years before and after El Nino events. Warm episode years and preceding and following years are refered to as WEY (0), WEY (-1) and WEY (1) respectively

384

averaged equatorial SST anomaly in the region of 5° N–5° S and 160° W–100° W (middle of Fig. 1), and the annual frequency of typhoon formation (bottom of Fig. 1).

The most prominent features of the composites in Fig. 1 are that the frequency of typhoon formation shows a maximum in WEY (2) and a minimum frequency is observed at WEY (0). An apparent correlation is also found in the sense that as the equatorial SST increase and the SOI decreases, the frequency of typhoon formation decreases, and vice versa.

Seasonal Variation of Correlation Patterns

Correlation coefficients between the annual frequency of typhoon formation and SST at each grid indicate 5° latitude, and longitude intervals for every season from winter to summer of the same year are computed in the first step. The correlations are also calculated for the period from winter to autumn for both the preceding year and two years before. In this way correlations are obtained for eleven seasons at every grid point.

In order to examine season-to-season variation of correlation between the frequency of typhoon formation and SST, the degree of correlation must first be defined. In this case, the number of samples total 30 (years), meaning the degree of freedom is 28. The correlation coefficient corresponding to a significant level of 5% is ± 0.36. The degree of correlation is defined as the sum of the positive or negative correlation coefficients, respectively, for the grid points which have a correlation coefficient whose absolute value is higher than 0.36. When the degree of correlation shows a high positive or negative value, the correlation between the frequency of typhoon formation and SST of the season concerned is high.

Figure 2 shows seasonal variation in the degree of correlation, which suggests the following interesting features.

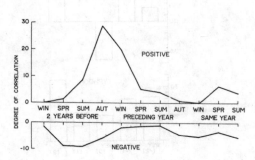

Fig. 2. Seasonal variations in degree of correlation between frequency of typhoon formation and sea surface temperature in the North Pacific. SPR, SUM, AUT and WIN indicate spring,

It should be noted that the degree of correlation between the frequency of typhoon formation and SST in the same year is unexpectedly low. It is the lagged correlation of two years before typhoon formation that shows a high degree of correlation.

Positive correlation shows large values during the period from the autumn of two years before typhoon formation to the winter of the preceding year. Few positive correlations are found from the autumn of the preceding year to the winter of the same year. Subsequently, the degree of correlation becomes somewhat larger in the typhoon season of the year concerned.

In general, the degree of negative correlation is much smaller as compared with that of positive correlation. The negative correlation between the frequency of typhoon formation and SST is not so significant. There is, however, a period when the degree of negative correlation becomes larger in the period from the spring to the autumn of two years before typhoon formation, and also in the autumn of the preceding year in addition to in every season of the same year.

Distribution of Correlation Coefficients

A typical example of the distribution of correlation coefficients between the frequency of typhoon formation and SST is shown in Fig. 3. This figure depicts the correlation pattern for each significant level of correlation coefficients exceeding 0.1%, 1%, and 5% by making use of areas shaded differently according to density.

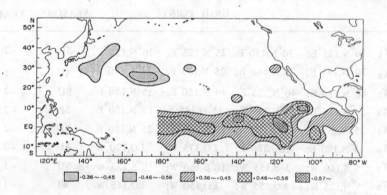

Fig. 3. Distribution of correlation coefficients between frequency of typhoon formation and sea surface temperature (autumn of two years before typhoon formation)

The North Pacific is divided into two regions showing positive and negative correlations for the autumn of two years before typhoon formation as shown in Fig. 3. Similar results were also presented in papers by Wang (1982), Xie et al. (1983) and Aoki and Yoshino (1984).

The banded region of positive correlation stretches from east to west in the southeastern part of the North Pacific, namely the eastern Equatorial Pacific, where the positive correlation region is wide and the correlation coefficients are significant at a 0.1% level at many grid points. The maximum value of the correlation coefficient is 0.63 at the grid point of $10°$ N and $105°$ W. The negative correlation region in the northwestern part of the North Pacific is narrow, and the maximum value of the correlation coefficient is low.

Development of Multiple Regression Equation

The following conditions are applied to select possible independent variables for multiple regression analysis;
1. The correlation coefficient is significant at a 5% level.
2. There exist more than four highly correlated grid points in adjacent positions.
3. When conditions (1) and (2) are satisfied in an area through several successive seasons, the season having the highest correlation is selected.

After this selection, four grid points are chosen in order of correlation for each area and the averaged SST on the four grid points is used as the possible independent variable. Thus, as shown in Table 1, twelve possible independent variables are obtained for the frequency of typhoon formation.

Table 1 List of possible independent variables. Averaged sea surface temperature on the four grid points is used in the development of the equation.

	GRID POINTS				SEASON*	YEAR**
X_1	$30°N,125°E;$	$30°N,130°E;$	$25°N,125°E;$	$20°N,125°E$	SU	-2
X_2	$30°N,135°E;$	$30°N,140°E;$	$25°N,135°E;$	$25°N,140°E$	AU	-1
X_3	$40°N,150°E;$	$40°N,155°E;$	$40°N,160°E;$	$35°N,150°E$	SU	-2
X_4	$20°N,145°E;$	$20°N,150°E;$	$15°N,145°E;$	$15°N,150°E$	SP	-2
X_5	$25 N,165°E;$	$25°N,170°E;$	$25°N,175°E;$	$25°N,180°E$	AU	-2
X_6	$5°N,180°E;$	$EQ,180°E;$	$EQ,175°W;$	$EQ,170°W$	AU	-2
X_7	$5°S,160°W;$	$5°S,155°W;$	$10°S,160°W;$	$10°S,155°W$	SU	-1
X_8	$5°N,155°W;$	$EQ,155°W;$	$EQ,150°W;$	$EQ,145°W$	WI	-1
X_9	$5°N,145°W;$	$5°N,140°W;$	$5°N,135°W;$	$5°N,130°W$	AU	-2
X_{10}	$10°N,120°W;$	$10°N,115°W;$	$10°N,110°W;$	$10°N,105°W$	WI	-1
X_{11}	$5°N,120°W;$	$5°N,115°W;$	$EQ,120°W;$	$EQ,115°W$	AU	-2
X_{12}	$5°S,115°W;$	$5°S,110°W;$	$5°S,105°W;$	$5°S,100°W$	AU	-2

* SP: spring; SU: summer; AU: autumn; WI: winter.
** -1: preceding year; -2: 2 years before

To predict frequency of typhoon formation, the equation is developed by applying a stepwise regression program using the possible independent variables mentioned above. The equation obtained is as follows for the frequency of typhoon formation,

$$Y_f = 41.50 - 2.970\,X_2 - 5.040\,X_4 + 4.963\,X_7 + 2.277\,X_{10}.$$

Multiple correlation coefficient and standard error of estimate by this equation based on development data are 0.85 and 2.9, respectively. Four years of independent data in 1951, 1952, 1983, and 1984 are available for testing purposes and the result is given in Table 2. The predictions appear to be acceptable by the current standard of forecasting technique.

Table 2 Predicted and observed frequencies of typhoon formation for independent data.

	1951	1952	1983	1984
prediction	23.2	25.0	24.0	27.9
observation	21	27	23	27

Thus, the regression equation is found to be effective. As independent variables in the regression equation are SST of the preceding year and two years before typhoon formation, this regression equation can be used for prediction of the frequency of typhoon formation in the following year.

References

Aoki, T, Yoshino M. M. (1984) *J Meteorol Soc Jpn,* **62**: 172–176
Horel, J. D Wallace J. M. (1981) *Mon Wea Rew.* **109**: 813–829
Rasmusson, E.M Carpenter T.H. (1982) *Mon Wea Rev* 110: 354–384
Rasmusson, E. M Carpenter T. H. (1983) *Mon Wea Rev* 111: 517–528
Tanaka, M (1982) *J Meteorol Soc Jpn* 60: 865–875
Wallace, J. M. Gutzler D. S. (1981) *Mon Wea Rev* 109: 784–812
Wang, J. Y. (1982) *Acta Oceanol Sin* 1: 40–46
Xie, S. M., Aoki T. Yoshino M. M. (1983) *Tenki* 30: 495–502 (in Japanese)

A Forecasting Model for the Upper Mixed Layer Depth of the Ocean

Jiang Jingzhong[1]

Abstract– In this paper, the similarity theory is applied to the forecasting of the depth of
the mixed layer (thermocline) during the warm seasons. The controlling processes are
assumed to be secular, nonadvective and nondivergent. Parameters include wind speed,
Coriolis effect, the coefficient of thermal expansion and heat storage within the mixed
layer. The concept of an universal function P(N) by Kitaigorodskii were used in the
forecasting model. The resulting forecast model consists mainly of two equations. The
constants in the equations were determined by using data from the East China Sea during
the warm seasons from 1975 to 1981. The tests apparently indicate good applicability of
this forecast method in the two specific regions in the East China Sea.

Introduction

In general it is difficult to develop a forecasting model for the thermal structure of the upper layer of the ocean in terms of dynamical approach. The reason
is that if all the physical processes are considered in the model, it would become
too complicated to obtain the solution. On the other hand, if too many assump-
tions are made to simplify the model, the model would be unrealistic.

Therefore, many authors have treated this problem based on empirical relations.
However, these relations are only valid for certain particular locations and for
limited time.

Another possible approach to developing a forecasting system is to use the
similarity theory. Kitaigorodskii (1960), Kitaigorodskii and Filushkin (1964)
investigated the application of the similarity theory to the forecasting of the ocean
thermal structure. The method used by Kitaigorodskii is also applied in this
chapter, with some modification of parameters, in an effort to obtain more realistic
results. The methods are used to develop a forecasting model which will be applied
to the predictions in two areas of the East China Sea. The values of all the
parameters are determined from the data in the East China Sea. In this paper, the
terms "the mixed layer depth (MLD)" and "depth of the thermocline" are
synonymous. Only those changes in the mixed layer depth which are secular,
nonadvective and nondivergent will be considered.

Parameters and Processes Affecting the Mixed Layer Depth During the Warm Seasons

There is general agreement among many investigators that during the warm seasons

---

1. Second Institute of Oceanography State Oceanic Administration, Hangzhou, China.

the mixed layer depth varied with (Kraus 1972; Fellor and Duobin 1975, Nihoul 1979; Kitaigorodskii 1960; Kitfaigooldskii and fiushkin 1964, Kraus and Turner 1967; Phillips 1977; Stevnson 1979) the following parameters:

1) wind;
2) the heat storage in the upper layer;
3) the net heat transfer downward across the air-sea interface;
4) Coriolis effect;
5) divergence in the upper layer of the ocean;
6) internal wave action;
7) advection.

Only the first four of the above processes will be considered in developing the present forecast model. The remainders are considered to contribute to the scatter of the results.

Selection of the Parameters Representing the Controlling Processes

In order to determine the representative parameters, it is usual to compare the magnitude of the terms in the relevant dynamical equations.

In this paper, an open region is considered. Assuming an unaccelerated flow in an unbounded and horizontally homogeneous ocean, the equations of motion in the XY-plane are reduced to:

$$0 = fv + \frac{1}{\rho} \frac{\partial \tau_{zx}}{\partial z} , \tag{1}$$

$$0 = -fu + \frac{1}{\rho} \frac{\partial \tau_{zy}}{\partial z}$$

where u and v are the components of velocity, τ_{zx} and τ_{zy} are the components of stress on the horizontal flow and f is the Coriolis parameter. At the interface, the stress is given by:

$$\tau = \rho v w^2 , \tag{2}$$

where τ is the stress on the water surface, w is the mean wind speed and v is a coefficient which is a function of wind speed and of the height where the wind speed is measured. Thus a measure of wind speed which is a representative of the stress effects should be introduced as a parameter in the forecasting model. However, the depth, to which wind effects reach, depends also on the Coriolis parameter as shown theoretically by Ekman. Thus, the Coriolis parameter must be considered in the forecasting model.

The maintenance of thermal structure depends on heat flow. The governing equation of temperature changes due to heat flow is:

$$\rho C_p \frac{\partial T}{\partial t} = -\frac{\partial}{\partial z} (k \frac{\partial T}{\partial z}) , \tag{3}$$

where ρ is the density of water, C_p is the specific heat of the water, T is the temperature of the water, and k is the eddy conductivity coefficient for heat. Equation (3) indicates that the local rate of change of temperature is a function of k as well as of the differences of temperature between layers.

It is difficult to calculate the value of k, because k depends upon the field of motion and the stability. Therefore, we will use another parameter, i.e., stability.

The density gradient, and thus the stability, can be represented by a combination of the coefficient of thermal expansion β and the temperature difference between layers; the following two parameters have been chosen to represent stability:
1) the coefficient of thermal expansion β;
2) the heat storage Q in the mixed layer.

Thus, besides the mixed layer depth (MLD) itself, four parameters have been used in the forecasting model:
1) the representative maximum wind (w);
2) the Coriolis parameter (f);
3) the coefficient of thermal expansion (β); and
4) the heat storage in the upper layer (Q).

Determination of Values of the Parameters

Wind Stress (w): It is necessary to find a value of wind speed which is representative of the influence of wind on the MLD over a period of time. The wind speed in the mixed layer depth model is expressed in the form of stress, and the stress coefficient γ is an increasing function of wind speed. Thus Equations (1) and (2) show that higher wind speed should be weighted more heavily in estimating the influence of wind stress upon the MLD. Therefore, one should avoid using linearly averaged wind in the formulation of a mixed layer depth model.

A "representative maximun wind" has been introduced and defined as the average of the five largest wind speeds of the eight usually reported during a 24-hour period (the least wind speed must be more than 4 m/sec in present paper). The period covers the interval, when the highest winds from 72 hours to 12 hours prior to the forecasting time present.

Coriolis Effect (Ω): The *Coriolis* parameter multiplied by 10^4 will be used to represent Coriolis effects:

$$\Omega = f \times 10^4 = 2\omega\sin\phi \times 10^4, \qquad (4)$$

where constant 10^4 was introduced for convenience.

The Heat Storage in the Mixed Layer (Q): One difficulty in the use of Q as a measure of stability is illustrated in Fig. 1. It shows that thermal structures with different stability may prevail with the same Q.

However, with the combination of Q, β, the MLD is able to provide a good approximation of the stability of the mixed layer. In this chapter, the MLD appears in the form of $P(N)$ in the forecast model which is initially determined by the physical parameters including mixed layer depth. The computation method of Q is as follows:

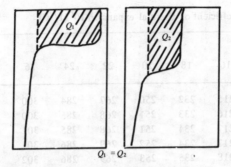

Fig. 1. Two possible thermal structures with same heat storage present in warm seasons

After the values of P and N are determined, the computation of Q will be carried out based on the average CTD, BT or Nansen cast. First, it is necessary to determine the "Area", depicted in Fig. 2. The dashed line in Fig. 2 is vertical to the sea surface, drawn from the point of maximun curvature of the temperature profile (roughly at the bottom of the thermocline).

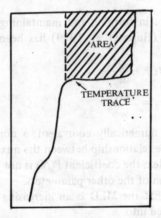

Fig. 2. Representation of the "Area" used in determining Q

Then given by

$$Q = \rho C_p \times \text{Area} \times 10^{-1} \quad (\text{Kg cal/cm}^2) \tag{5}$$

where the dimensions are cal/$°$C. cm^3 (ρC_p) and m.$°$C (Area) respectively.

The Coefficient of Thermal Expansion (β): Values of β used in this paper were interpolated from those given by Sverdrup (Sverdrup et al. 1964). These values are tabulated in Table 1.

The Forecasting Model

The Pairs of $P(N)$. As above, four parameters have been chosen as representative of

Table 1. Values of the coefficient of thermal expansion $10^6 (1/°C)$

S‰	T(°C) 14	16	18	20	22	24	26	28	30
30.0	196	215	232	250	267	284	300	316	332
30.5	197	216	233	251	268	285	300	316	332
31.0	198	217	234	251	268	285	301	317	332
31.5	199	217	234	252	269	286	301	317	332
32.0	200	218	235	253	269	286	302	317	333
32.5	201	219	236	254	270	287	302	318	333
33.0	201	220	237	254	271	287	302	318	333
33.5	202	220	238	255	271	288	303	318	333
34.0	203	221	238	256	272	288	303	318	334
34.5	204	222	239	256	272	289	304	319	334
35.0	205	223	240	257	273	289	304	319	334

the factors which are effective in forming and maintaining a mixed layer in the warm season. The π theorem (Hugo et al. 1979) has been applied to these four parameters and results in:

$$\pi_1 = \frac{\beta Q \Omega}{w}, \tag{6}$$

$$\pi_2 = \frac{H\Omega}{w} \tag{7}$$

In order for $\pi 1$, π_2 to be numerically equivalent, a nondimension coefficient $P(N)$ is introduced. Because the relationship between the mixed layer depth and the other parameters is very complex, the coefficient $P(N)$ is not , in general, a constant and must be found as a function of the other parameters.

In this case, it is known that the MLD is an increasing function of the wind, and π_1 and π_2 may be combined into

$$\pi_2 = P\frac{1}{\pi_1}. \tag{8}$$

Substituting for π_1 and π_2 and solving for the MLD gives:

$$MLD = P \times \frac{w^2}{\beta Q \Omega^2}. \tag{9}$$

Use of the π theorem, as before, yields another nondimension N of these parameters:

$$N = \frac{\beta Q \Omega}{w}. \tag{10}$$

Substituting both P and N into the equations with the same data and plotting the corresponding pairs of P and N give a function $P(N)$ in its graphical form. If the parameters are truly characteristic of the controlling processes, the plot of P versus N should have little scatter. If the parameters used are not truly representative, then large scatter results.

Based on the observation data from the East China Sea, two types of thermal structure are considered, type I occurs in the area to the southwest of the Cheju Island region, and type II occurs between the continental shelf and Kuroshio (Fig. 3). Because the two types of thermal structure are quite different in the physical processes, they will be treated separately.

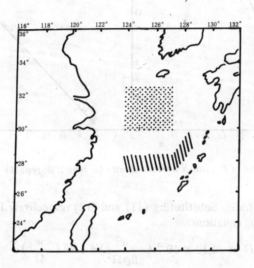

Fig. 3. The distribution of two types of thermal structure (Dotted: type I; Hatched: type II)

As regards type I, the thermocline is formed mainly due to increase of the sea surface temperature during warm seasons. The MLD is less than 30 meters, and temperature gradient is very large (more than 0.5°C/m). For type II the thermocline is formed because of piling the warm mixing water in the continental shelf above the cold Kuroshio water. The MLD of type II is 50 meters deeper than that of type I, while its thermocline is weaker.

In the sea area near Cheju Island, 300 data sets of temperature profiles from 21 stations have been used to compute the values of the parameters, W, Q, H, β, and to determine the pairs of P and N. These data were taken from June to August during the year 1975 to 1981. In order to remove the effects of internal wave, each pair of P and N are averaged for ten (or more) observations.

As regards type II in the Kuroshio region, data from more stations than those for type I have been used.

The equations for the best-fit line by least-squares technique are:

Type I $\qquad P = -4.12 \times 10^{-4} + 4.44N$, $\qquad$ (11)

Type II $\qquad P = -3.97 \times 10^{-3} + 10.9N$. $\qquad$ (12)

Fig. 4 shows the best-fit line of P versus N by least-squares.

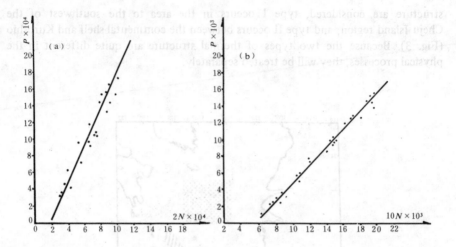

Fig. 4. The best-fit line of P versus N by least-squares (a: type I; B: type II)

The Forecasting Model. Substituting (11) and (12) respectively into Equation (9) gives the forecasting equations:

Type I $\qquad MLD = -4.12 \times 10^{-4} \left(\dfrac{w^2}{\beta \varrho \Omega^2} \right) + 4.44 \left(\dfrac{w}{\Omega} \right),$ (13)

Type II $\qquad MLD = -3.97 \times 10^{-3} \left(\dfrac{w^2}{\beta \varrho \Omega^2} \right) + 10.9 \left(\dfrac{w}{\Omega} \right).$ (14)

Test of the Forecasting Equations

Two examples of the applicaiton of Equations (13) and (14) to determine MLD are as follows.

Example 1. Forecast to be made for 7 August, 1979 based on data observed on 6 July 1979 at the south-west of the Cheju Island.
1) From the temperature profiles taken on 6 July 1979, Q_T was determined from equation (5).
2) Using the climatological data to compute the average net heat teansfer downward across the air-sea interface Q' (Kg cal/cm^2 day). Then the heat storage is given by

$$Q = Q_T + Q'.$$ (15)

3) Values of β, Ω, and w were determined by the formula in section 4.
4) Equation (13) provided a forecast MLD of 28.66 meters. The survey MLD was at 26.61 meters on 6 July, 1979.

Example 2. Forecast of the mixed layer depth of 21 August 1979 was based on data taken on 7 August, 1979 at the continental shelf. Steps 1,2 and 3 are the same as those in example 1. Equation (14) was used to forecast the MLD. The forecast MLD is 63.70 meters and observational value is 68.08 meters.

Equations (13) and (14) can be used to forecast MLD over any length of time for which the parameters can be accurately predicted. Twenty four tests of the equation were made using independent data for forecast periods from 15 to 30 days. Among them, seventeen tests belong to type I and seven to type II. The statistical results of forecasting are summarized in Table 2.

Table 2. Statistical results of predictions

Type of the Thermal Structure	I	II
Observed Mean *MLD* (m)	26.14	66.20
Algebraic Mean of the Errors (m)	−3.14	5.04
Mean of Absolute Errors (m)	5.34	7.24
Standard Deviation (m)	5.63	6.35
Percentage Within 1 Standard Deviation	65	57
Percentage Within 2 Standard Deviation	94	71

Conclusions

In this paper, the concept of a universal function $P(N)$ used by the famous oceanographer Kitaigorodskii has been generalized to develop a forecasting model of the mixed layer depth of the ocean. The tests showed: in the specific sea areas, equations (13) and (14) appear to be useful in forecasting the mixed layer depth (thermocline) at a reasonable degree of accuracy.

This forecasting model can be applied to a specific sea area not only for a single station, as Kitaigorodskii did. Therefore the applicability of universal function $P(N)$, as proposed by Kitaigorodskii, is greatly generalized.

References

Hugo, B.F. et al, (1979) *Mixing in inland and coastal waters.* Academic Press, London, pp 23–29

396

Kitaigorodskii, S. A., (1960) *Izv Acad, Sci USSR Geophys Ser* 3: 284–287 (English ed)

Kitaigorodskii S.A. Filushkin, B. N., (1964) Application of the similarity theory to the analysis of the observation in the upper ocean. *Oceanol Stu* 13, Izd-vo. Nauka, M

Kraus, E.B. (1972) *Atmosphere-ocean interaction.* Oxford Univ Press (Clarendon) pp 151–212

Kraus, E. B., Turner, J. B., (1967) *Tellus* 19: 98–106

Mellor, G.I. Duobin, P. A., (1975) *J Phys Oceanogr* 5: 718–728

Nihoul, C. J. ed., (1979) *Marine forecasting.* Elsevier Amsterdam, pp 1–33

Phillips, O. M., (1977) Entrainment. In: Kraus E. B. (ed) *Modelling and prediction of the upper layers in the ocean* pp 92–101

Stevenson, J. W., (1979) *J Phys Oceanogr* 9: 57–64

Sverdrup, H. U., et al, (1946) The oceans, Chap 3: Prentice-Hall, Inc. New York, pp 44–77

Prediction of Indian Monsoon Variability: Evaluation and Prospects Including Development of a New Model

V. Thapliyal[1]

Abstract – The interannual variability of the summer monsoon rains has a profound impact on the socio-economic environment of India. For nearly a century, several attempts have therefore been made to develop techniques for predicting the variability of seasonal monsoon rainfall over the country. In the first decade of this century, India (Walker 1924) introduced the concept of correlation in long-range prediction and since then this method is being used in some form or the other. Initially, the predictors used were essentially surface observations from different parts of the world, but during the last two decades attempts have been made to identify some other predictors related to the upper air, slow varying boundary conditions (like sea surface temperature; snow cover etc.), long period oscillations and Quasi Biennial Oscillations etc. A few encouraging results have emerged from these studies such as, for example, the recently identified relationship between the winter lower stratospheric circulation feature and the Indian monsoon rainfall has succeeded in predicting the large monsoon failure over India. In this chapter an attempt is made to discuss the evaluation of long-range forecast techniques as they were developed in India. The attempt will also be made to discuss the possible physical linkages which may be operative for generating the large seasonal variations of the monsoon.

For long-term predictions the general circulation models have not yet produced satisfactory techniques. However, India (Thapliyal 1981) has recently introduced a new concept in the field of long-range prediction and has succeeded in developing stochastic dynamic forecast models. Conceptually, the new models promise to be superior to the existing operational long-range prediction techniques, as they utilize the transfer dynamics of the atmosphere as well as of the stochastic processes for the long-range prediction of monsoon rainfall variability over India. In addition, the operational results from these models have been found very encouraging as compared with other models.

An attempt has been made to develop these models further by incorporating in them a few other oscillations like QBO, southern oscillation etc. The exact mathematical expression of the new model suggests that the input of the previous four years contributes significantly to the monsoon variability. In addition, the atmospheric oscillations of 2, 4, 6, and 11 years, also influence the monsoon variability. Further analysis of the model suggests that the atmosphere has a much longer memory than is presently assumed by the scientists. It appears that until ocean-atmosphere-interactive-general circulation models are fully developed, the dynamic stochastic transfer models may provide a useful tool for predicting the year-to-year variation of the climate.

Introduction

Despite recent advances in science and technology, the economy of India depends critically on the summer monsoon (June to September) rainfall which accounts around 75 percent of the annual rain over most parts of the country. As the variation in the monsoon rainfall has a profound socio-economic impact over the coun-

---

1. India Meteorological Department Pune 411 005, India

398

try, there has always been a great demand for the Long Range Prediction (LRP) of seasonal monsoon rainfall (or year-to-year variability of the rainfall) from different sections of the society. In fact, a stage has reached where various users starting from the common man to the National Government are demanding various kinds of LRP of weather for carrying out effective planning in different fields of activity. Realizing this, the Indian Weather scientists have started serious attempts to develop the adequate LRP techniques which could provide precise forecasts for variability of monsoon rainfall over India.

India has a long history of issuing operational LRP of monsoon rainfall over different regions of the country. Based mainly on Himalayan snow, the first official LRP of monsoon rainfall covering the whole of India and Burma was issued by Blanford, the then reporter of the India Meteorological Department on June 4, 1886. The next important development took place in first decade of this century when Sir Gilbert Walker, the then Director General of the Indian Meteorological Department, introduced the concept of correlation in the field of LRP and succeeded in removing the subjectivity from the earlier forecast methods. Since then, the correlation technique is being widely used by many countries, including advanced ones for preparing their operational long range forecasts. Recently, the introduction of a new concept of the dynamic stochastic transfer of the atmosphere in the field of LRP of monsoon rainfall in India has resulted in the development of a superior technique which is being operationally used to predict the monsoon rainfall over two broad subdivisions of India, namely Peninsula and Northwest India (Fig. 1). In this paper, a critical review on the LRP of monsoon rainfall in India has been presented. Details regarding the regression and the dynamic-stochastic transfer models have also been presented.

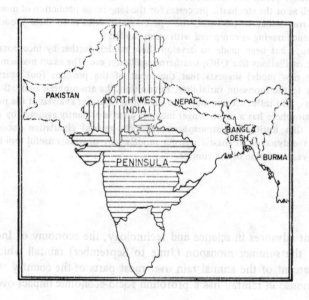

Fig. 1. Forecast subdivisions of India

Early Developments
,

India was the first country which started, a century ago, the systematic work of developing the techniques for LRP of seasonal monsoon (June to September) rainfall over India. After India experienced a serious famine in 1877, a year of highly deficient monsoon rainfall over several parts of the country, the Government of India's anxiety for the earliest possible information about the progress of the monsoon grew substantially and it called upon Blanford (1884) who established the Indian Meteorological Department in 1875, to prepare monsoon forecasts. At that time so little was known of meteorology that hardly any recognized method of Long Range Prediction (LRP) existed for preparing this type of forecast. Blanford noted an association between excessive winter and spring snowfalls in the Himalayas and subsequent drought over India and he used this relationship to issue tentative forecasts from 1882 to 1885. The success achieved infused greater confidence and the first of the regular series of operational LRP covering the whole of India and Burma was issued on June 4, 1886. Since then the long-range forecasts of monsoon rainfall have been issued by the Indian Meteorological Department regularly every year. Sir John, Eliot who succeeded Blanford in 1889, utilized weather conditions over the whole of India and surrounding regions to prepare too elaborate (30 pages) forecast of monsoon rainfall. The forecasts after 1895 were based mainly on (i) Himalayan snow, October to May, (ii) the local peculiarities of weather in India immediately antecedent to advance of monsoon across the coasts of Bombay and Bengal, and (iii) local peculiarities in Indian Ocean and Australia.

Realizing the complexities of the LRP problem Sir Gilbert Walker (1908, 1918, 1923) initiated studies of worldwide variation of weather elements such as pressure, temperature, rainfall etc., even though the main aim was to develop objective LRP method for predicting monsoon rainfall over India. While searching for the potential predictors, Walker described and coined the word 'Southern Oscillation' and the two 'Northern Oscillations' (North Atlantic and North Pacific). He also recognized that the variation of monsoon rainfall in India from the average conditions occurs on so large a scale that it would be remarkable if they were not preceded and followed by abnormal conditions in areas some distance away from the region over which the precipitation actually occurs.

Walker (1923, 1924a) utilized the concept of correlation to test whether preceding events anywhere in the world have a significant relationship with subsequent monsoon (June to September) rainfall over India. For developing the forecast formulae, Walker used those weather factors which showed significant correlation with the predictant (Indian monsoon rainfall). Among several factors, he finally chose those which had least inter-correlation coefficients among themselves and showed maximum multiple correlation coefficient. The first forecast by using the regression equation with multiple correlation coefficient (cc) equal to 0.58 was issued in 1909. The details of the formula are given below:

$$\{R_{IB}\} = 0.30\{F_1\} + 0.53\{F_2\} - 0.35\{F_3\} - 0.04\{F_4\}, \tag{1}$$

where R_{IB} is seasonal monsoon rainfall over India and Burma, F_1 is Himalayan snow accumulation at the end of May, F_2 is South American pressure (0.25 March + 0.50 April + 0.25 May), F_3 is Mauritius pressure (May), F_4 is Zanzibar district rain (April + May) and $\{\ \}$ indicates the departure of the factor from its normal.

Thus Walker succeeded in introducing the concept of correlation in the field of LRP. As mentioned above, this approach is still being very widely used for operational LRP of the weather.

The above equation was modified and a new regression equation having multiple cc equal to 0.69 was introduced in 1916. The details of the formula are given below:

$$\{R_I\} = -0.44\{F_1\} + 1.61\{F_2\} - 23.83\{F_3\} - 0.18\{F_4\} - 0.17\{F_5\}, \tag{2}$$

where R_I is predicted monsoon rainfall over India, and F_5 is Ceylon rainfall for the month of May.

Forecasts for India as a whole were prepared up to 1916. In subsequent years the experimental monsoon rainfall forecasts were issued for different homogeneous regions of India whose final form came to exist from 1924.

Forecasting Monsoon Rainfall for Homogeneous Regions of India

By dividing the country into three homogeneous regions, Walker (1924) was able to develop better formulae for forecasting than those when India as a whole was considered. Based on correlation, Walker divided the country in 1914 into 4 homogeneous regions, which were used up to 1922. Subsequently he again divided the country into 3 main homogeneous areas namely (i) Northeast India, (ii) Peninsula and (iii) Northwest India. These regions were slightly modified in 1949 and 1961 and their current forms, as referred to in the latest monsoon forecasts, are shown in Fig. 1.

The monsoon rainfall for a homogeneous area is weighted seasonal rainfall for those meteorological subdivisions which constituted the forecast region. A set of three regression formulae each containing 6 factors was developed by Walker (1924b) for forecasting monsoon rainfall over the different homogeneous subdivisions of India namely NE India, Peninsula and NW India. Details regarding these forecast formulae are given below.

Forecast for NE India

The first regression formula for forecasting monsoon rainfall in NE India was developed by Walker in 1922 and is given below:

$$[R_{NEI}] = 0.38[F_6] - 0.26[F_7], \tag{3}$$

where R_{NEI} is monsoon rainfall in NE India and F_6 and F_7 are respectively, Seychelles May wind velocity and rainfall. The brackets [ ] denote the proportional departures i. e. the ratio of the departures to their standard deviations.

The above formula did not have satisfactory multiple cc (0.52) and provided inaccurate forecasts for subsequent years. Besides, the variability of monsoon rainfall in NE India is small and the coefficient of variation is only 8 percent of the normal rainfall, it is possible to issue the forecast based on actual distribution of rainfall in the past 4 years and without using any regression formula by stating that rainfall will be within ± 10 percent of the normal on 80 percent of the occasions. In addition, subsequent studies were unable to identify significant predictors which could help to prepare more accurate forecasts than those obtained from the climatology alone. In view of these considerations, the issue of separate forecasts for NE India was discontinued in 1935 and the same situation continues even today.

Forecast for the Peninsula

The first regression formula for forecasting monsoon rainfall in the Peninsula was developed by Walker in 1922 and is given below:

$$\{R_P\} = 1.61\{F_8\} - 0.29\{F_9\} - 0.02\{F_{10}\} - 77.30\{F_{11}\} - 0.21\{F_{12}\}$$
$$- 0.35\{F_{13}\} , \tag{4}$$

where R_P is monsoon rainfall in the Peninsula, F_8 is South American pressure (0.5 April + 0.5 May), F_9 is Zanzibar rain (May), F_{10} is Java rain (October to February), F_{11} is Cape Town pressure (September to November), F_{12} is South Rhodosian rain (October to April) and F_{13} is Dutch Harbour temperature (March + April).

The above formula had multiple cc equal to 0.76 and was used for preparing operational forecasts from 1924 to 1930. Subsequently, it was revised on 8 occasions (1931, 1934, 1935, 1941, 1954, 1961, 1970, and 1977) on the basis of correlation coefficient of individual factors as well as the multiple cc of the formula. Details regarding the latest formula which emerged after 1977 revision are given below:

$$\{R_P\} = -0.296\{F_8\} + 2.242\{F_{14}\} + 1.127\{F_{15}\} - 58.1595 , \tag{5}$$

where F_{14} is Indian March Minimum temperature in degrees Celcius (average of Calcutta, Jaisalmer, and Jaipur), and F_{15} is the Indian ridge (the mean April subtropical ridge position in degrees of latitude along 75° E longitude).

On comparing the first formula [Eq. (4)] with the latest one [Eq. (5)] it is noted that south American pressure (F_8) finds place in both formulae. On studying all the forecast formulae developed for the Peninsula, it has been found that the factor south American pressure (F_8) was used as predictor in all of them. This suggests that the south American pressure has survived eight revisions and has provided useful indications regarding subsequent monsoon over the Peninsula during the last 60 years. Rao (1964) has calculated decadal ccs for F_8 and has noticed that the south American pressure changed its relationship from positive to negative in the 1940's. However, after examining the correlation coefficients for periods ranging from 20 years (1901 to 1920) to 82 years (1901 to 1982) the author has noted that the factor has never changed its basic (long term) association with the monsoon rainfall (Fig. 2) since 1901. It thus appears that the correlation (ρ) can be divided into two parts as given below:

402

$$\rho = \bar{\rho} + \rho', \tag{6}$$

where ρ is total correlation coefficient, $\bar{\rho}$ is the long-term or basic relationship i.e., ccs for long periods, say 50 years or more, and ρ' is the perturbation relationship such as 10-year running cc. On this basis it can be concluded that the basic relationship between south American pressure and monsoon rainfall has not changed during the last century; however, the perturbation component of the relationship changed its sign in the 1940's. Thus the factor has maintained its basic association with monsoon rainfall during the last 82 years (Fig. 2) and appears to have some physical linkage with the Indian summer monsoon. With the present state of the art, the Southern Oscillation, which recently emerged again a powerful tool for providing physical basis for LRP, may be one of the agencies responsible for creating the physical linkage between south American pressure and monsoon rainfall over India. After conducting his world weather studies, probably Walker (1923, 1924a) himself realied this fact.

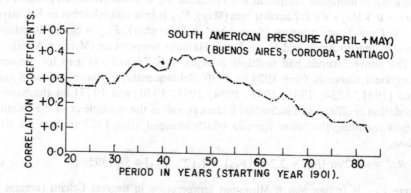

Fig. 2. Correlation coefficients between monsoon rainfall in Peninsula and Predictor

In the current formula two more factors, namely Indian temperature and April ridge location are also used. Probably, these factors are connected with the heat sources distribution in the pre-monsoon months and may have some physical links with the coming events. However, the agency responsible for creating the physical links with the future monsoon behavior is not yet clear.

Forecast for NW India

The first regression formula for forecasting the monsoon rainfall in NW India (Fig.

1) was developed by Walker in 1922 and is given below:

$$\{R_{NWI}\} = -0.95 \{F_1\} + 0.29 \{F_8\} - 44.50 \{F_{11}\} - 0.36 \{F_{12}\} - 0.53 \{F_{13}\}$$
$$-17.0 \{F_{17}\} , \tag{7}$$

where F_{17} is Equatorial pressure (January to May) and other factors have been defined earlier.

The above formula had a multiple cc equal to 0.57 and was used for forecasting monsoon rainfall in NW India from 1924 to 1930. The formula was revised six times (1931, 1941, 1946, 1965, 1972 and 1977) on the basis of correlation and multiple correlation. Details of the latest formula having multiple cc equal to 0.73 are given below:

$$\{R_{NWI}\} = 25.863 \{F_{17}\} - 1.433 \{F_{18}\} - 0.129 \{F_{19}\} + 0.759 \{F_{15}\}$$
$$-11.682 , \tag{8}$$

where F_{18} is Argentina pressure (April), F_{17} is Ludhiana mean temperature (April) and other factors have been defined earlier.

On comparing the first forecast formula (7) with the latest (8), it is noted that two factors, namely Equatorial pressure (F_{17}) and south American pressure (as Argentina pressure F_{18} is a part of south American pressure, F_8), are common in the two formulae. After studying all the formulae developed up to 1982, it has been noted that these two factors were retained in all the six revisions (1931, 1941; 1946, 1955, 1972 and 1977) carried out so far. Regarding Equatorial pressure (F_{17}) Rao (1964) has pointed out that during the 1920's and 1930's, the decadal ccs between the pressure and the monsoon rainfall have changed their signs. As mentioned in an earlier section, we consider them as a perturbation part of the association (ρ') and our study reveals that the basic association ($\bar{\rho}$) between these factors have not yet changed the sign. To support this view, the ccs for different periods ranging from 20 to 82 years have been calculated and are plotted in Fig. 3. It is seen from the figure that the decrease of cc was fast up to 1940 and very slow in subsequent increasing periods. The value of cc for 82 year (1901–1982) period is little over − 0.4, which is significant at the 0.1 percent level. Thus, the basic association ($\bar{\rho}$) between the Equatorial pressure and rainfall is remarkably significant. This suggests the possibility of some physical links between the Equatorial pressure and the monsoon rainfall over India. Southern Oscillation seems to be one of the agency for providing the physical links between the south American pressure (or Argentina pressure) and the Equatorial pressure with the monsoon rainfall over India. The other two factors used in the current formula are the Indian April ridge and the Ludhiana mean temperature. These parameters are related with pre-monsoon conditions and a possible agency for their physical linkage with the subsequent monsoon is not yet clear.

Assessment of Regression Forecasts

Every year at the beginning of the monsoon season, the long-range forecasts a: prepared and published by the India Meteorological Department. After the season

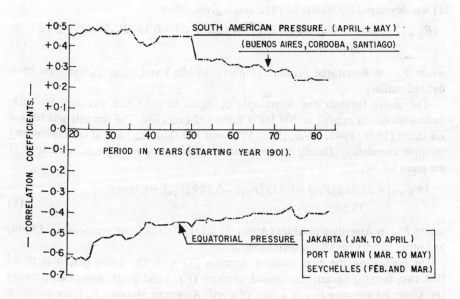

Fig. 3. Correlation coefficients between monsoon rainfall in northwest India and different predictors

over the monsoon rainfall amounts from various observatories of the country become available and the long-range forecasts are routinely verified. Statements showing actual rainfall and the comparison of all the operational forecasts with the actual rainfall are published.

Montgomary (1940) verified the forecasts from Walker's formulae with data up to 1936 and concluded that in spite of its early success, the formula has broken down in 16 years (1921–1936). Jagannathan (1960) reviewed the seasonal forecasting techniques in India and during 1932 to 1960 he found that the percentage of correct monsoon forecast of Peninsula and NW India were 72 and 76 percent respectively. We have verified forecasts for subsequent years (1961 to 1982) and the relevant information for operational forecasts issued so far is given below.

Table 1. Verification of forecasts issued during 1924 to 1982

Monsoon Rainfall	Percentage of Correct Forecasts from		Skill Score
Forecast for	Formula	Climatology	(S)
Peninsula	64.4	49.0	0.30
Northwest India	62.7	38.0	0.46

It is seen from the above table that the accuracy of the forecasts is nearly

65 percent. It may, however, be mentioned that the forecasts issued during the different years, even though they were not based on the same formula, have been pooled together. As such the inferences drawn from the analysis carried out here, relate to the regression forecasting technique rather than the individual formula.

It may be worthwhile to examine whether the forecast technique has given more information than the climatological knowledge. For issuing the forecast in popular language, seven categories for monsoon rainfall, namely normal (± 10 percent of the normal), slight excess/defect (± 11 to ± 25), moderate excess/defect (± 26 to ± 50) and large excess/defect (≳ 50) are used. For calculating the climatological probabilities of the different category of the rainfall the data for the past 100 years (1883 to 1982) have been analyed. The study has revealed that the probabilities of occurrence of the normal monsoon rainfall over both the subdivisions, namely Peninsula and NW India, are maximum and are shown in Table 1. If the forecasts are to be issued based solely on climatology, it would be advisable to predict normal monsoon rainfall every year, as this category has the highest percentage of success. Any forecast technique, having some skill score should therefore provide more accurate forecasts than those obtained from the above method. To examine this aspect the skill score has been defined as follows

$$S = (R - E) / (T - E),$$ (9)

where S is the skill score, R is the percentage of correct forecast, T is the total number of forecasts (100 in this case) and E is the percentage of forecasts expected to be correct on the basis of climatology.

In the case of a negative value of S, the forecast technique's skill score is lower than that of climatology. On the other hand, if the value of S is positive, the forecast technique's skill is higher as compared to that of the climatology. It is seen from Table 1 that the skill scores of the forecasts for Peninsula and NW India are not only positive but also considerably higher. It can, therefore, be concluded that the regression forecast technique operationally used in India for predicting the monsoon rainfall over different homogeneous regions of the country has provided more accurate forecasts than those obtained from climatology.

Remarks on Correlation Technique

On examining the first set of the formulae [Eqs. (3, 4, 6)] developed by Walker in 1922 and the latest formulae [Eqs. (5, 7)] developed in 1977, it is noted that the philosophy of the presumed influence of meteorological parameters on subsequent monsoon has changed over the years. Walker probably felt that monsoon is mainly influenced by the extra-Indian factors and that is why all his predictors except the Himalayan snow (Eqs. 3, 4, 6) were of extra-Indian origin. On the other hand, the evolution of the forecast formulae during the last 60 years has changed this idea, and current formulae (Eqs. 4, 6) indicate that both Indian and extra-Indian pre-monsoon conditions affect the subsequent monsoon over India.

The most discouraging feature of the regression technique is that the correlation coefficients vary with time and even change their sign. This reduces the accuracy of the forecasts which in turn limits the confidence in the regression technique. However, the idea that the regression equation fails in all extreme deficient years does not seem to hold good, as the regression formulae (Eqs. 5 and 8) correctly predicted the deficient monsoon rainfall in 1979 and 1982. Nevertheless, the percentage of forecast failure is much higher in extreme years as compared to the normal rainfall years. In spite of all these limitations, if minor considerations like replacing the factor in proper time are taken care of, the accuracy of the forecast can be increased. For example, the forecast foumulae were not revised in the 1940's and 1950's although many consecutive forecasts failed. It is interesting to note that the monsoon had near normal rainfall during 1940 to 1960, probably the best period in the past century, and still the forecast failures were remarkably high during this period. If the forecast formulae had been revised after 3 or 4 consecutive failures, the accuracy of the forecasts might have been higher. Indian experience with the regression technique indicates that after a formula is developed, it generally gives reliable indications for 4 to 5 years, because some factor or the other either changes its sign or becomes unimportant.

The accuracy of the forecasts can probably be further imporoved if the unstable factors like Argentina pressure (in NW India formula) are replaced by the more stable factors like South American pressure (Figs. 3, 4). The long-term association of Argentina pressure has changed the sign, while that of South American pressure still has positive cc with the rainfall. Similarly, inclusion of the Himalayan snow cover, a factor used from 1882 to 1955, into the regression formula for NW India may increase the accuracy of the forecast. After noticing the change in its short period (10 year) ccs the factor was considered to have lost its association with the monsoon rainfall. We have, however, examined the long period ccs between the Himalayan snow and the rainfall and have found that the relationship has not changed up to 1982 (Fig. 4). The cc of the Himalayan snow for 82 year period is very close to -0.2 which is significant at the 5 percent level. It appears from the figure that the basic relationship of the Himalayan snow with the rainfall almost reaches an asymptotic form. In fact this factor has maintained its relationship with the monsoon rainfall during the past century and appears to have some physical linkage with the Indian monsoon rainfall. This view gains indirect support from some general ciculation studies (Charney and Shukla 1981) which have suggested that slowly changing boundary conditions like snow cover, sea surface temperature etc. provide some physical basis for long-range monsoon prediction in low latitudes. To increas the accuracy of the forecast, it may probably be advisable to develop forecast fomulae which use predictors connected with the southern oscillation, and slow changing boundary conditions such as snow cover, sea surface temperature etc. Further studies in this direction may produce useful results, as the southern oscillation seems to be the single most important feature of the atmospheric circulation whose perod is large enough to be of practical value for long-range forecast of highly periodic phenomena like summer monsoon of Southeast Asia.

Dynamic Stochastic Transfer Model for Long-Range Prediction of Rainfall

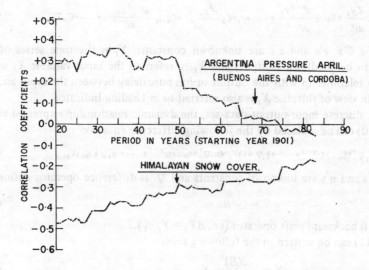

Fig. 4. Correlation coefficients between monsoon rainfall in north-west India and different predictors

The general circulation models, the alternate hope of the forecasters, have not yet attained the ability to produce operational long-range forecast of rainfall. Due to inherent limitations of regression and multiple regression techniques, the percentage of accuracy of their forecasts cannot be increased beyond a certain limit. In view of these difficulties, the author (Thapliyal 1982 a, b, 1983, 1984) has recently developed a few dynamic-stochastic transfer models which utilize the dynamic transfer property of the atmosphere for predicting the monsoon rainfall in two broad subdivisions of India, namely, the Peninsula and NW India. During the past two decades, these models have provided more accurate forecasts than those obtained from other conventional techniques like regression, multiple regression, etc. Thus, the dynamic-stochastic transfer model is superior to other conventional techniques not only in concept but also in performance. It appears that the technique has the potential to play a dominant role in developing better operational LRP models in the years to come. In view of these considerations, the concept and the method employed for developing the dynamic-stochastic transfer models have been briefly described below.

Formulation of the Problem

The atmosphere is a complex dynamic system and it transforms various inputs into numerous outputs. Let us assume that the atmosphere is a linear dynamic system and it converts input series X into output series Y. For such an atmosphere, the transfer dynamic relationship between the continuous input X_t and output

Y_t can be represented by a linear differential equation (Box and Jenkins 1976; Thapliyal 1981, 1984) of the form given as

$$K_r \frac{d^r y_t}{dt^r} + K_{r-1} \frac{d^{r-1} y_t}{dt^{r-1}} + \cdots + y_t = C_s \frac{d^s x_{t-b}}{dt^s} + C_{s-1} \frac{d^{s-1} x_{t-b}}{dt^{s-1}} + \cdots + C_o X_{t-b} \qquad (10)$$

where K's C's, r's and s's are unknown constants, Y_t is the time series of the variable to be predicted (output) and X_{t-b} refers to the input variable X_t which has lag b relationship with the output or b is pure delay between the input and the output. In view of this, the X_t is also referred to as Leading Indicator.

For a discrete input-output data set, the dynamic relationship expressed above in Eq. (10) can be replaced by the following difference equation

$$(\xi_\gamma \nabla^\gamma + \xi_{\gamma-1} \nabla^{\gamma-1} + \cdots + \xi_1 \nabla + 1) Y_t = (\eta_s \nabla^s + \eta_{s-1} \nabla^{s-1} + \cdots + \eta_1 \nabla + \eta_0) X_{t-b} , \qquad (11)$$

where ξ's and η's are unknown constants and ∇ is difference operator defined as follows

$$\nabla = 1 - B \qquad (12)$$

where B is backward shift operator (i.e. $BY_t = Y_{t-1}$).

Eq. (11) can be written in the following form

$$Y_t = \frac{\omega(B)}{\delta(B)} X_{t-b} , \qquad (13)$$

where

$$\omega(B) = \omega_0 - \omega_1 B - \omega_2 B^2 - \cdots - \omega_s B^s ,$$
$$\delta(B) = 1 - \delta_1 B - \delta_2 B^2 - \cdots - \delta_r B^r , \qquad (14)$$

B is backward shift operator defined in Eq. (12) and ω's are δ's are unknown constants.

Eq. (13) represents a linear dynamic transfer relationship between the input and the output (say rainfall over a particular homogeneous area) of the atmosphere and is generally referred to as "Dynamic Transfer Model". From this model, it is possible to forecast the output (rainfall) time series Y_t provided the values of r, s, δ and ω figuring in Eqs. (13) and (14) are known. It may be noted that these parameters express a linear dynamic transfer process of the atmosphere through which the input X_t has been transformed into the output series Y_t.

It is possible to develop a dynamic transfer model similar to Eq. (13) which has a finite number of inputs in it and can be used to forecast one particular output of the atmosphere, say rainfall. But in actual practice the development of such a model becomes difficult, as the identification of all the possible input parameters of the atmosphere is not feasible. By using the dynamics of the Auto Regressive Integrated Moving Average (ARIMA) process it is, however, statistically possible to include in the model given in Eq. (13) the resultant effects of other inputs, outputs, and feedback processes of the atmosphere. Let us assume that the inputs other than X_t, outputs other than Y_t and the feedback processes of the atmosphere are infecting the predicted output from the dynamic transfer model [Eq. (13)] by an amount N_t, according to

$$Y_t = \frac{\omega(B)}{\delta(B)} X_t + N_t , \qquad (15)$$

where N_t is the 'noise and is supposed to be independent of X_t.

For the sake of generality it is assumed that the input, output and noise series are nonstationary, and by differencing them D times they become stationary, so that

$$\nabla^D Y_t = y_t , \qquad \nabla^D X_t = x_t , \quad \text{and} \quad \nabla^D N_t = n_t , \qquad (16)$$

where ∇ represents backward difference operator and is defined in Eq. (12).

By using Eqs. (16) into (15), we obtain

$$y_t = \frac{\omega(B)}{\delta(B)} x_{t-b} + n_t \qquad (17)$$

In the absence of any control on the atmosphere, the noise n_t is modeled by using the ARIMA process, which postulates that a time series can be represented as an output from a dynamic system to which the input is a series of random shocks (or white noise) and for which the transfer function can be parsimoniously expressed as the ratio of two polynomials in B, the backward shift operator (Box and Jenkins 1976, Thapliyal 1981, 1984) that is

$$n_t = \frac{\phi(B)}{\theta(B)} a_t , \qquad (18)$$

where

$$\phi(B) = 1 - \phi_1 B - \phi_2 B^2 - \cdots - \phi_p B^p$$
$$\theta(B) = 1 - \theta_1 B - \theta_2 B^2 - \cdots - \theta_q B^q , \qquad (19)$$

ϕ's, θ's, p and q are unknown constants, and n'_t are random shocks. It may be noted here that the values of p and q indicate the order of Auto-Regressive and Moving Average processes, respectively.

On substituting the value of n_t from Eq. (18) into Eq. (17), we obtain

$$y_t = \frac{\omega(B)}{\delta(B)} x_{t-b} + \frac{\phi(B)}{\theta(B)} a_t ; \quad b \geqslant 0 \qquad (20)$$

The above equation gives the general form of the Dynamic Stochastic or Dynamic Shock Transfer (DST) model.

The first and second functions on the right hand side of Eq. (20) are respectively known as dynamic and random shocks transfer function. In other words, the model takes into account the dynamic transfer relationship not only between the input and the output but between the noise and the random shocks also. Thus, an optimal output forecast of an atmospheric variable y_t, say monsoon rainfall, can be obtained from the DST-equation (20) provided a suitable lead parameter to the atmosphere is known.

Model Development for LRP of Monsoon Rainfall over Peninsula

In this section, an attempt has been made to develop an exact mathematical expression for Dynamic-Stochastic Transfer (DST) model [Eq. (20)] so that long-range forecasts of seasonal monsoon rainfall (June to September) over one of the major homogeneous subdivisions of India, namely the Peninsula (Fig. 1) can be prepared and made available to the users. For building the forecast model let us consider that the atmosphere is a linear dynamic system which converts a suitable lead input, X into the output, Y, say monsoon rainfall over Peninsular India. Recently, the author (Thapliyal 1981) has studied various parameters and has shown that the location of the 500 hPa April subtropical ridge along 75°E longitude over India can be used as a lead input parameter for developing a suitable long-range forecast DST model for the Peninsula.

For building up and testing the DST forecast model, the input (500 hPa April ridge position over India) and the output (monsoon rainfall in Peninsular India) data for recent 44 years (1939 to 1982) have been collected from two recent publications (Banerjee et al. 1978; Thapliyal 1982) and used in this study. The data have been utilized to identify the dynamic [Eq. (13)] and stochastic [Eq. (18)] transfer functions which constitute the model [Eq. (20)]. A detailed procedure for obtaining an exact mathematical expression for the dynamic transfer function has been described elsewhere (Box and Jenkins 1976; Thapliyal 1981). It may however, be mentioned that by regarding the output a linear aggregate of a series of superimposed impulse response funcitons, scaled by the input, it has been possible to identify the exact form of the dynamic transfer function as given below

$$y_t = [\{\omega_0 + \sum_{i=1}^{4}\omega_i B^i\}/ \{1+\sum_{i=1}^{4}\delta_i B^i\}]x_t , \qquad (21)$$

where

$$y_t = \log(R_t/R_{t-1}) ,$$
$$x_t = X_t - X_{t-1} , \qquad (22)$$

where R_t is the monsoon rainfall in centimeters over Peninsula, X_t is the location of 500 hPa April subtropical ridge in degress of latitude, along 75°E longitude and t is year like 1980, 1981, etc. The values of δ's and ω's are given below in Table 2.

Table 2. Values of δ's and ω's

i	0	1	2	3	4
δ_i		−0.8986	−0.5319	−0.2459	−0.2007
ω_i	0.0554	−0.0520	−0.0385	−0.0197	−0.0056

The usual procedure for identifying the stochastic (shock) transfer function utilizes the dynamics of the ARIMA process which postulates that a time series can be represented as an output from a dynamic system to which the inputs are random shocks. By following the identification procedure described by the author elsewhere (Thapliyal 1981), the following stochastic transfer function has been obtained

$$a_t = N_t + \sum_{i=1}^{13} \theta_i a_{t-i} , \tag{23}$$

$$N_t = (y_t)_A - (y_t)_F , \tag{24}$$

where $(y_t)_A$ is actual rainfll index defined in Eq. (22), $(y_t)_F$ is forecast rainfall index obtained from Eq. (21) and values of θ's are given below in Table 3.

Table 3. Values of θ's

i	θ_i	i	θ_i	i	θ_i	i	θ_i
1	0.530	5	−0.050	9	−0.300	13	0.335
2	0.095	6	0.715	10	−0.250	14	−
3	−0.095	7	0.170	11	−0.555	15	−
4	0.650	8	0.015	12	0.005	16	−

From Eqs. (20), (21) and (23), the DST model can be expressed as follows

$$y_t = \omega_0 x_t + \sum_{i=1}^{4} (\delta_i y_{t-i} - \omega_i x_{t-i}) + \hat{a}_t - \sum_{i=1}^{13} \theta_i a_{t-i} , \tag{25}$$

where $\hat{a}_t$ represents residuals.

On substituting the values of x and y from Eq. (21) into the above equation and neglecting the residual terms, i.e., $a_t = 0$, the final form of the DST model is given below

$$R_t = R_{t-1} \left[\left\{ \prod_{i=1}^{4} (R_{t-i}/R_{t-i-1})^{\delta_i} \right\} \cdot \exp \left\{ \omega_0 (X_t - X_{t-1}) \right. \right.$$
$$\left. \left. - \sum_{i=1}^{4} \omega_i (X_{t-i} - X_{t-i-1}) - \sum_{i=1}^{13} \theta_i a_{t-i} \right\} \right] , \tag{26}$$

where R_t and X_t are defined in Eq. (22) and the values of δ, ω and θ are given in Tables 2 and 3 respectively.

From Eq. (26) it is thus possible to obtain the forecast for R_t, i.e., subsequent monsoon rainfall (June to September) over Peninsula, at the end of April when the lead input X_t becomes available.

Model Interpretation

It is seen from the right hand side of Eq. (25) that all the terms contributing to the forecast R_t multiply each other. This suggests that for LRP of monsoon rainfall the DST model utilizes some sort of nonlinear interactions among all the parameters of its three main constituents viz. output, input, and random shocks. A careful examination of the above equation indicates that the memory of the atmospheric transfer dynamics under these nonlinear interactions is much longer than is generally believed. In the present form of the model, the input (April ridge pattern) and the output (seasonal monsoon rainfall) of previous 5 years, and the random shocks of previous 13 years, dominate the nonlinear interations of the atmosphere and thereby significantly contribute to the forecast rainfall, R_t. The first part of the model, namely dynamic transfer function when considered separately, indicates that the nonlinear interactions among the past 5 years, inputs and the outputs are probably responsible for providing 5-year memory to the atmospheric transfer dynamics. On the other hand, when the second part of the model (stochastic transfer function) is added along to the first one, it is noted that the nonlinear interactions among the past values of the three, viz the input, the output, and the random shocks, are probably responsible for providing 13-year memory to the stochastic-dynamic transfer processes of the atmosphere. This suggests that the atmospheric transfer dynamics has a much longer memory than is considered at present.

By examining the magnitude of the constants, used in the model like δ, ω and θ, it is possible to have some ideal regarding the amount of relative contribution that each one of the past values of the input, the output, and the random shocks makes to the forecast. The magnitude of δ decreases almost exponentially from δ_1 to δ_4 (Table 2). This indicates that the contribution from past rainfall values as can be seen from Eq. (26), decreases exponentially from $(R_{t-1}/R_{t-2})^{\delta_1}$ to $(R_{t-4}/R_{t-5})^{\delta_4}$. Thus the successive past 5-year rainfall values $(R_{t-1}$ to $R_{t-5})$ contribute to the forecast. Similarly, the immediate successive past 5-year values [Eq. (26)] of the input parameter (April ridge) also contribute to the forecast. In this case also the contribution to R_t almost decreases exponentially from $(X_{t-1} -X_{t-2})$ ω_1 to $(X_{t-4} - X_{t-5})$ ω_4. As the successive past 5-year values of the input (April ridge) and the output (rainfall) contribute to the forecast, it is possible that the dynamic transfer processes involving input and output of the atmosphere has a longer memory of the order of 5-years.

As seen above, the contribution from the input (April ridge) and the output (rainfall) time series to the forecast has a trend that decreases with the successive past values. This type of trend is not found with the contributions that come from the past successive random shocks. If we examine Table 3 we find that the magnitudes of the θ do not show any regular trend, but indicate some oscillatory pattern. As compared with others, the magnitudes of θ_1, θ_4, θ_6, and θ_{11} are much larger, of which θ_6 has maximum value followed in decreasing order by θ_4, θ_{11}, and θ_1. This suggests that maximum contribution to the current year rainfall forecast comes from the random shocks of the past sixth year followed in decreasing order, by the random shocks of the past fourth, eleventh, and first year. It appears that maximum contribtuion to the forecast comes from the southern oscillation (θ_4 and θ_6). A significant contribution also comes from the eleventh year oscillation, which may be either a multiple of the southern oscillation or connected with the sunspot cycle. Although small, all the successive past 13 year's values of the random shocks make some contribution to the forecasts. This suggests that the dynamic stochastic transfer memory of the atmosphere is of the order of 13 years.

Model Forecast Verification

The input (April ridge) and the output (rainfall) data up to 1976 have been used for building up the DST forecast model. After the completion of the developmental work in 1976, the model is being used as one of the forecast tools by the Indian

Meteorological Department for preparing the operational long-range forecasts of monsoon rainfall in the Peninsula. By utiliing the input data of the past 6 years (1977 to 1982), the forecasts for each year have been obtained from Eq. (26) and are presented together with the actual rainfall amounts in Fig. 5. It is seen from the figure that the predicted rainfall amounts are very close to the realized ones. The modle has correctly predicted not only the droughts of 1979 and 1982, but good monsoon rainfall in other years also. Even in two consecutive contrasting rainfall years, like 1979 (a drought year) and 1980 (a good rainfall year), the closeness of the predicted and the realized values stands out remarkably well. Thus the performance of the model during the verification period (1977–1982) has been found excellent.

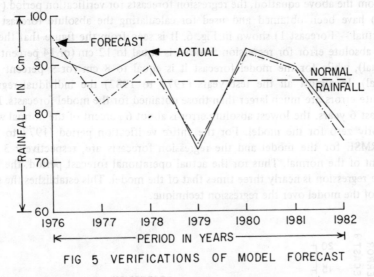

FIG 5 VERIFICATIONS OF MODEL FORECAST

Fig. 5. Verifications of model forecast

The preciseness of the model forecast can be judged by examining the amount of variance explained by the model and also the Root Mean Square Error (RMSE) of the model. During the operational period (1977–1982) the model predictions have explained 99 percent of the variance, which is very high. Even the RMSE for the operational forecasts is equal to 3 cm which is just 3 percent of the normal rianfall (87.4 cm) and less than one fourth of the standard deviation. it can, therefore, be concluded that during recent years the DST model has provided excellent operational long-range forecasts of the rainfall in Peninsula and can be used with a high degree of confidence for preparing forecasts of seasonal monsoon (June to September) rainfall about one month prior to the start of the season over India.

Comparison of Best Operational LRP Techniques with the DST Model

The regression method of LRP is still considered to be one of the best operational models. However, correlation coefficients between the predictor and the predictant vary with time and seriously affect the accuracy of regression forecasts, particularly in extreme years. In order to compare the performance of the DST model against the regression, the same amount (1939–1976) of data for the input (April ridge) and the output (monsoon rainfall) which were earlier used for building the new model have been utilized here for formulating the following regression equation

$$R = 17.7454 + 4.6086X, \tag{27}$$

where R is amount of monsoon rainfall in cm, over Peninsula and X is the mean April 500 hPa ridge position along $75°$E longitude.

From the above equation, the regression forecasts for verification period (1977–1982) have been obtained and used for calculating the absolute forecast errors (1 Actual − Forecast 1) shown in Fig. 6. It is seen from the figure that the maximum absolute error for regression forecast is equal to 12 cm (or 14 percent of the mormal), while for the model forecast it is equal to 8 cm (or 9 percent of the normal). In almost all the test years (1977 to 1982) the individual regression absolute errors are much larger than those obtained for the model forecasts. During the past 6 years, the lowest absolute error is about 6 percent of the normal while it is nearly zero for the model. For the entire verification period (1977 to 1982), the RMSE for the model and the regression forecasts are, respectively 3 and 8 percent of the normal. Thus for the actual operational forecast period, the RMSE of the regression is nearly three times that of the model. This establishes the superiority of the model over the regression technique.

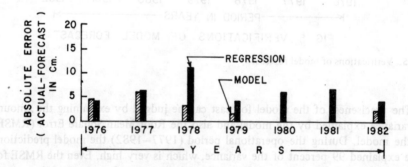

Fig. 6. Year-to-year errors for model and regression forecasts

For comparing the performance of the model with the multiple regression technique, the forecasts obtained from these methods have been analyzed. During the verification period (1977 to 1982), the India, Meteorological Department has used the multiple regression equation, given in Eq. (5), for preparing LRP of monsoon rainfall over the Peninsula. The multiple regression forecasts are expressed as mentioned earlier in seven categories of rainfall like large excess, moderate excess, slight excess, normal, slight defect, moderate defact, and large defect. By expressing the model forecasts in the above seven categories, it has been found that during the verificaiton period (1977 to 1982) the forecasts from the model and the multiple regression formula proved correct on 100 and 83 percent occasions respectively. This establishes the superiority of the DST model over the multiple regression also.

Remarks on the DST Model

The Dynamic-Stochastic-Transfer model developed in this section has provided more encouraging results than those obtained from the regression and the multiple regression techniques, which as considered to be superior to many of the conventional LRP methods, currently being used in the world. Such encouraging results are expected from DST models developed for other regions of the world also, because for long-range prediction of weather the new model utilizes the transfer-dynamic-relationship not only between the lead input and the output, but also between the random shocks and the output of the atmosphere. Thus, the DST model not only provides more accurate LRP, but also attempts to introduce a new concept in the field of long-range forecasting. It appears that the development of similar DST models for other regions and the meteorological elements would considerably increase the reliability of the operational LRP of weather.

Acknowledgements. The author is thankful to the Director General of Meteorology, India Meteorological Department for providing the necessary facilities to carry out the work. He is also thankful to Shri H.M. Chaudhury, Additional Director General of Meteorology (Research) and Shri Nootan Das, Deputy Director General of Meteorology (Climatology and Geophysics) for encouragement and useful suggestions, to the members of staff of Long Range Forecast Unit, particularly Shri R.W. Roberts for assistance in computations, and to Shri S. Gurumurthy for typing the manuscript.

References

Banerjee, A. K., Sen P.N. Raman C. R. V. (1978) *Ind J Met Hydl Geophys* 29: 425–431
Blanford, H.F. (1884) *Proc R Soc* Lond 37: 3
Box, G.E.P, Jenkins G.M. (1976) *Time series analysis, forecasting and control.* Holdenday
Charney, J.G. Shukla J. (1981) *Monsoon dynamics.* Cambridge Univ Press
Jagannathan, P. (1960) *Seasonal forecasting in India – a review.* Meteorol Office, Poona
Montgomery R.B. (1940) *Report on the work of G.T. Walker.* Washington-MWR Suppl No. 39, pp 1–26
Rao, K.N. (1964) *Seasonal forecasting – India WMO Tech* Note No 66, pp 17–30
Thapliyal, V. (1981) ARIMA model for long range prediction of Monsoon Rainfall in Peninsular India. *Meteorol Monogr Climatol* No 12 pp 1–22
Thapliyal, V. (1982a) vol 33. *Mausam* pp 399–404
Thapliyal, V. (1982b) *In Proc Symp Hydrol Aspects Mountainous Watersheds,* November 4–6, vol 1. Univ Roorkee, India, pp I-45 to I-50
Thapliyal, V. (1983) Dynamic-Noise transfer model for long range prediction of Monsoon rainfall in Peninsula. *Nat Symp Climate Dynam Long Range Prediction,* February PRL-Ahmadabad, India
Thapliyal V (1984) Long range prediction of rainfall (Indian Monsoon) communicated to

Vayu Mandal

Walker, G.T. (1908) Correlation in seasonal variation of climate (Introduction) *Memoir India Meteorol Dep* (IMD Mem) vol 20 part VI, pp 117–124

Walker, G.T. (1918) *Q J Roy Met Soc* 43: 218–219

Walker, G.T. (1923) Correlation in seasonal variation of weather VIII. A preliminary study of world weather. *IMD Mem* vol 24 part IV pp 75–131

Walker, G.T. (1924a), Correlation in seasonal variation of weather IX. A further study of world weather. *IMD Mem* vol 24 part IX pp 275–332

Walker, G.T. (1924b) Correlation in seasonal variation of weather X Applications to seasonal forecasting in India. *IMD Mem* part X, pp 333–345

Limit Cycle in a Simple Nonlinear Low-Order Air-Sea Interaction Model

Jin Feifei[1]

Abstrace – In this paper, a barotropic model with topographic forcing derived by Charney (1979) and the equation derived by Pedlosky (1975) are used. A long-term coupled oscillation can exist in the non-linear atmosphere-ocean interaction model. Its amplitude depends on the height of topography, the wave length, the friction and the thermal exchangerate. Its period depends on the amplitude of the oscillation and the time scale of advection in the ocean. Beyond the period of the oscillation, the solution is asymmetrical. Forced wave and zonal flow of the atmosphere have both fast and slow varying processes, but the amplitude of anomalous SST always varies slowly. It seems possible that this kind of oscillation has some contribution to the seasonal and annual variation of atmospheric circulation and SST.

Model

$$\frac{\partial}{\partial t}\nabla\psi + J(\psi,\nabla\psi + f_0\frac{h}{H}) + \beta\frac{\partial}{\partial x}\psi = -\gamma\nabla(\psi - \psi^*) + \mu(a_0\psi - T_s)$$

ψ —— stream function

ψ^* —— thermal forcing

T_s —— SST

h —— orography

This quasi-geostrophic barotropic model with orographic and thermal forcing is similar to Charney's (1979) model. The difference is that here we introduct T_s as a variable heating source:

In order to close the system, we use the following SST equation

$$\frac{d}{dt}T_s \doteq \frac{\partial}{\partial t}T_s + V_s\frac{\partial T_s}{\partial y} = \lambda(a_0\psi - T_s)$$

$$V_s = \alpha_1\psi$$

These equations are similar to that of Pedlosky (1981) for upper layer SST evolution. Here we take the hypothesis that large-scale sea flow is nearly Sverdrup flow so we can use Sverdrup relation to represent this sea flow.

In this model, we include two kinds of basic air-sea interaction processes, i.e., T_s as a forcing source changes atmospheric circulation and changed circulation in turn redistributes SST by Sverdrup sea flow.

Nondimensionalizing these equations and then decomposing ψ, T_s as

$$\psi = -v_0(y - y_0) + \psi'$$

$$T_s = -v_0(y - y_0) + T_s'$$

---

1. Institute of Atmospheric Physics, Academic, Sinica, Beijing, China

and highly truncating ψ', T_s' by spectral expanding

$$\psi' = \phi_A \cos y + \phi_n \sin y\, e^{inx} + (\ast) + \cdots$$

$$T_s' = (\theta_A \cos y + \cdots)\,\delta(x)$$

$$\delta(x) = \begin{array}{ll} 1 & \text{on ocean} \\ 0 & \text{on land} \end{array}$$

$$f_0\,\frac{h}{H} = h_0 \sin y\, e^{inx} + (\ast) + \cdots$$

$$\psi^* = \phi_A{}^* \cos y + \cdots$$

(where we only consider one zonal SST component, one zonal and one wave component of the circulation), then we have

$$\frac{d}{dt}\phi_A = -\gamma^*(\phi_A - \phi_A^*) - \mu^*(\phi_A - \theta_A/2) + \alpha_{n_1} I_m(\phi_n{}^* h_0)$$

$$\frac{d}{dt}\phi_n = -(\gamma^* + \mu_n{}^*)\phi_n - ib_{n_1}\phi_n + ih_0(nv_0 + \alpha_{n_2}\phi_A)$$

$$\frac{d}{dt}\theta_A = -\alpha_0{}^* U_0 \phi_A - \lambda^*(\theta_A - \phi_A).$$

This is a highly truncated model. In this model, (mechanically and symetrically thermally, topography and zonal heating) forced components of the circulation $\phi n, \phi A$, and θA, the zonal component of T_s, interact on each other. The two basic interaction processes metioned above still remain in this highly truncated model.

For convenience, we only consider weak nonlinear problem. Using the multiple scale method, we simplify the system once more. and take following simplification. Let

$$U_0 = \beta/(1 + n^2) + |\Delta|\eta, \qquad |\Delta| \ll 1 \tag{1}$$

$$\theta(\mu^*) = o(r^*) = o(|\Delta|), \quad o(\lambda^*) = o(\alpha^*) = o(|\Delta|^{1+p}) \tag{2}$$

$$o(h_0) = |\Delta|^{\frac{1}{2}}(\phi_n^{(1)} + |\Delta|^{\frac{1}{2}}\phi_n^{(2)} + \cdots) \tag{3}$$

$$o(h_0) = o(|\Delta|^{2/3}) \tag{4}$$

$$\phi_A = |\Delta|^{\frac{1}{2}}(\phi_A^{(1)} + |\Delta|^{\frac{1}{2}}\phi_A^{(2)} + \cdots)$$
$$\theta_A = |\Delta|^{\frac{1}{2}}(\theta_A^{(1)} + |\Delta|^{\frac{1}{2}}\theta_A^{(2)} + \cdots)$$
$$T = |\Delta|t \tag{5}$$

$$\frac{d}{dt} = \frac{\partial}{\partial t} + |\Delta|\frac{\partial}{\partial T}.$$

The first one is near resonant condition. The second is weak friction condition. The third represents weak mechanical forcing. The forth shows that ϕn, ϕA, θA are expanded in the asymptotic series. We introduce T as a slow time scale. By analyzing each order equation, we can finally obtain

$$\frac{d}{dT} A = -R_n A - i(\alpha_{n2}v + n\eta)A + inU_0 h_0$$

$$\frac{d}{dT} V = -b\left[\frac{d}{dT}|A|^2 + 2R_n|A|^2\right] - R_\mu V + H_0 + Rw_s$$

$$\frac{d}{dT} W_s = -|\Delta|^P \alpha_0^* [v - R_s(v - w_s)],$$

where $A = \phi_n^{(1)}$, $V = \phi_A^{(2)}$ $w_s = \theta_A^{(2)}$. It can be shown that both $\phi_A^{(1)}$ and $\theta_A^{(1)}$ equal to zero. In this system, we can find that there is a small parameter $|\Delta|^P \alpha_0^*$. We denote it as ε. This small parameter is important for further discussion.

Decomposition of the System and Qualitative Analysis of the Phase Space

Let $\qquad \varepsilon = |\Delta|^P \alpha_0^*$, $\qquad \tau = \quad \tau = \quad T$

Because ε is much smaller than one, T is a faster time scale than τ. We decompose the system (2) into the following two systems

$$- R_n A - i(\alpha_{n2}v + n\eta)A + inU_0 h_0 = 0$$

$$- 2bR_n|\Delta|^2 - R_\mu v + H_0 + Rw_s = 0$$

$$\frac{d}{d\tau} w_s = -v + R_s(v - w_s)$$

and

$$\frac{d}{dT} A = -R_n A - i(\alpha_{n2}v + n\eta)A + inU_0 h_0$$

$$\frac{d}{dT} V = -b\left[\frac{d}{dT}|A|^2 + 2R_n|A|^2\right] - R_\mu v + H_0 + Rw_s$$

$$\frac{d}{dT} w_s = 0,$$

where system (3) is a slowly varying subsystem. In this system, the components of atmospheric circulation A, v *are described* by balanced equations, and the SST component w_s varies in time scale τ . A and v change with w_s. This subsystem is qualitatively similar to Chao's long-range weather forecast model (Chao et al 1977). System (4) is a faster varying subsystem. In this system, the SST component remains constant. The atmospheric circulation is then adjusted to external.

Using these two subsystem, we can analyze the qualitative vector field in the phase space. On the projected phase plane (w_s, v), the quasi-balanced state curve is determined by the following equation

$$Rw_s = -H_0 + R_\mu v + 2bR_n \frac{n^2 U_0^2 |h_0|^2}{R_n^2 + (\alpha_{n2}v + n\eta)^2} .$$

It is deduced from the first two equations of system (3). Because $dw_S/dt = 0$, so away from the curve, the vector field on the phase plane is horizontal. In this chapter we have shown that the states on the right and left branches are stable and the states on the middel branch. are unstable. The phase point moves to stable branches and goes away from the unstable middle branch. So we get the horizontal arrows shown in Fig. 1.

On and near the curve, the states are described by system (3). When dw_S/dt is positive on the left branch and negative on the right branch, the arrows are shown in Fig. 1.

From the qualitative vector field as shown in Fig. 1, we can easily find that there is a close trajectory $P_2Q_1Q_2P_1P_2$ on the phase plane. In other words, there exists a limit cycle. It consists of four segments of phase trajectories. On P_2Q_1 and Q_2P_1, the phase point moves slowly in time scale . On the P_1P_2 and Q_1Q_2 the phase point moves faster in time scale T. The period of the cycle depends mainly on the time used in the P_2Q_1 and Q_2P_1 processes.

This limit cycle forms an oscillation in the air-sea model. The physical processes of the oscillation are clear.

In Fig. 1, when the phase point is located at the left branch of the curve, it shows that there is easterly zonal wind modification because V is negative. This zonal wind produces Sverdrup flow V_s, as shown in Fig. 2. Therefore, there is warm SST advection in the southern part of the ocean and cold SST advection in the northern part of the ocean, which amplifies the zonal SST anomalous. So, w_s increases.

On the other hand, when the phase point is located at the right branch of the curve, it shows that there is westerly zonal wind modification, which makes Sverdrup flow and SST advection reversed. So, w_s decreases.

When the phase point reaches critical states Q_1 or P_1 as shown in Fig. 1 it jumps from one branch to another because of instability. The rapidly varying processes P_1P_2 and Q_1Q_2 take place and the oscillation forms.

Numerical Results

We use the R–K method to integrate system (2). One example is shown in Fig.3.

Numerical result of the trajectory in phase space is similar to the qualitative one in Fig. 1. Figure 3 also shows the solution of $|A|$. φ (wave phase) and V, w_s over a period of thirteen months. In the oscillation the forced wave moves back-and forward in a definite geographical region. The amplitude of the forced wave decreases when the wave moves to ocean and increases when it moves to land. Zonal wind modification is positive when wave amplitude is smaller and it becomes negative when wave amplitude is larger. These features are qualitatively similar to the annual variation of the subtropical circulation of upper layer troposphere.

Conclusions

(a) A long-term coupled oscillation can exist in the nonlinear air-sea model. The period depends on its amplitude and the time scale of advection in the ocean.

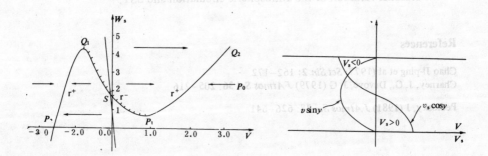

Fig. 1. Qualitation vector field on phase plane

Fig. 2. Distribution of zonal wind modification and sea flow

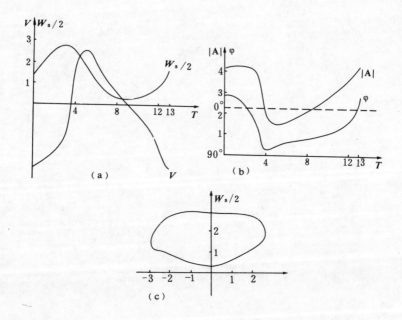

Fig. 3. (a) and (b): the solutions of W, V, A over one period; (c): phase trajectory on phase plane

period depends on its amplitude and the time scale of advection in the ocean.

(b) Over the period of the oscillation, the solution is asymmetrical. Forced wave and zonal flow of the circulation have both rapid and slow variation processes, but the amplitude of SST component always varies slowly.

(c) It seems possible that this kind of oscillation makes some contribution to the seasonal variation of the atmospheric circulation and SST.

References

Chao Ji-ping et al (1977) *Sci Sin* **2**: 162–172
Charney, J. G., Devove, J. G (1979) *J Atmos Sci* **36**: 205–216

Pedlosky, J (1981) *J Atmos Sci* **38**: 626–641

Stochastic Dynamic Climate Prediction Model with Nonlinear Feedback

Shi Yongnian[1]

Abstract — In this paper, stochastic-dynamic climate prediction equations are obtained on the basis of the abstract dynamic equations with quadratic feedback terms. The problem of interaction between first-order moments and second-order moments is discussed. Some examples are given.

Introduction

It has been proved by theory and practice that it is completely impossible to forecast the weather in detail over half a month or longer by means of a method based on the short-range numerical forecasting model, and of simply extending the procedure to a long period. The lack of predictability is due to the presence of dynamical instabilities and nonlinear interactions (feedbacks). However, if the object which we forecast is the mean value of meteorological elements, the average meteorological fields or any other statistics, there will be a certain degree of predictability. The physical background of this predictability is the relative stability of the ultralarge scale motion.

As the object of this kind of prediction is climatological statistics, we call it climate prediction. Naturally, our method should be established on a foundation connecting dynamics with the theory of probability. Namely, we have to solve stochastic-dynamic differential equations.

Basic Equations

Let $Y = Y (y_1, y_2 \ldots \ldots y_n)$ be an n-dimensional vector representing the state of a closed climate system. Generally speaking, the closed equation system which we use to forecast the climate system may be written as

$$\frac{dy_1}{dt} = F_1 (y_1, y_2, \ldots \ldots, y_n)$$

$$\frac{dy_2}{dt} = F_2 (y_1, y_2, \ldots \ldots, y_n)$$

$$\ldots \ldots \ldots$$

$$\frac{dy_n}{dt} = F_n (y_1, y_2, \ldots \ldots, y_n)$$

(1)

---

1. Department of Meteorology, Nanjing University, Nanjing, China

424

or, for simplicity, be stated as

$$\frac{d\mathbf{Y}}{dt} = \mathbf{F}(\mathbf{Y}) \tag{2}$$

We may call it the abstract dynamic equations.

For a pure determinate problem, the so-called forecasting is the integration of Eqs. (1) or (2) started from the initial point, i.e.,

$$\mathbf{Y}(t) = \mathbf{Y}_0 + \int_{t_0}^{t} \mathbf{F}(\mathbf{Y}) \, dt , \tag{3}$$

where $\mathbf{Y}_0 = \mathbf{Y}(t=t_0)$. Thus, the endpoint of n-dimensional vector $\mathbf{Y}$ will move along a definite trajectory in n-dimensional phase space. But from the viewpoint of stochastic dynamics, it is doubtful whether there must be a definite trajectory. Because of the errors in observation and interpolation to data-void areas, the initial conditions should be uncertain. This may be represented by a random vector or a set of random variables. In addition, the formulations of the dynamical equations and the parameterizations of the physical processes are only approximate. Then the endpoint of a state vector cannot moves along a definite trajectory in the phase space. As a result, the predictand must also be a random vector. The solution which we endeavor to find should be the probability distribution function or the probability density function.

The general equation to solve the probability density function is the so-called kinetic equation,

$$\frac{\partial \varphi}{\partial t} = \sum_{n=1}^{\infty} \sum_{i=1}^{n} \frac{(-1)^n}{n!} \frac{\partial^n}{\partial y^n} [\alpha_{in}(y_i ; t)\varphi], \tag{4}$$

where φ is the probability density function, and

$$\alpha_{in}(y_i , t) = \lim_{\Delta t \to 0} (\frac{1}{\Delta t}) E\left\{ [\eta_i(t + \Delta t) - \eta_i(t)]^n \mid \eta_i(t) = y_i \right\}$$

are called the derivate moments of the random vector. Theoratically, a_{in} can be derived or calculated from dynamic equations. But in practice, we can only derive the derivate moments from a few kinds of very simple dynamic equation. Moreover, any calculation of φ by direct numerical integration needs a huge amount of machine time, so, it is difficult to do it by this method

Epstein (1969) suggested an approximate method to solve the stochastic-dynamic problem. This is the moment method.

If the abstract dynamic equations are linear feedback, then the prediction equations for first-order moments are independent of any other order moments, as are the equations for second and other order moments. Thus, if a random vector is of Gaussian type, namely it is in normal probability distribution, then the stochastic-dynamic problem can be solved completely, because a Gaussian random vector is characterized by the first two moments. If the random vector is not Gaussian, we can obtain the information about the mean value and the uncertainty around this mean value by means of the first two moments.

If the abstract dynamic equations are quadratic feedback, they may be written as,

$$\frac{dy_i}{dt} = \sum_{j=1}^{n} \sum_{k=1}^{n} a_{ijk}\, y_j y_k - \sum_{j=1}^{n} b_{ij}\, y_j + C_i \tag{5}$$

$$(\,i=1,2,......,n)$$

From these equations, we can obtain the prognostic equations of first-order original moments μ_i

$$\frac{d\mu_i}{dt} = \sum_{j=1}^{n} \sum_{k=1}^{n} \left(a_{ijk}\, (\mu_j\mu_k + \sigma_{jk}) \right) - \sum_{j=1}^{n} b_{ij}\,\mu_j + C_i \tag{6}$$

$$(\,i=1,2,......,n),$$

where σ_{jk} is the second-order central moment of η_j and η_k, and, i.e., $\sigma_{jk} = E\,(\eta_j - \mu_j)(\eta_k - \mu_k)$. Thus, if we want to predict the first moments, we must simultaneously predict the second-order moments. The prognostic equations of second-order central moments are

$$\frac{d\sigma_{ij}}{dt} = \sum_{k=1}^{n} \sum_{l=1}^{n} \left[a_{jkl}\, (\mu_k\,\sigma_{il} + \mu_l\,\sigma_{ik}) + a_{jkl}(\mu_k\,\sigma_{jl} + \mu_l\,\sigma_{jk}) \right]$$

$$- \sum_{k=1}^{n} (b_{ik}\,\sigma_{jk} + b_{jk}\,\sigma_{ik}) + \sum_{k=1}^{n} \sum_{l=1}^{n} (a_{jkl}\,\tau_{ikl} + a_{ikl}\tau_{jkl}) \tag{7}$$

$$(\,i=1,2,......,n;\ j=1,2,......,n),$$

where τ_{ikl} and τ_{jkl} are third-order central moments,

$$\tau_{ikl} = E\left\{ (\eta_i - \mu_i)(\eta_k - \mu_k)(\eta_l - \mu_l) \right\}$$

$$\tau_{jkl} = E\left\{ (\eta_j - \mu_j)(\eta_k - \mu_k)(\eta_l - \mu_l) \right\}$$

Just as in the linear feedback model, if we only consider the first and second-order moments, then the stochastic-dynamic prediction equations consist of n equations to predict first-order moments and $n+n(n-1)/2$ equations to predict second order moments (i.e., n equations to predict variances, $n(n-1)/2$ equations to predict covariances) . Hence, the equations form a closed set.

Examples

We can write the prognostic equations of first moments of u, v and w derived from primitive equations,

426

$$\frac{\partial \bar{u}}{\partial t} + \left(\frac{\partial}{\partial x} (\overline{u^2} + \overline{u'^2}) + \frac{\partial}{\partial y} (\bar{u}\bar{v} + \overline{u'v'}) + \frac{\partial}{\partial z} (\bar{u}\bar{w} + \overline{u'w'}) \right)$$

$$= - \frac{1}{\rho} \cdot \frac{\partial \bar{p}}{\partial x} + 2\Omega (\bar{v} \sin \varphi - \bar{w} \cos \varphi) + \overline{F}_x \qquad (8a)$$

$$\frac{\partial \bar{v}}{\partial t} + \frac{\partial}{\partial x} (\bar{u}\bar{v} + \overline{u'v'}) + \frac{\partial}{\partial y} (\overline{v^2} + \overline{v'^2}) + \frac{\partial}{\partial z} (\bar{v}\bar{w} + \overline{v'w'})$$

$$= - \frac{1}{\rho} \frac{\partial \bar{p}}{\partial y} - 2\Omega \bar{u} \cos \varphi + \overline{F}_y \qquad (8b)$$

$$\frac{\partial \bar{w}}{\partial t} + \frac{\partial}{\partial x} (\bar{u}\bar{w} + \overline{u'w'}) + \frac{\partial}{\partial y} (\bar{v}\bar{w} + \overline{v'w'}) + \frac{\partial}{\partial z} (\overline{w^2} + \overline{w'^2})$$

$$= - \frac{1}{\rho} \frac{\partial \bar{p}}{\partial z} + g + 2\Omega \bar{u} \cos \varphi + \overline{F}_z \qquad (8c)$$

$$\frac{\partial \bar{u}}{\partial x} + \frac{\partial \bar{v}}{\partial y} + \frac{\partial \bar{w}}{\partial z} = 0 \qquad (8d)$$

For the prediction of second-order moments, we can obtain the equations as the following type,

$$\frac{\partial}{\partial t} \overline{u'v'} + \frac{\partial}{\partial x} (\bar{u}\,\overline{u'v'}) + \frac{\partial}{\partial y} (\bar{v}\,\overline{u'v'}) + \frac{\partial}{\partial z} (\bar{w}\,\overline{u'v'})$$

$$+ \frac{\partial}{\partial x} (\overline{u'^2 v'}) + \frac{\partial}{\partial y} (\overline{u'v'^2}) + \frac{\partial}{\partial z} (\overline{u'v'w'}) \qquad (9)$$

$$= -2\Omega \overline{v'w'} \cos \varphi \,,$$

where u, v, w are the first-order mometns, $\overline{u'^2}$ etc. are variances, $\overline{u'v'}$ etc. are covariances, $\overline{u'^2 v'}$, $\overline{u'v'^2}$, $\overline{u'v'm'}$ are third-order central moments.

Take the maximum simplified dynamic equations by Lorenz (1960) as the fundamental equations,

$$\begin{cases} \dfrac{dA}{dt} = -\dfrac{1}{2} \left(\dfrac{1}{k^2} - \dfrac{1}{k^2 + l^2} \right) k l \, F \, G \\[3mm] \dfrac{dF}{dt} = \dfrac{1}{2} \left(\dfrac{1}{l^2} - \dfrac{1}{k^2 + l^2} \right) k l \, A \, G \\[3mm] \dfrac{dG}{dt} = -\left(\dfrac{1}{l^2} - \dfrac{1}{k^2} \right) k l \, A \, F \,, \end{cases} \qquad (10)$$

where **A**, **F**, and **G** are the coefficients of the low-order expansion of vorticity,

$$\nabla^2 \psi = A \cos ly + F \cos kx + G \sin ly \sin kx$$

the corresponding stochastic-dynamic equations with quadratic feedback terms, truncated at third order moments, can be derived as (Epstein 1969)

$$
\begin{cases}
\dfrac{d\mu_A}{dt} = C_1 \left(\mu_F \mu_G + \sigma_{FG} \right) \\[2em]
\dfrac{d\mu_F}{dt} = C_2 \left(\mu_A \mu_G + \sigma_{AG} \right) \\[2em]
\dfrac{d\mu_G}{dt} = C_3 \left(\mu_A \mu_F + \sigma_{AF} \right)
\end{cases}
\tag{11}
$$

$$
\begin{cases}
\dfrac{d\sigma_{AA}}{dt} = 2C_1 \left(\mu_F \sigma_{AG} + \mu_G \sigma_{AF} \right) \\[2em]
\dfrac{d\sigma_{FF}}{dt} = 2C_2 \left(\mu_A \sigma_{FG} + \mu_G \sigma_{AF} \right) \\[2em]
\dfrac{d\sigma_{GG}}{dt} = 2C_3 \left(\mu_A \sigma_{FG} + \mu_F \sigma_{AG} \right) \\[2em]
\dfrac{d\sigma_{AF}}{dt} = C_1 \left(\mu_G \sigma_{FF} + \mu_F \sigma_{FG} \right) + C_2 \left(\mu_G \sigma_{AA} + \mu_A \sigma_{AG} \right) \\[2em]
\dfrac{d\sigma_{AG}}{dt} = C_1 \left(\mu_F \sigma_{GG} + \mu_G \sigma_{FG} \right) + C_3 \left(\mu_F \sigma_{AA} + \mu_A \sigma_{AF} \right) \\[2em]
\dfrac{d\sigma_{FG}}{dt} = C_2 \left(\mu_A \sigma_{GG} + \mu_G \sigma_{AG} \right) + C_3 \left(\mu_A \sigma_{FF} + \mu_F \sigma_{AF} \right)
\end{cases}
\tag{12}
$$

In the above equations, $\mu_x = E\{x\}$, $\sigma_{xy} = E\{(x - \mu_x)(y - \mu_y)\}$

$$C_1 = -\left[2\alpha(1 + \alpha^2) \right]^{-1}, \qquad C_2 = \alpha^3 / \left[2(\alpha^2 + 1) \right],$$

$$C_3 = -(\alpha^2 - 1)/\alpha, \qquad \alpha = k/l .$$

The point in the nine-dimensional phase space at which the nine variables (μ_A, μ_F, μ_G, σ_{AA}, σ_{FF}, σ_{GG}, σ_{FG}, σ_{AG}, σ_{AF}) are all zero is apparently an equilibrium point. But there are many equilibrium points beside this point. For example, for

$\mu_A = \Sigma_{AF} = \Sigma_{AG} = 0$, and other six variables can satisfy the following equations,

$$\mu_F \, \mu_G = -\sigma_{FG}$$

$$\mu_G = \pm \sqrt{\sigma_{GG} + C_3 \, \sigma_{AA}/C_1}$$

$$\mu_F = \pm \sqrt{\sigma_{FF} + C_2 \, \sigma_{AA}/C_1}$$

it is equilibrium point as well. For $\mu_F = \sigma_{AF} = \sigma_{FG} = 0$ or $\mu_G = \sigma_{AG} = \sigma_{FG} = 0$, there are also three equations to determine equilibrium point respectively.

Set the coordinates of equilibrium points into the following matrix,

$$
\begin{pmatrix}
0 & C_1\mu_G & C_1\mu_F & 0 & 0 & 0 & C_1 & 0 & 0 \\
C_2\mu_G & 0 & C_2\mu_A & 0 & 0 & 0 & 0 & C_2 & 0 \\
C_3\mu_F & C_3\mu_A & 0 & 0 & 0 & 0 & 0 & 0 & C_3 \\
0 & 2C_1\sigma_{AG} & 2C_1\sigma_{AF} & 0 & 0 & 0 & 0 & 2C_1\mu_F & 2C_1\mu_G \\
2C_2\sigma_{FG} & 0 & 2C_2\sigma_{AF} & 0 & 0 & 0 & 2C_2\mu_A & 0 & 2C_2\mu_G \\
2C_3\sigma_{FG} & 2C_3\sigma_{AF} & 0 & 0 & 0 & 0 & 2C_3\mu_A & 2C_3\mu_F & 0 \\
C_2\sigma_{GG}+C_3\sigma_{FF} & C_3\sigma_{AF} & C_2\sigma_{AF} & 0 & C_3\mu_A & C_2\mu_A & 0 & C_2\mu_G & C_3\mu_F \\
C_3\sigma_{AF} & C_3\sigma_{AA}+C_1\sigma_{GG} & C_1\sigma_{FG} & C_3\mu_F & 0 & C_1\mu_F & C_1\mu_G & 0 & C_3\mu_A \\
C_2\sigma_{AF} & C_1\sigma_{FG} & C_1\sigma_{FF}+C_2\sigma_{AA} & C_2\mu_G & C_1\mu_G & 0 & C_1\mu_F & C_2\mu_A & 0
\end{pmatrix}
$$

and calculate the eigenvalues of this matrix, we can discuss the stability of these equilibrium points.

Upon taking the parameter $a = 2$, initial condition $\mu_A = 0.12$, $\mu_F = 0.24$, $\mu_G = 0$, $\sigma_{ji} = 0$ $(i \neq j)$, but $\sigma_{ii}(i = A, F, G) = 0.0001, 0.001, 0.01, 0.1$ or 1.0 respectively. We integrate these equations numerically by the Runge-Kutta method. The corresponding results of μ_A, μ_F and μ_G are depicted in Fig. 1.

From the figures we can clearly see that the larger the initial variances are, the more the variability in mean value with time will be. This result implies that the low order spectral coefficients of vorticity are ergodic. This is the very nature which we can use to estimate the population averages, variances, or covariances. Even if the initial average streamline fields are the same, the final fields may be very different due to different initial variances.

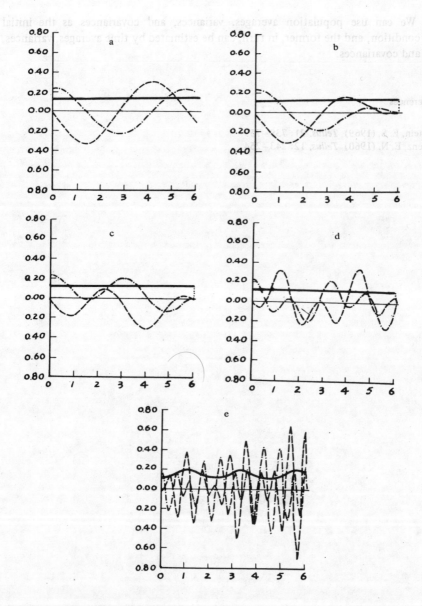

Fig. 1. The corresponding results of μ_a (——), μ_F (—·—·—) and μ_G (— — —) for $\sigma\mu=0.0001$ (a), 0.001 (b), 0.01 (c), 0.1 (d) and 1 (e) respectively

Conclusions

(1) The stochastic-dynamic climate model is a possible approach to the predictions of averages, deviations and other statistics of meteorological variables.

430

(2) We can use population averages, variances, and covariances as the initial condition, and the former, in turn, can be estimated by time averages, variances, and covariances.

References

Epstein, E. S, (1969) *Tellus,* 21: 739–759
Lorenz, E. N. (1960) *Tellus,* 12: 243–254.

A Two-Dimensional Radiation-Turbulence Climate Model: Sensitivity to Cirrus Radiative Properties

Szu-cheng S. Ou[1] and Kuo-nan Liou[1]

Abstract

Based on the thermodynamic energy balance between radiation and vertical plus horizontal dynamic transports, a two-dimensional radiation-turbulence climate model is developed. This model consists of a broadband solar and IR radiation transfer scheme previously presented by the authors and vertical and horizontal dynamic eddy transports utilizing the elementary turbulent theory. In the model, three kinds of feedback mechanism are considered: the humidity feedback via the constant relative humidity assumption, the ice-albedo feedback via a preliminary correlation between the surface albedo and the surface temperature, and the dynamic transport feedback through the parameterization of eddy transports and the prescribed mean wind field. A standard temperature field, which differs from the climatological data by no more than 0.1 °C, is first obtained by solving the coupled thermodynamic and surface flux budget equations using climatological distributions of H_2O, CO_2, O_3, surface albedo and cloud properties. The model-derived atmospheric radiation budget, surface energy balance and horizontal transport patterns compare reasonably well with available observational data. Further validation of the model includes sensitivity studies on the effects of doubling of CO_2 and a 2% increase in the solar constant. The temperature changes relative to the standard field on these experiments agree closely with those presented by Manabe and Wetherald utilizing a general circulation model. To investigate the two-dimensional cirrus/radiation interaction, a relationship between the cirrus IR emissivity and solar reflectance (and transmittance) is established based on parameterization equations. On the basis of a number of experiments involving various couplings and feedbacks, we find that (1) the humidity and albedo feedbacks are most active in the tropics and arctic area, respectively, (2) the dynamic transport is a negative feedback in the equatorial and polar regions but a positive one in mid-latitudes, and (3) the relative importance of each feedback depends only slightly on the radiative properties of cirrus. Finally, we demonstrate that slight variations on the cirrus IR emissivity lead to significant temperature perturbations in the arctic surface and tropical troposphere.

1. Department of Meteorology, University of Utah, Salt Lake City, Utah 84112 USA
* For details see the *Journal of the Atmospheric Science,* August, 1984.

A Hybrid Coastal Wave Climate Model — Past, Present and Future

Joseph P. S.[1]

Abstract — For nations surrounded by coastlines, the coastal wave climate is essential for shoreline development, fishing, navigation, and for offshore engineering etc. A simple, accurate, and economically cheap wave climate model is discussed in this paper which is based on the wave growth equation of Toba (1978) and on the 3/2 power law of wave heights and periods. The model's ability to give the details of wave climate such as spatial and temporal distribution of wave properties in the ocean and along the coastline, design wave heights and spectral properties is tested for severe situations like the typhoons of the Japan Sea, severe cyclones of the Atlantic Ocean, Arabian Sea and Bay of Bengal. The results of the model studies are in good agreement with field measurements and observations and are discussed in detail.

Introduction

The limited availability of wave data for design applications (extreme wave conditions) and for wave climate studies in an adequate format has been a major problem. Observational wave data in the form of ship reports are limited in accuracy by biases in measurements and locations which render such data useless for some applications, while long time series of reliably observed wave data are scarce. However, sufficiently long time series (statistically significant time period) of meteorological records (wind and pressure) are available to establish wave climatology and the wave hindcast approach employs wave models as a transfer function between the wave histories at one location and meteorological records, with the limitations in the meteorological time series data providing a boundary on the ability to forecast or hindcast the ocean wave field.

For a given sea surface wind, the problem of wind wave prediction can be reduced to three steps: estimation of surface momentum fulx, determination of the fraction of momentum transferred to the wave field and finally, the solution of an appropriate wave momentum balance equation. For the first part, that is essentially the determination of the surface drag coefficient, the present state of our knowledge is taken to be adequate for wave prediction purposes.

The second step, that is the determination of the fraction of momentum retained as wave momentum or otherwise termed the wave growth mechanism of wind waves, present knowledge is very limited. The combined mechanism of Philips (1958) and Miles (1957) was considered to be a final solution to the growth mechanisms of wind waves for about 10 years following its proposition but it was seen to be ineffective with the gradual accumulation of observational as well as experimental results. As the available wave growth mechanisms are ineffective in

---

1. Centre for Earth Science Studies, Regional Centre, Cochin-18, India

giving the actual growth rates for wind waves, the wave prediction problem has to depend to a great extent on some empirical formulas which describe wind wave growth in terms of other parameters.

Although our knowledge about mechanisms of wind wave growth is limited it has been recognized that there is self-similarity structure in growing wind waves (e.g., Tokuda and Toba 1981), presumably due to the strong nonlinearity inherent in windwaves under the action of wind, as studied by Okuda (1980). The similarity in spectral shape has been considered in many ways by many investigators since the proposition of $g^2 f^{-5}$ type spectral from by Phillips (1958), up to recent concepts of self-stabilizing nature and rapid adjustment of spectrum after a change of wind as presented in Hasselmann et al. (1973, 1976). These similarities contribute greatly to the reduction of complexity in wave prediction models.

Numerical Scheme of a Parametric Model with Basic Equations: The 3/2 power law

Toba (1972) proposed a power law

$$H^* = BT^{* \, 2/3} \tag{1}$$

where $B = 0.062$, $H^* = \frac{gH}{u^{*2}}$, $T^* = \frac{gT}{u^*}$ with u^* as friction velocity, H =significant wave height, T = significant wave period and g acceleration due to gravity. This power law makes it possible to represent the wind wave field in terms of a single variable, for example, significant height or period or wave energy per unit area of the sea surface. This power law is considered to have been substantiated by wave data in laboratory as well as in field (e.g., Kawai et al. 1977; Mitsuyasu et al. 1980).

The 3/2 power law forms the basis for the single parameter representation of the growing wind wave field (Toba 1978) after its transformation to:

$$E^* = B_f \, f_p^{*-3}, \quad B_f = 2.1 \times 10^{-4} \tag{2}$$

where $E^* \equiv g^2 E/u_*^4$, $f_p^* \equiv u_* f_p/g$, f_p the peak frequency,

where $f_p = (1.05 \, T_s)^{-1}$ \tag{3}

after Mitsuyasu (1968) and Toba (1973) and

$$E = \int_0^\infty \phi \, (f) \, df = H_s^2/16 \tag{4}$$

after Longuest-Higgins (1952) with $\phi \, (f)$ as the spectral density.

By the use of the power law (1) along with

$$\widetilde{T}_s/2\pi = 1.37 \, [1- \{ \, 1+ 0.004\widetilde{x} \, \}^{-5}] \quad \text{wilson (1965)}, \tag{5}$$

with a representative value of drag coefficient $C_D = 1.2 \times 10^{-3}$ a growth equation for wind waves in terms of non-dimension total energy E^* with a source function having the form of a simple stochastic error function was expressed by Toba (1978) as

$$\frac{\partial E^{*2/3}}{\partial t^*} - \frac{E^{*1/2}}{a} \frac{\partial (E^{*2/3})}{\partial x^*} = G_0 R [1- \text{erf} \, (bE^{*1/3})] \tag{6}$$

where erf $(x) \equiv 2\pi^{-1/2} \int_0^x \exp(-t^2)\, dt$

and $a = 0.74$, $G_0 R = 2.4 \times 10^{14}$ and $b = 0.12$. It should be noted that in (6)

$$G_0 R \equiv G_0 \ [1 - \text{erf}\ (bE^{*1/3})] \tag{7}$$

represents the proportion of momentum retained as the wave momentum to the total momentum transferred from the wind to the sea.

The selection of 3/2 power law as the basis of the present wave prediction groups this model into the parametric type. Though it belongs to such a class, once a representative property of the wave field is predicted, the spectral quantities can also be estimated, due to the presence of the similarity structure in growing wind waves and of mutually transformable relationships among representative wave properties as discussed by Toba (1978).

The merit of the prediction model using Toba's (1978) growth equation (6) among others using significant wave properties seems to be considerable. In most of the traditional methods, the significant wave height and wave period are separately estimated from the use of two separate fetch graphs. But in the present case, the prediction of a single parameter of the wave field, for example the wave energy, is sufficient for the estimation of any other property of the wave field through the similarity relations. In addition, the predicted heights and periods from two different fetch graphs by traditional methods do not necessarily satisfy the similarity relation between height and period.

Although simplicity and superiority are present in the prediction of the wind wave part of the wave field, the prediction equation (6) cannot be extended to the swell part of the wave field, as the swells have no similarity in structure. For swell components, the spectral treatment seems to be the best, as each of the swell component is free to propagate as approximately independent water waves without any significant influence from other swell components. Thus, for the swell part of the wave field, spectral concepts are used in the present model. The simplified treatment of the wind wave part with a single parameter and the treatment of the swell part with spectral concepts give a hybrid nature to the presently used model. A similar hybrid treatment can be found in Gunther et al. (1979) and in Weare and Worthington (1978).

Prediction Scheme for the Wind Waves

The present prediction equation for the wind waves is a first-order differential equation which can be easily integrated numerically. A flow chart of wind wave prediction along with swell formation conditions is shown in Fig. 1. The conditions of zero initial energy for the entire prediction region and of no energy flow into the prediction region from outside through the numerical boundaries, as in Kawai (1979) and in Joseph et al. (1981), are used in the present chart also. The only input data required for running this model is the time series of wind velocity over the ferecast/hindcast region. The basic equation is integrated with the iteration scheme in Kawai et al. (1979), and the grid interpolation method in Joseph et al. (1981) is also used in the present model without any modification. Since the

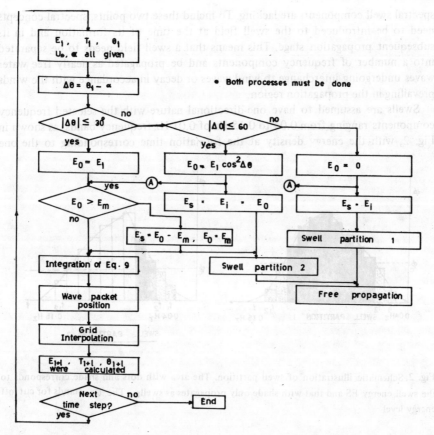

Fig. 1. Main scheme of the model. The X represents wind direction. θ – the wave direction, Es' – the total swell energy and Em the saturated wind wave energy

present model is for deep water waves, for obtaining shallow water wave conditions, wave refraction techniques need to be incorporated with this model. Simple wave refraction models are readily available at present. Though the swell formation by the wind wave saturation or by wind direction shifts remains the same in the above two references, the treatment of the swell field is different and this will be described in detail in the next section.

Prediction Scheme for Swells

Once swell formation occurs, swells need to be propagated in the prediction region with some suitable scheme. For simplicity and with economical considerations it can be done with available empirical relations which predict the gross nature of a swell field. In the empirical treatment used in Kawai (1979) and Joseph (1981), any treatment of possible interchanges to wind waves during their propagation through strong wind region and an energy partition of the original swell field into spectral swell compónents are lacking. To includ these two points, spectral concepts need to be introduced to the swell field at the time of its formation and in its subsequent propagation stage. This means that a swell field needs to be separated into a number of frequency components and be propagated as nearly free water waves undergoing interchange to wind waves or decay in accordance with the winds prevailing in the propagation region.

Swells are assumed to have one-directional nature with the selected frequency components ranging from 0.04 to 0.15 Hz of 0.01 Hz frequency bands, as shown in Fig. 2, with the enerɤv density at the formation time corresponding to the one

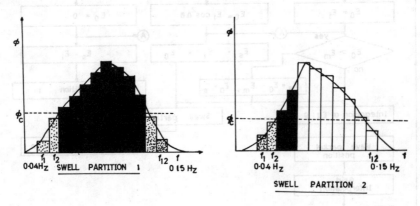

Fig. 2. Schematic illustration of swell partition. The area with dots and shade corresponds to the swell energy ES and that with shade only propagates as swells. The ϕc stands for cut off energy level

directional spectrum of wind waves. For frequencies greater than 0.15 Hz, the energy is cut off, since the energy density is small. This limitation in the number of frequency bands was introduced by the limitation of the computer memory. According to the formation process of the swell, the energy partition into the above frequency bins is given in different ways. The swell partition 2, shown in Fig. 2, corresponds to swell formation by wind wave energy saturation or by wind direction change $|\Delta\theta| \leqslant 60°$ ($\Delta\theta$ = wind direction − wind wave direction). The swell partition 1 shown in Fig. 2, corresponds to swell formation due to wind direction change $|\Delta\theta|$ greater than 60°. In both the partitions, ϕ_c is an energy cut off level to neglect minor swell packets. As to the value of ϕ_c' it will be reasonable to adopt a value of the peak energy level of saturated wind waves with a peak frequency at the uppermost swell band. In the swell partition 1, only the dark area is considered. In swell partition 2, the dark area's energy only propagated as swells, the white area remains as the energy of the wind waves. The dotted portion's energy which remains below the cut-off energy level is neglected.

The scheme for free propagation of swell is shown as a flow chart in Fig. 3. The subscript j is the frequency number. After the phase speed c is estimated, readjustment is made by use of the wind at the nearest grid to the swell's newly attained position. Namely, for an adverse wind of $|\Delta\theta| > 90°$, swells propagate in the original direction by decaying by Inoue's (1967) formula:

$$B(f, u^*) = 0.00139 \exp\left[-7000\left\{(u_*/c) - 0.031\right\}^2\right] + 0.725(u_*/c)^2$$
$$\exp\left[-0.0004(c/u_*)^2\right] f \tag{8}$$

where f is the frequency. For $/C\theta/$ between $60°$ and $90°$, swells propagate as free waves, for $/\Delta\theta/$ smaller than $60°$ the energy of the components is again added to the wind wave energy, otherwise freely propagated.

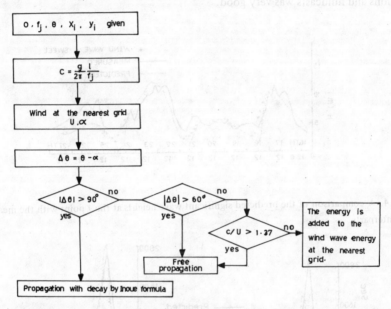

Fig. 3. Scheme for free propagation of swells. The (xi, yi) stands for the position of the swell component

At the stage of the output of swell packets, 12 directional bins of $30°$ are used, and for a particular bin the swell component having maximum energy in each frequency band is adopted and it is assumed to take the mid-angle of the bin.

This hybrid model now makes it possible to provide the entire spectrum of wave field at any grid point by a superposition of the combined energy of a swell component from all directions with the available energy of the wind wave component of same frequency. A directional spectrum may also be possible if a directional spreading function is applied to the wind wave spectrum and then by the addition of the frequency-directional swell components to the wind wave directional spectrum.

Model Validation

The best way to determine the accuracy of a particular wave model is to compare wave conditions computed by the model to measured wave conditions. In par-

438

ticular, knowledge of frequency spectra, directional spectra, and extreme waves are useful. Unfortunately, few suitable wave data are available and few detailed comparisons have been made. Many model results are compared with the measured wave data of the N. Atlantic cyclone of December 1959. Figure 4 shows the significant wave height time series which shows that difference between hindcast measurements is not great. The reason for the few differences is described in detail in Joseph et al. (1981). The spectral features of hindcasted waves are shown in Fig. 5 for the time intervals of peak wave development. The agreement between predictions and hindcasts was very good.

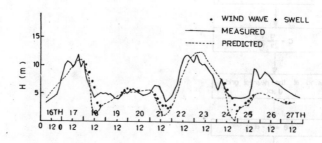

Fig. 4. A comparison of the predicted significant wave heights at the J point with the measured counterparts

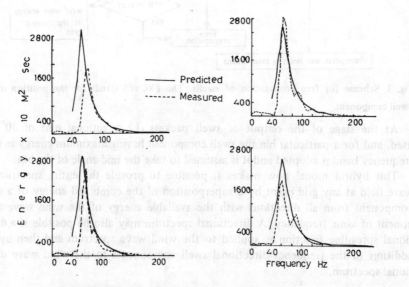

Fig. 5. Comparison of hindcasted wave spectra with their measured counterparts

As a second test for the model, an enclosed sea which is free from the entrance of swells through numerical boundaries during wave prediction interval is preferred. The Japan sea seems to satisfy this condition quite well and the interval from 1st April to 7th April 1978 was selected for hindcast studies. The reasons for this selection and the details of the meteorological conditions are given in detail in Joseph et al. (1981). The spatial distribution of the wave heights are shown in Fig. 6. The agreement between hindcast and measurement seems to be remarkable for April 4th and 7th. The reasons for the inaccuracies for the predictions on 5th and 6th depends on the inaccurate wind field deduced from pressure charts and these are described in detail in Joseph et al. (1981). No time series measurements of significant height or spectral measurements were available, and hence no comparison was attempted with hindcasts.

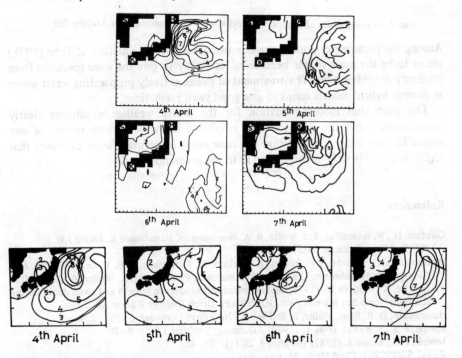

Fig. 6. Comparison of predicted significant heights with measured counterparts for Japan Sea

This model was also tested for the waves in the Arabian Sea. The cyclonic situation of in June 1976 is taken for it since wave measurements are available for it. The hindcasted and measured waves are compared in Fig. 7. The agreement seems to be good and the slight deviations at the time of the peak wave development may be due to the inaccuracy of wind field used in the prediction.

Conclusions

The superiority of the parametric wave prediction models over spectral models is indicated from the consideration of its simplicity and less computational expense.

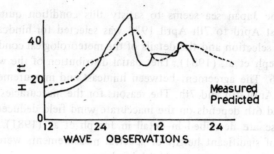

Fig. 7. A comparison of hindcasted heights with measurements for Arabian Sea

Among the parametric models, the one using the growth equation of Toba (1978) seems to be the most simple because of its ability to give the wave spectrum from similarity considerations. The treatment of swells as freely propagating water waves as done in hybrid models seems to give good swell predictions.

The prediction model validation for the severe weather conditions clearly demonstrated that the model can be employed for wave climate studies of any region for any time, provided that the time series of accurate wind data over that region is available to be incorporated in to the prediction model.

References

Gunther, H., W. Rosenthal, T. J. Weare, B. A. Worthington, Hasselmann K. Ewing J.W. (1979) *J Phys Oceanogr* 84:5727–5738

Hassalmann, K., Barnett, T.P. Bouws, E. Carlson, H. Cartwright, D.E. Enke, K. Ewing, J.A. Gienapp, H. Hassalmann, D.E. Sell, W. Kruseman, P. Meerburg, A. Muller, P. Olbers, D.J. Richter, K. Waldern W. (1973) Measurements of wind-wave growth and swell decay during the Joint North Sea Wave Project (JONSWAP). *Dtsch Hydrogr Z Supp* A8: 12

Hasselmann, D. B. Rose, Muller, P. Sell W. (1976) *J Phys Oceanogr*, 6: 200–228

Joseph, P. S., S. Kawai, Toba Y. (1980) *J Oceanogr Soci Jpn* 37 (1): 9–20

Joseph, P. S., Kawai S. (1981) *Geophys J* 28 (1): 27–45

Kawai, S. (1979) J Fluid Mech, 93: 661–703

Kawai, S. K. Okada, Toba Y. (1977) *J Oceanogr Soc Jpn* 33: 137–50

Kawai S., Joseph P.S. Toba Y. (1976) *J Oceanogr Soc Jpn* 35: 151–167

Longuest – Higgins M S (1952) *J Mar Res* 11: 245–266

Miles, J. W. (1957) *J Fluid Mech* 31: 185–204

Mitsuyasu H (1968) On the growth of the spectrum of wind generated waves (I) *Rep Inst Appl Mech Kyshu Uni* 16: 459–482

Mitsuyasu H. F. Tasai, Suhara, T. Mizuno, S. Ohkusa, M. Honda, T. Rikishi K. (1980) *J Phys Oceanogr* 10: 286–296

Okuda K. (1980) Study on the internal structure of wind waves. Ph D Thesis, Division of Science, Tohoku Univ 145 pp

Phillips, O. M., (1958) *J Fluid Mech* 2: 417–445

Toba, Y. (1972) *J Oceanogr Soc Jpn* 28: 109–121

Toba Y. (1973) *J Phys Oceanogr Soc Jpn* 29: 209–220

Toba Y., (1978) *J Phys Oceanogr* 8: 494–507

Tokuda, M. Toba Y. (1981) *J Oceanogr Soc Jpn* (in Press)

Wilson, B. W. (1965) *Dtsch Hydrogr Z* **18**: 114–130

Weare, J. and B. A. Worthington (1978): A numerical model hindcast of severe wave conditions for the North Sea. *Proc. of the NATO symposium on Turbulent Fluzez through the sea. Surface-Wave Dynamics and Prediction*, Ile de Bendor, New York.